THE BIG BOOK OF
OUTSIDE SUDOKU
400++ PUZZLES & VARIANTS

DJAP

I0446479

The Big Book of Outside Sudoku

FIRST EDITION: DECEMBER 2023

ISBN 979-8-87099-462-8

Introduction and How to Solve Outside Sudoku

Welcome! The puzzles in this book will challenge you in unique ways, while still being based on the basic rule of sudoku: Fill in the boxes so that the numbers 1–9 appear once (and only once) in every row, column, and 3×3 box. It is strongly recommended to become proficient at solving regular sudoku before attempting to solve the puzzles in this book.

In Outside Sudoku, the starting clues have been thrown outside of the grid. The numbers outside the grid are to be placed in **one of the nearest 3 squares** of the corresponding row or column, in some order. Consequently, **the 6 squares** past those first 3 squares **do *not* contain** any of those outside numbers, a fact which will often be **very important** to remember. For any given row or column, there can be 1, 2, or 3 numbers given outside, or sometimes none at all! Even when only 1 number is given, the rule is the same: that 1 number must be placed in one of the first 3 nearest squares. Please carefully study this sample puzzle and the corresponding solution. It will help you understand the rules. Start by looking for two outside clues for the same number which correspond to the same 3x3 box, such as the 5 in the upper left corner (or 9 in bottom-left, or 8 in upper-right, or 6 in bottom-right).

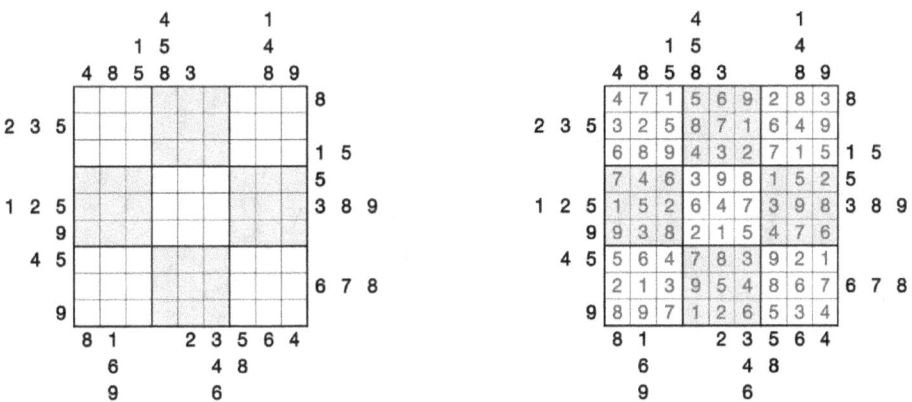

Think about the clues this way: each outside clue will be placed somewhere in one of the 3 nearest cells. Imagine it's already placed there (pick any of the 3). Now, because of the basic rules of Sudoku, this means that the same number cannot be placed anywhere else in the corresponding 3x3 box, nor can it be placed anywhere in the remainder of the corresponding row/column! **Got it?**

Consecutive and Non-Consecutive Sudoku

In **Consecutive** Sudoku, the grid is sprinkled with horizontal and vertical bars. The digits on either side of each bar are consecutive, though which number is greater is not indicated. Example: a bar adjacent to a 4 means that the digit on the other side could only be a 3 or a 5. All possible bars have been placed; the lack of a bar between any two digits tells you that they are *not* consecutive. Here's a sample **Outside Consecutive puzzle** and answer.

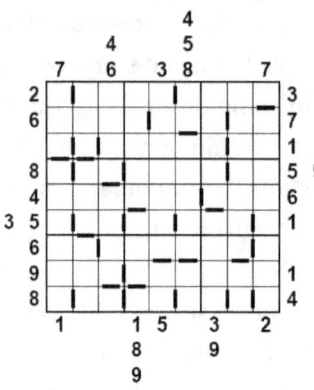

In the 6th column, clues 4-5-8 on top correspond to the first 3 cells. There is a bar indicating that two of those three must be consecutive numbers – which could be either 4|5 or 5|4, and hence the last one of the three must be 8! **Easy?** Now, have a look at this one:

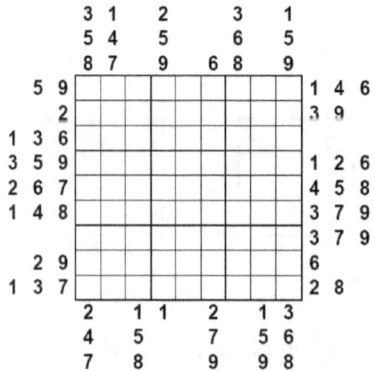

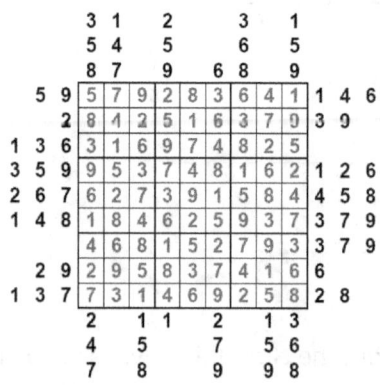

It looks like a regular Outside Sudoku, but, in fact, this is a **Non-Consecutive Outside Sudoku,** in which two **consecutive numbers must never be placed next to each other** horizontally or vertically in the 9x9 grid. Puzzles #201-#300 in this book are all Non-Consecutive, and this is always written next to each puzzle.

Jigsaw Outside Sudoku

Puzzles #101 to #200 are **Jigsaw Outside Sudoku**. The "outside sudoku" **rules remain the same**, as they correspond to rows and columns. However, the 3x3 boxes have been reshaped into "jigsaw nonets" - strings of 9 cells marked with thick lines. Each jigsaw nonet must contain one of each digit 1 to 9. That's all.

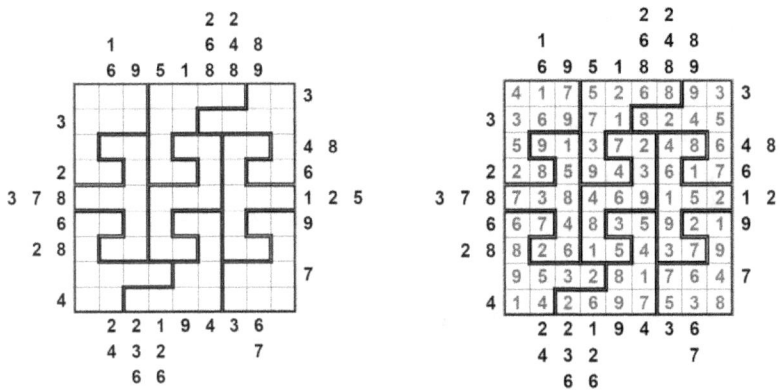

Frame Sudoku

Finally, bonus puzzles are called Frame Sudoku. The clues represent **the sum of the 3 nearest numbers** in that row/column! And that's all you are given to work with. And yet it is still possible to solve the puzzle **without guessing**!

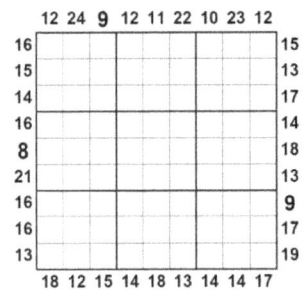

	12	24	9	12	11	22	10	23	12	
16	3	7	6	2	4	8	5	9	1	15
15	5	8	2	7	1	9	3	6	4	13
14	4	9	1	3	6	5	2	8	7	17
16	6	3	7	9	5	1	8	4	2	14
8	1	2	5	4	8	7	6	3	9	18
21	8	4	9	6	3	2	7	1	5	13
16	7	5	4	8	9	3	1	2	6	9
16	2	6	8	1	7	4	9	5	3	17
13	9	1	3	5	2	6	4	7	8	19
	18	12	15	14	18	13	14	14	17	

If you like my puzzles, please visit my store `amazon.com/djape` for more.

To keep the printing cost down and to save some trees, the solutions are not printed in the book, but they can easily be downloaded from this link:

djape.net/outside400

Happy solving and enjoy!   —Djape

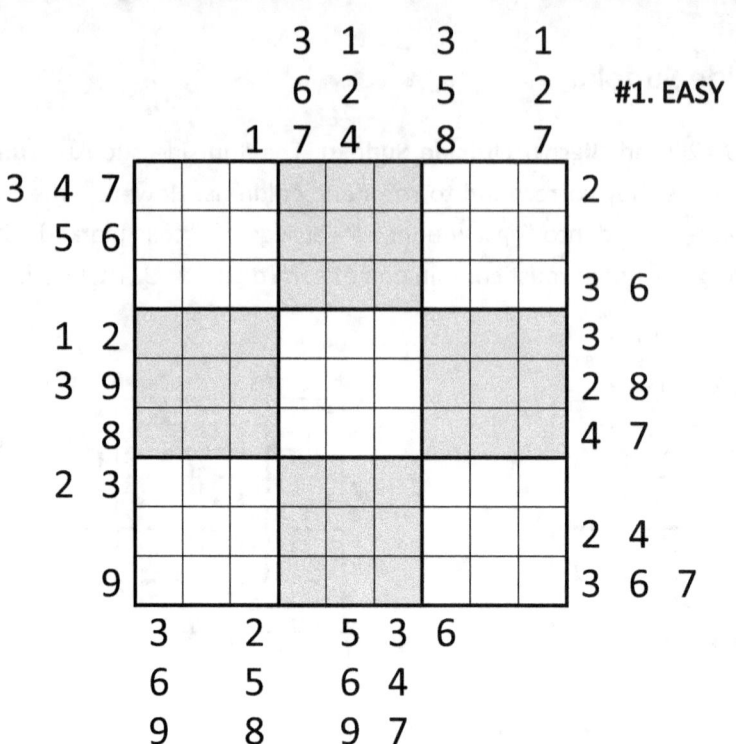

#1. EASY

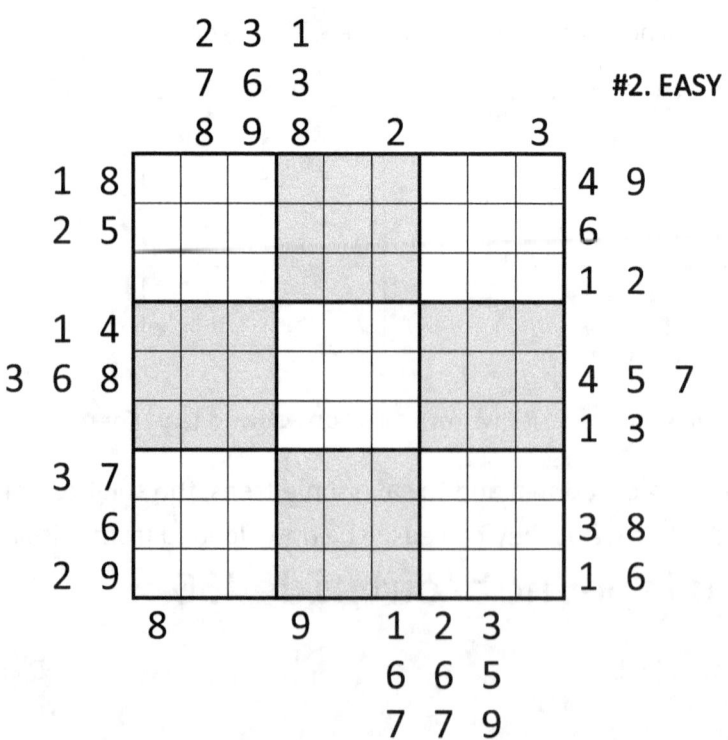

#2. EASY

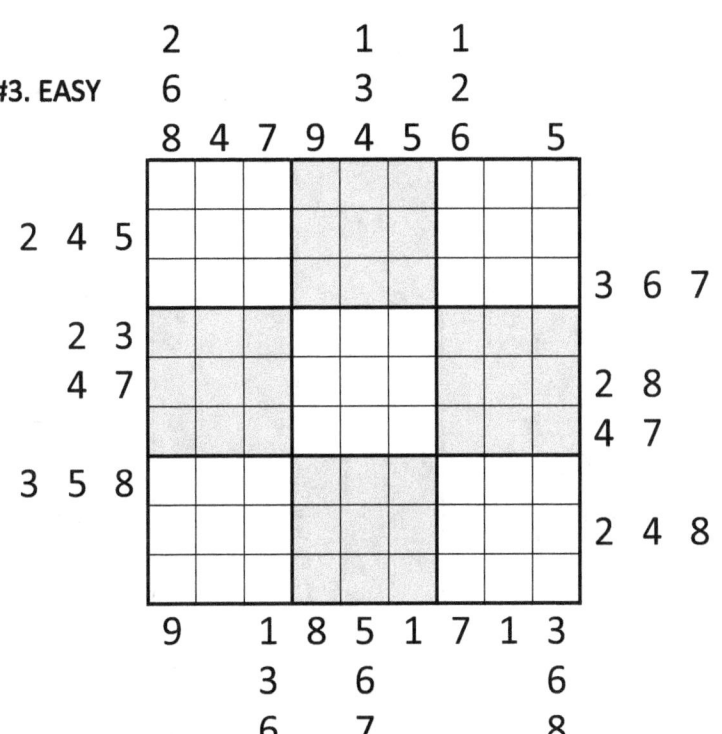

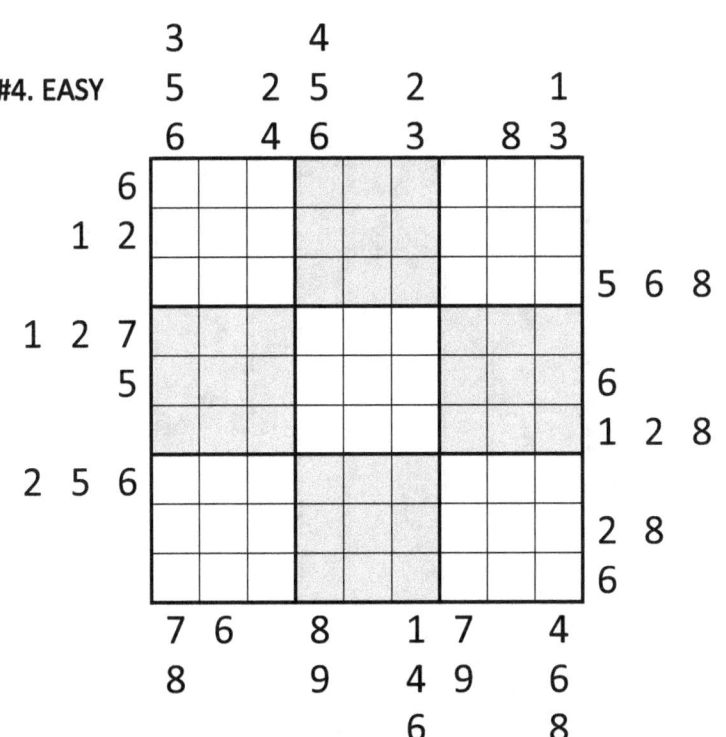

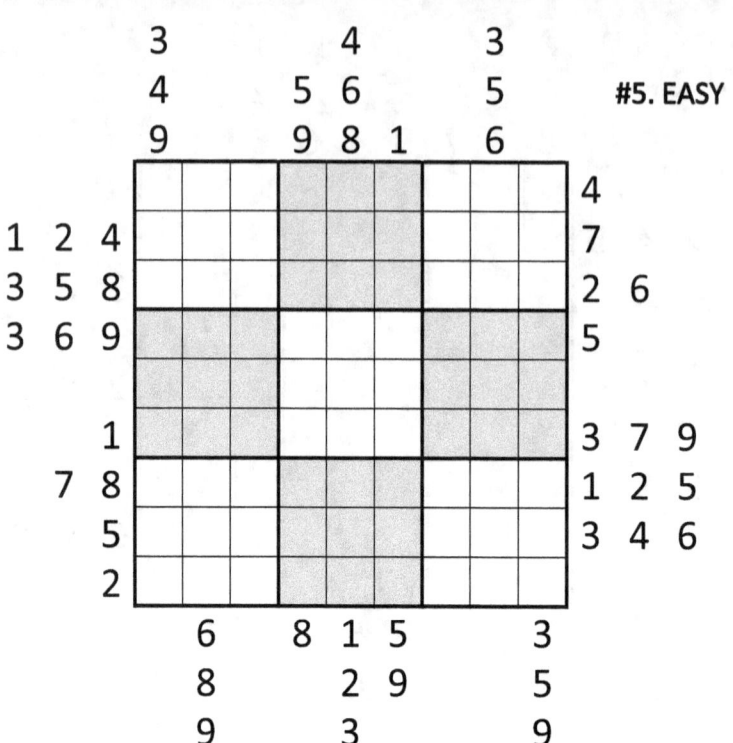

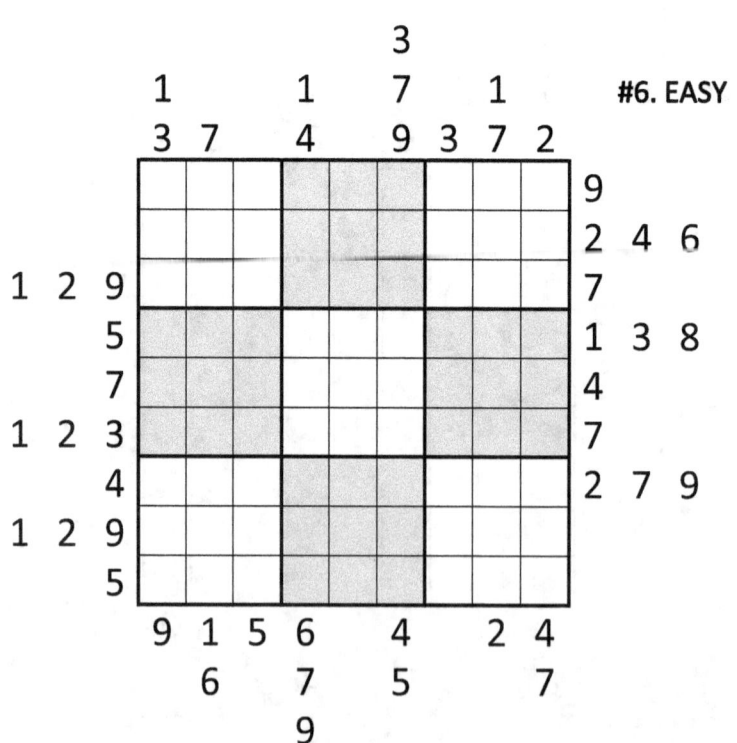

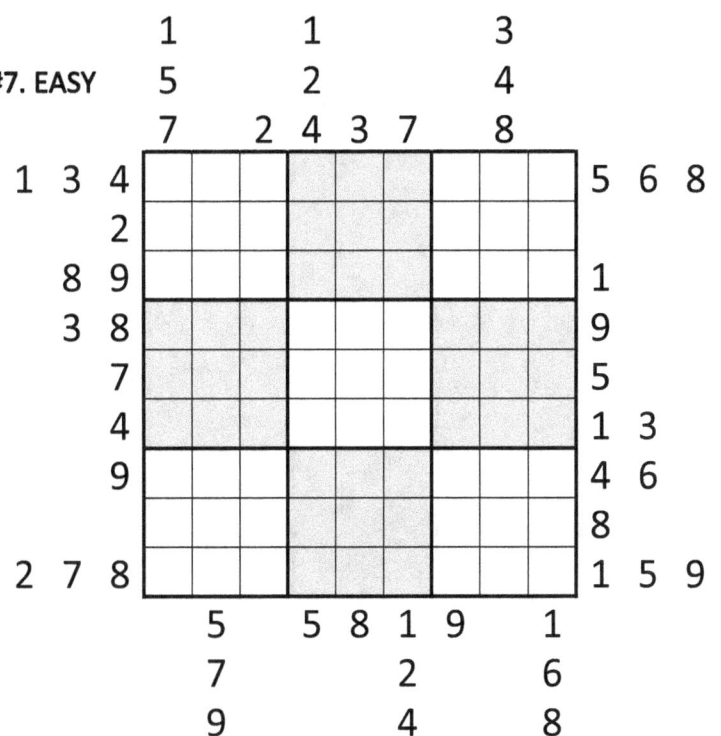

#7. EASY

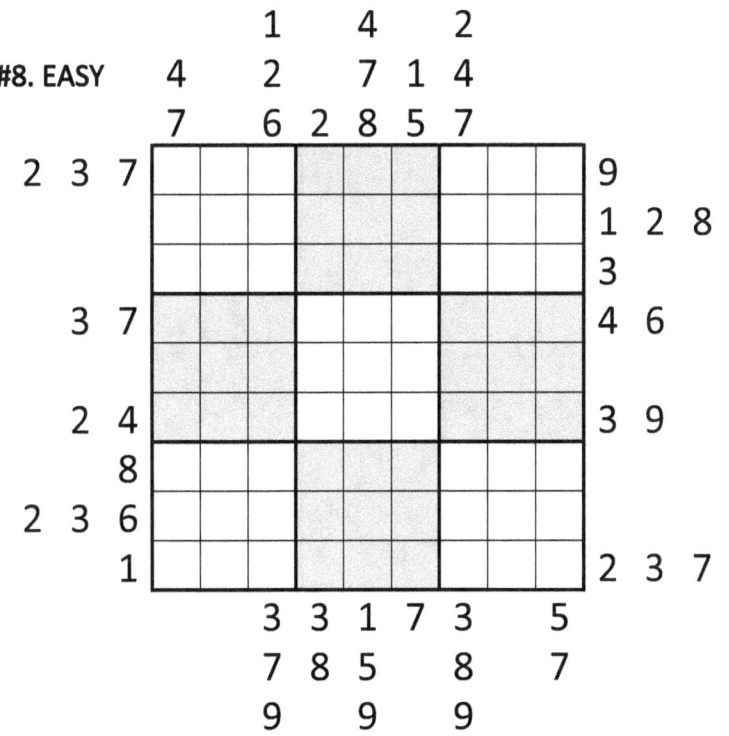

#8. EASY

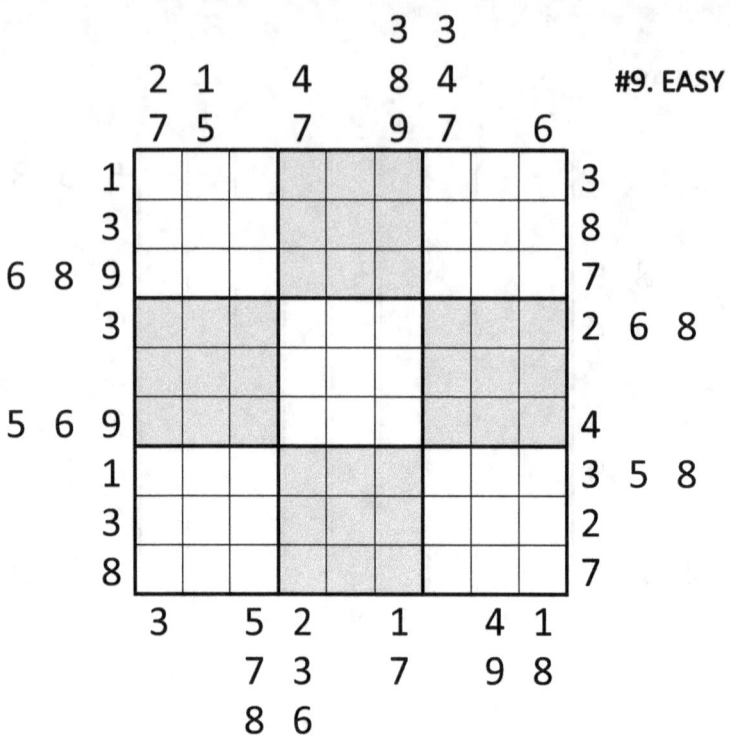

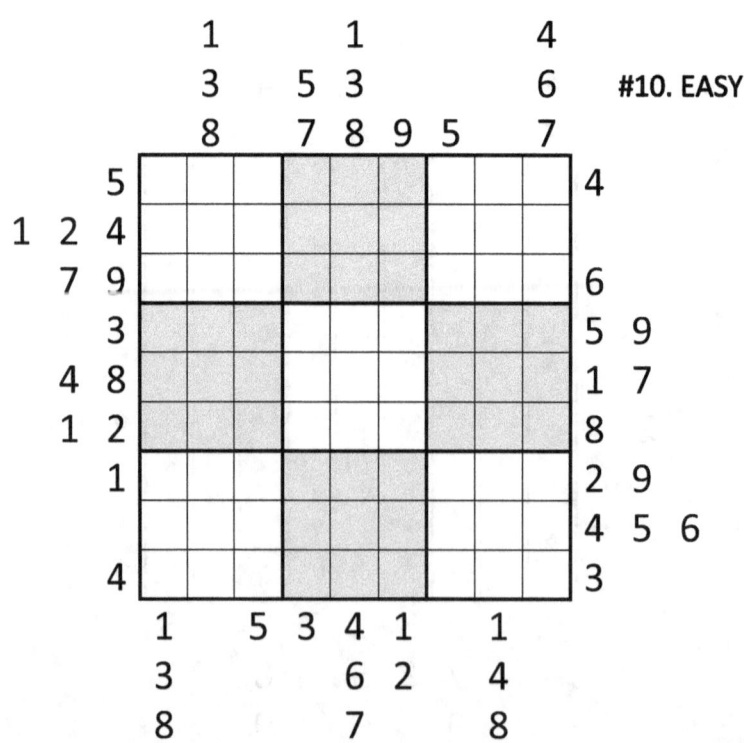

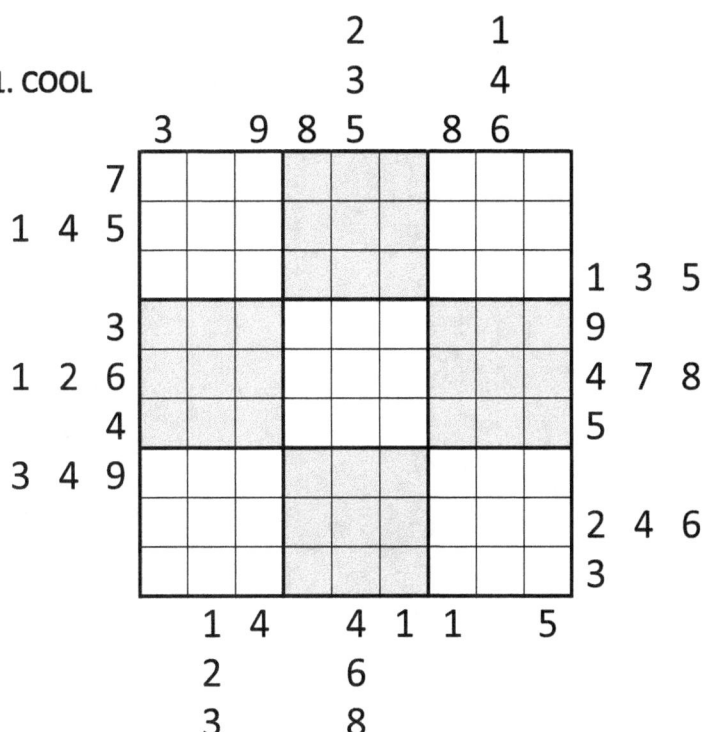

#11. COOL

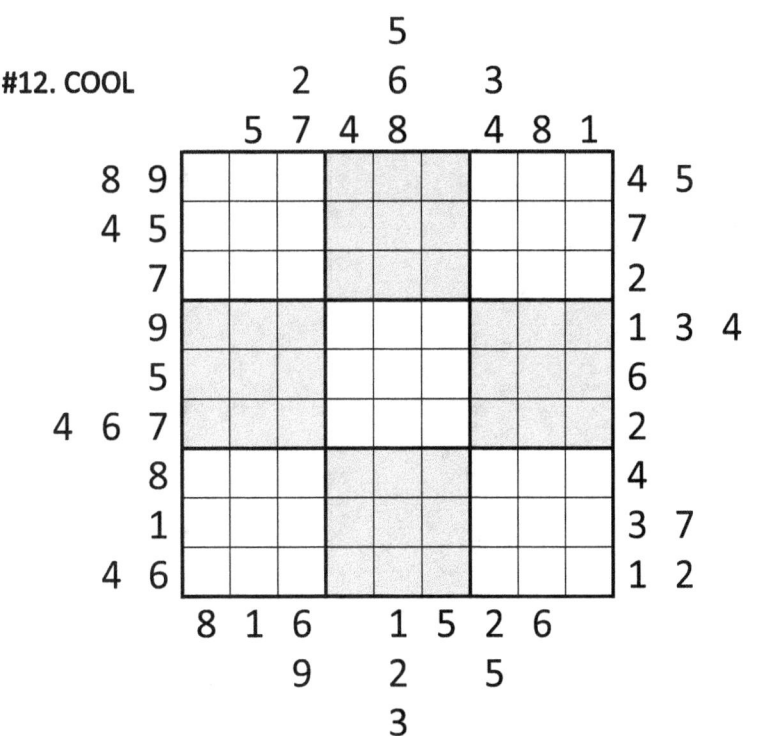

#12. COOL

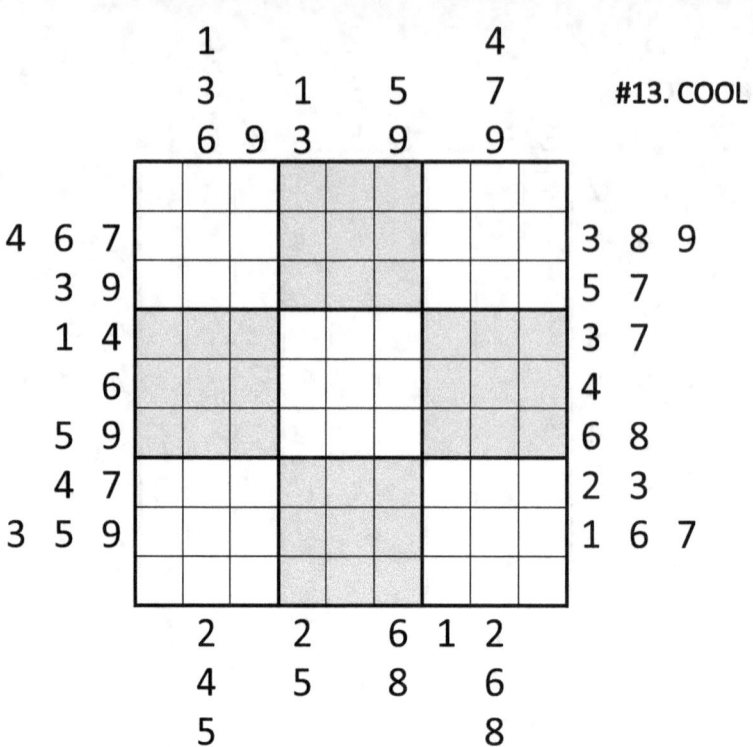

#13. COOL

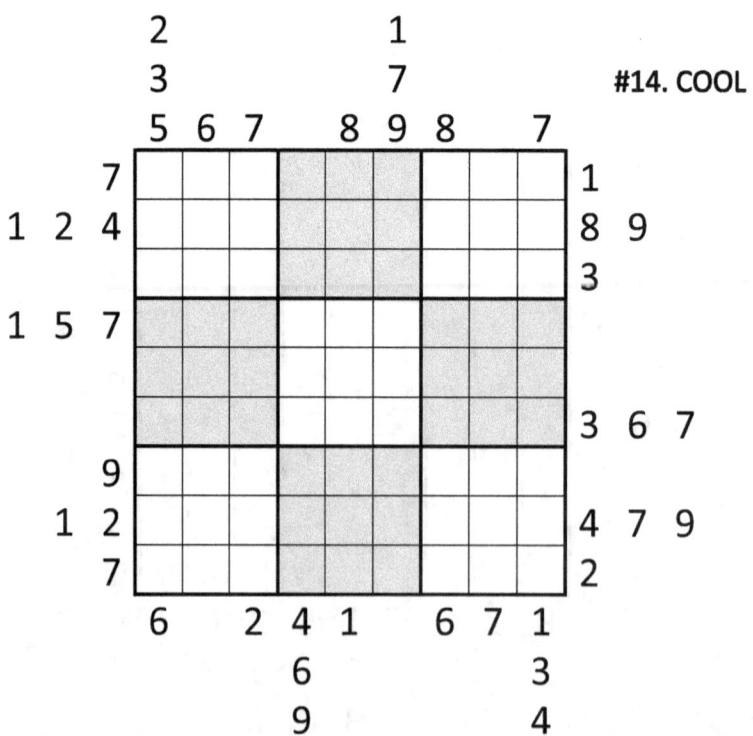

#14. COOL

#15. COOL

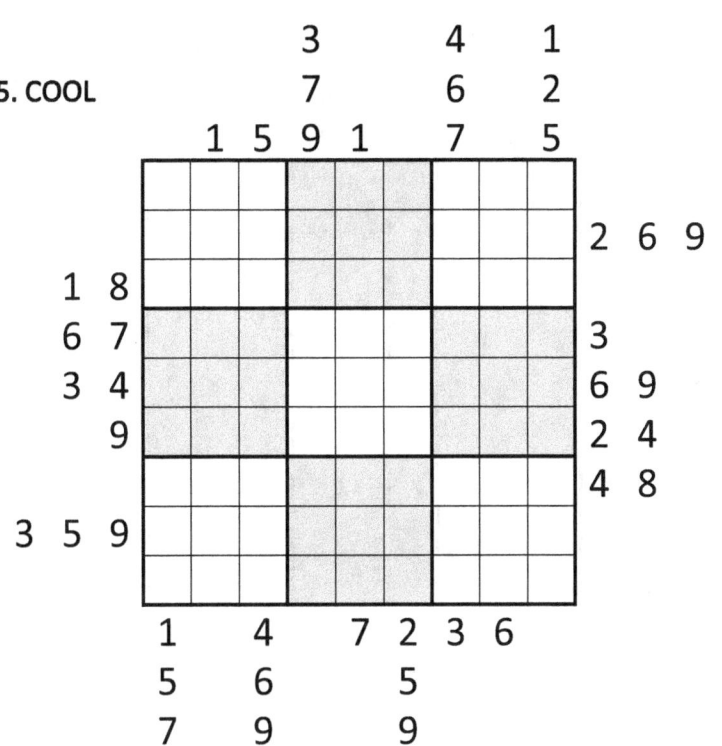

#16. COOL

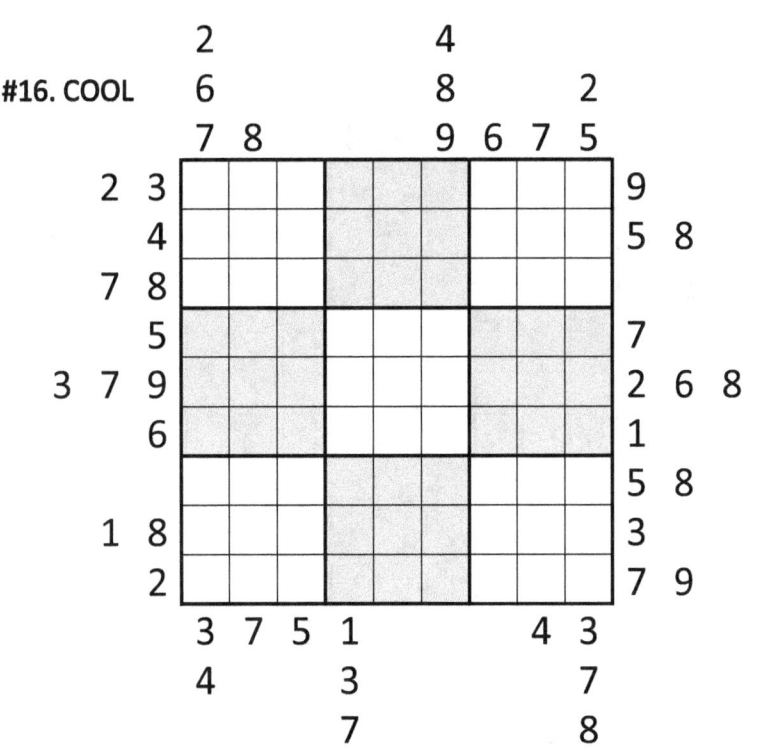

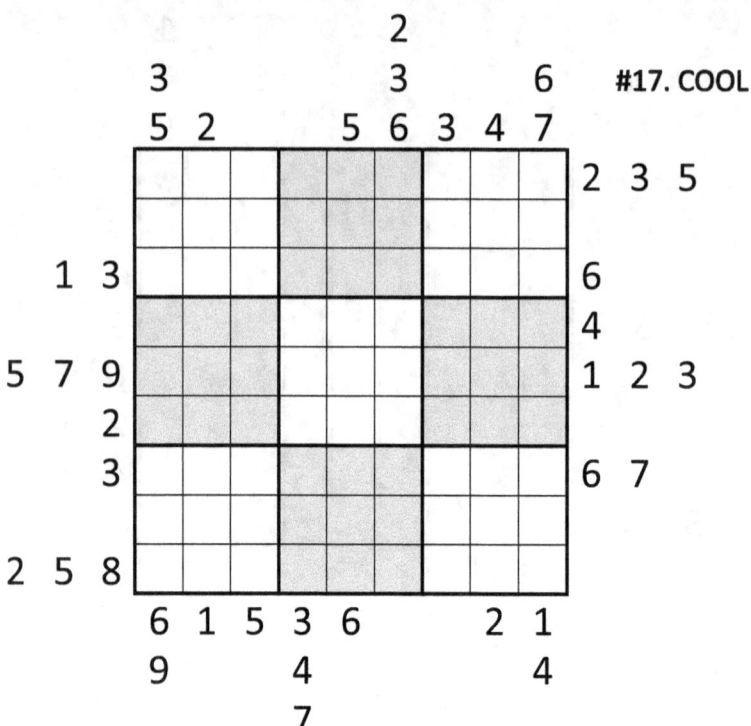

#17. COOL

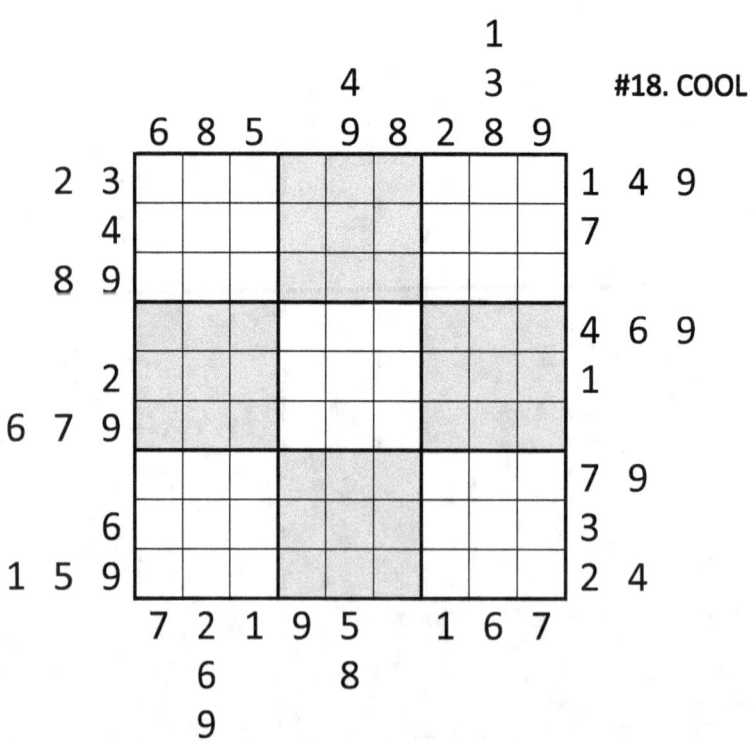

#18. COOL

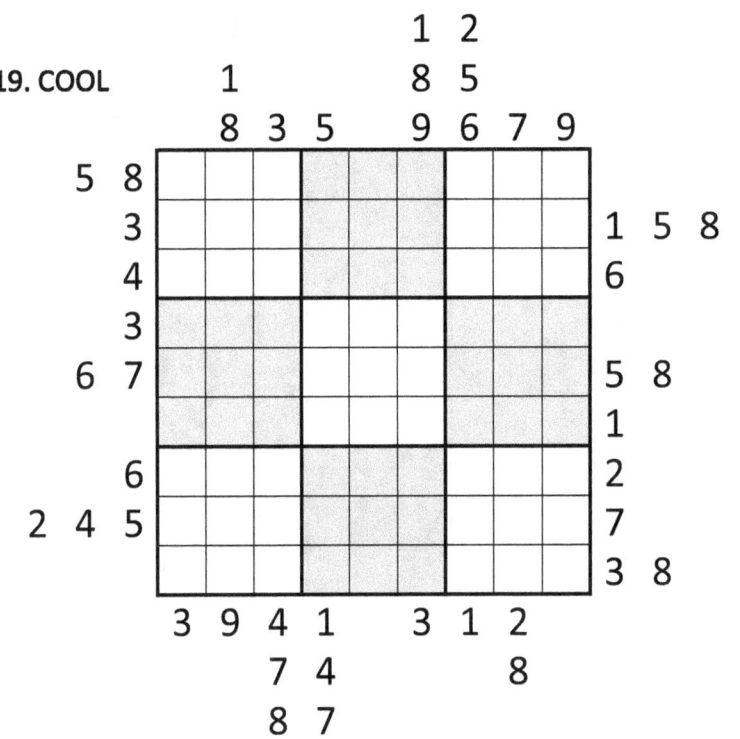

#19. COOL

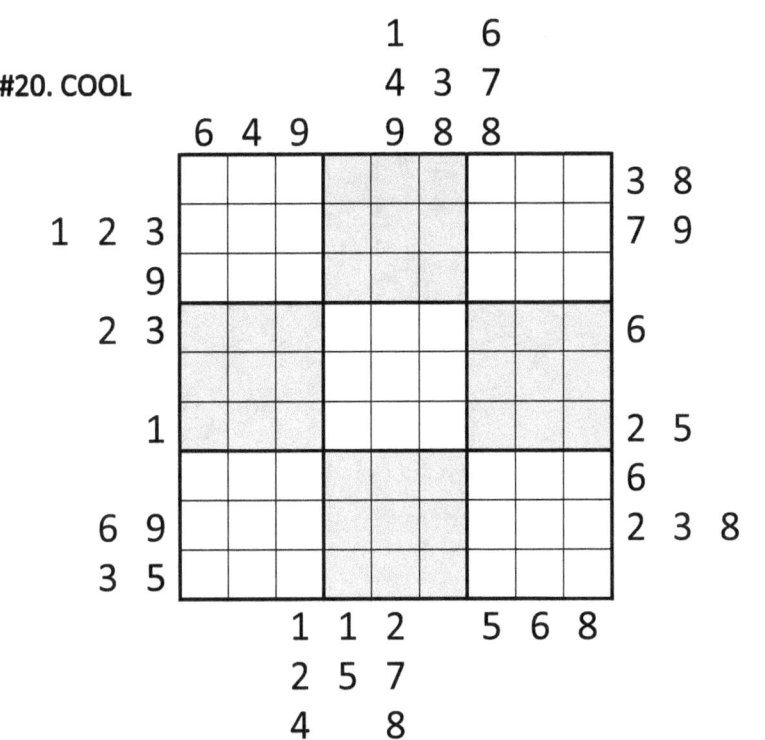

#20. COOL

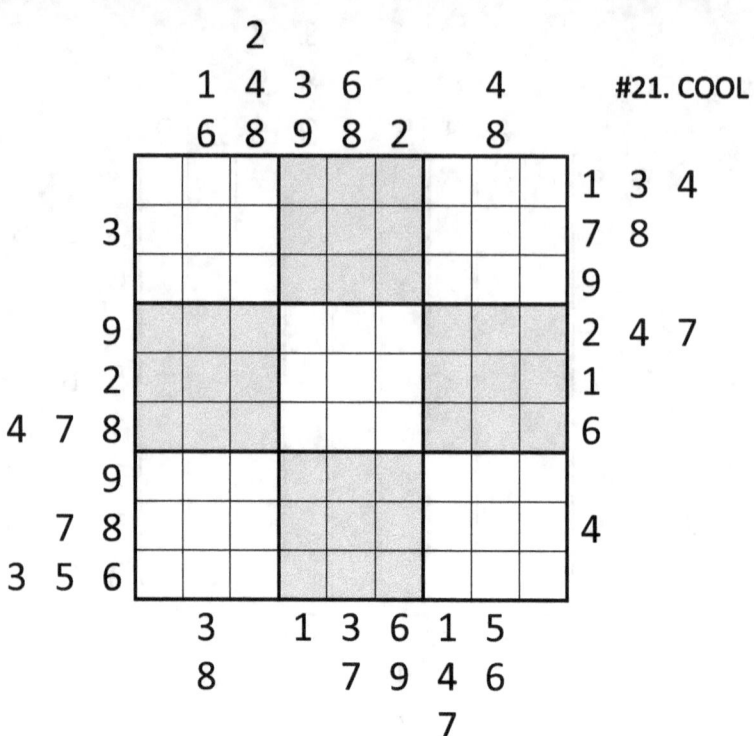

#21. COOL

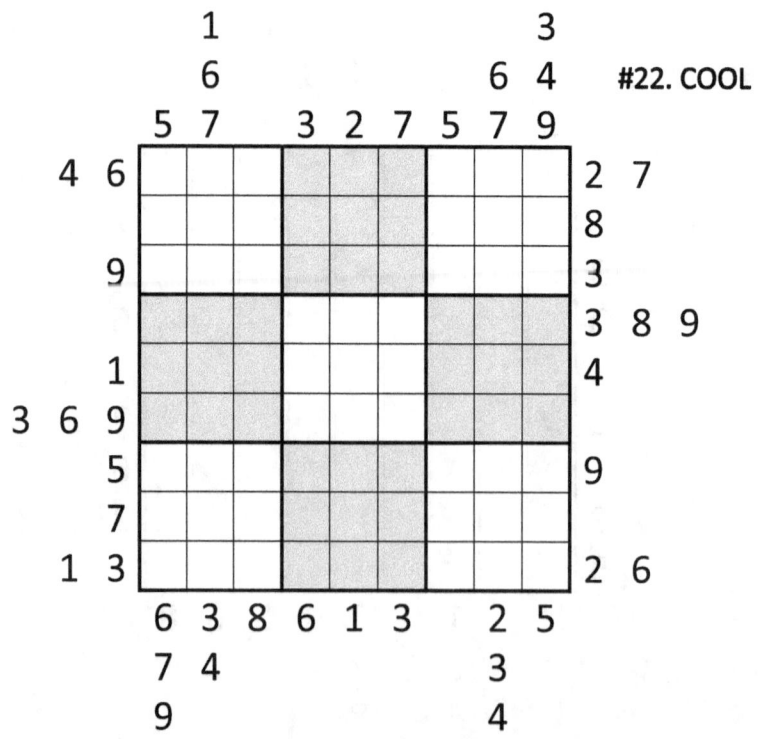

#22. COOL

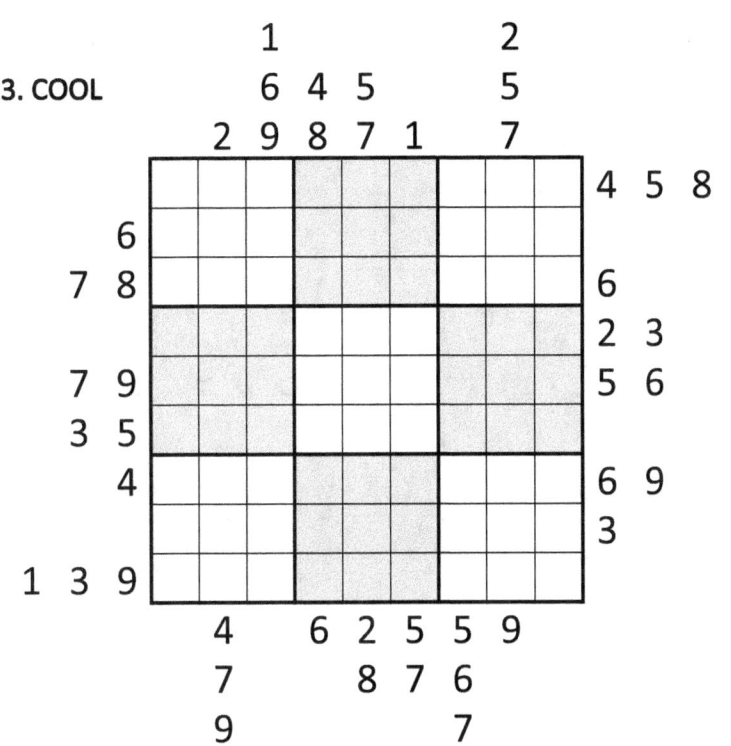

#23. COOL

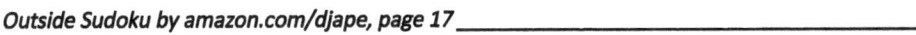

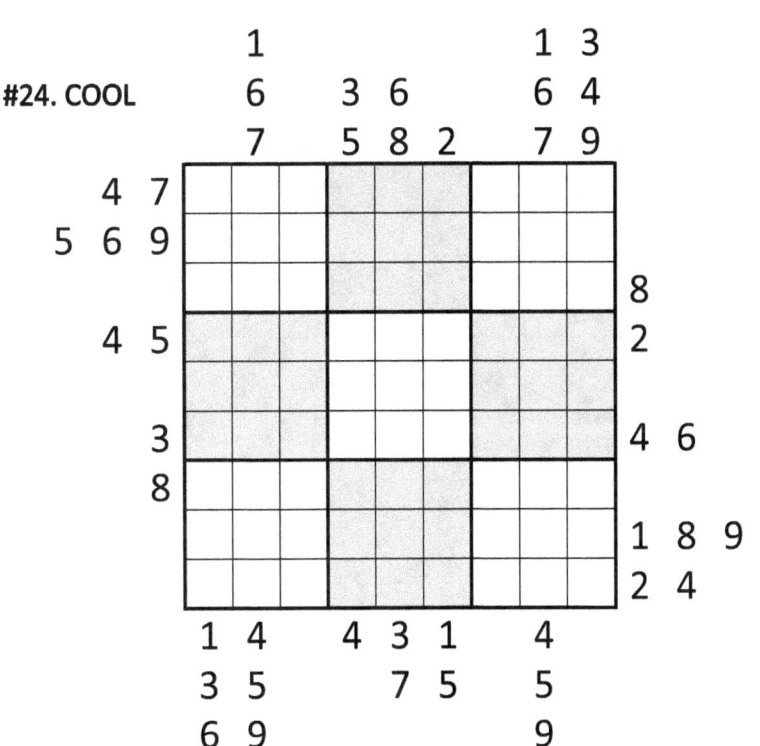

#24. COOL

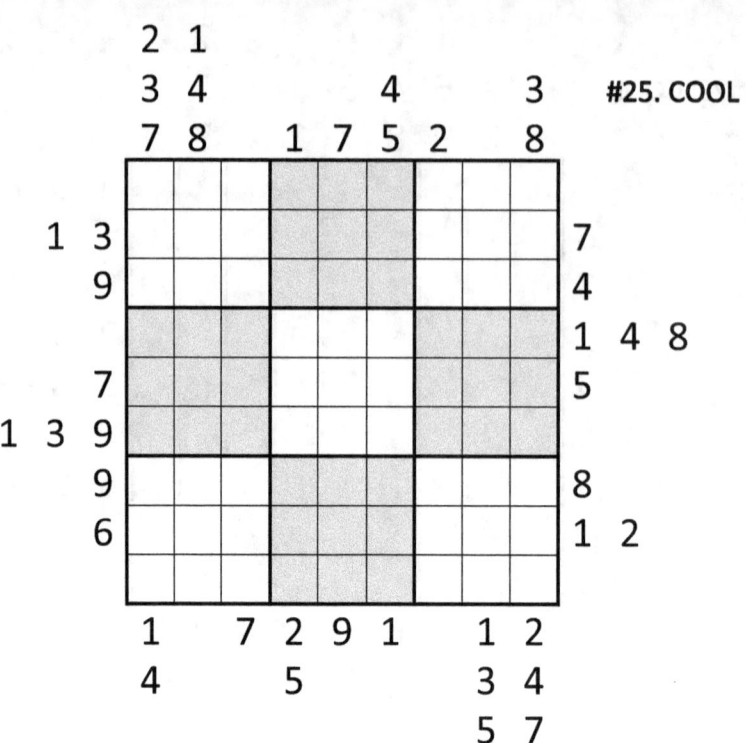

#25. COOL

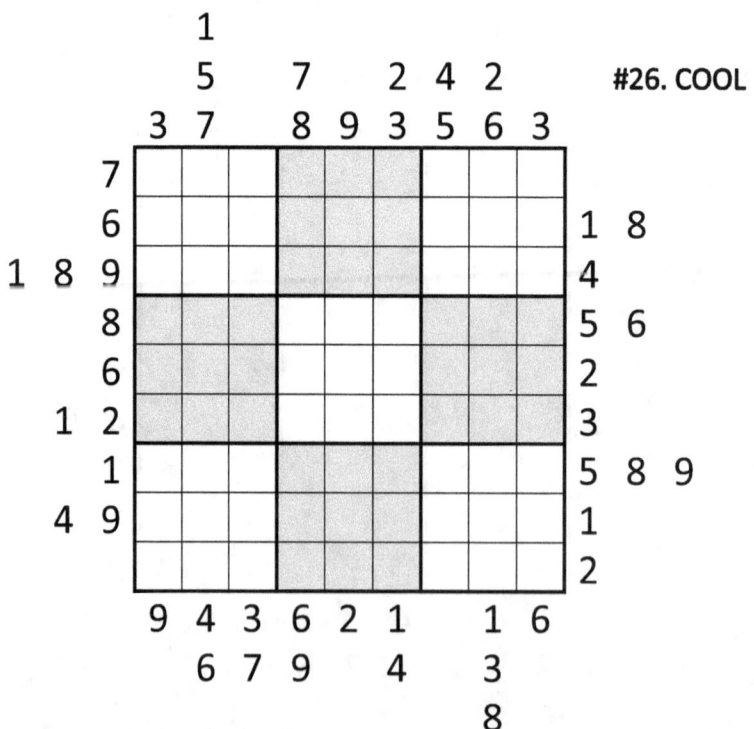

#26. COOL

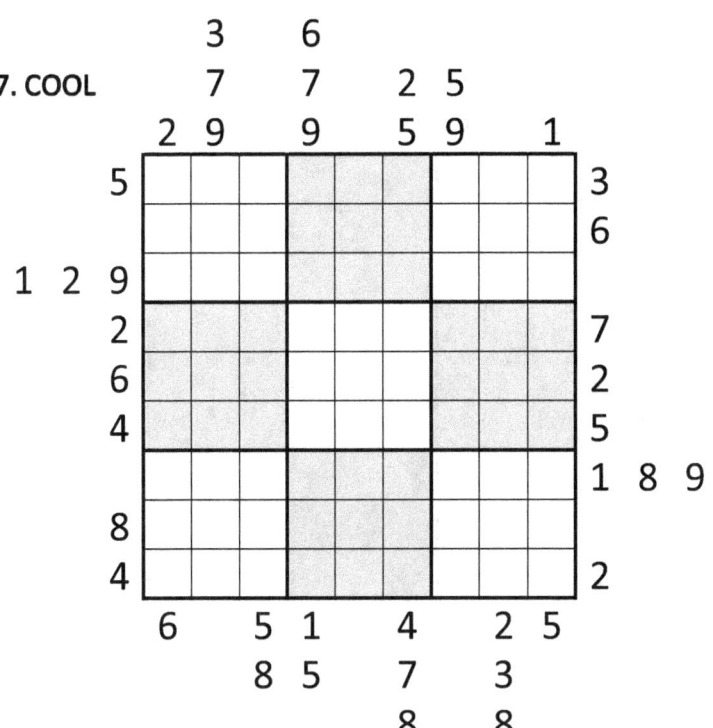

#27. COOL

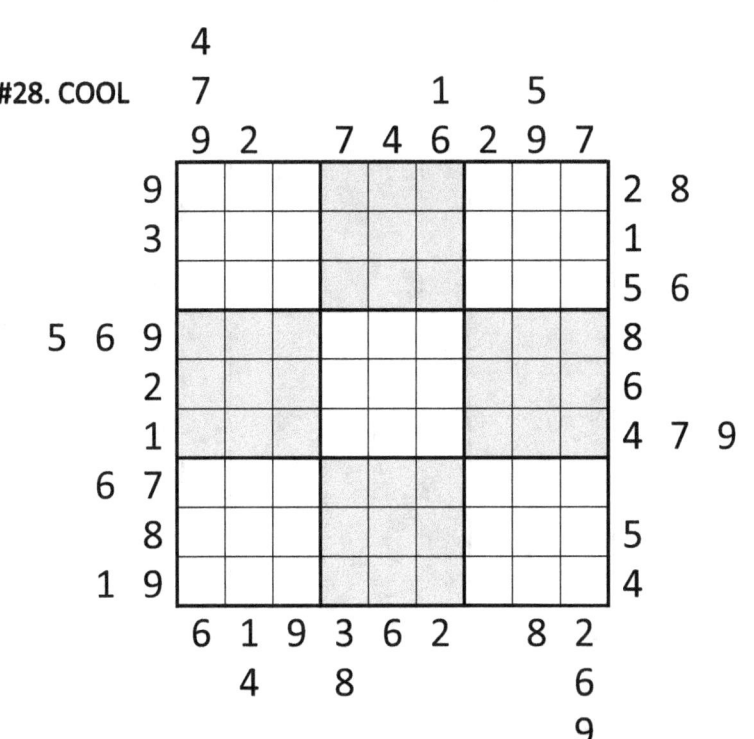

#28. COOL

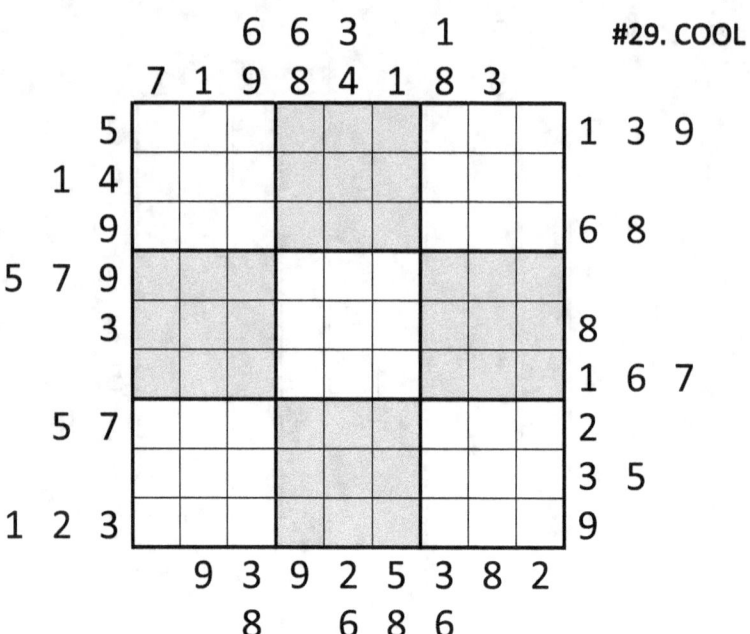

#29. COOL

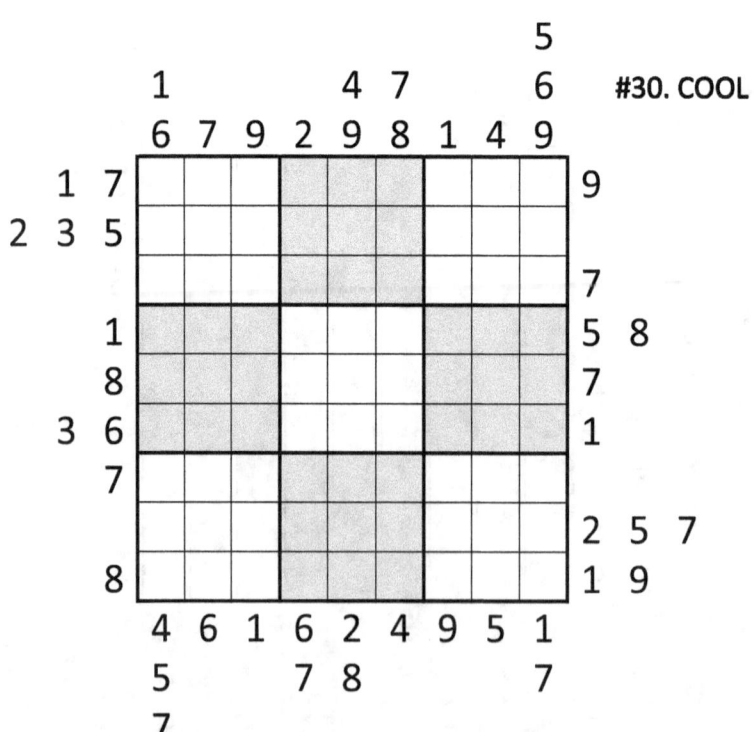

#30. COOL

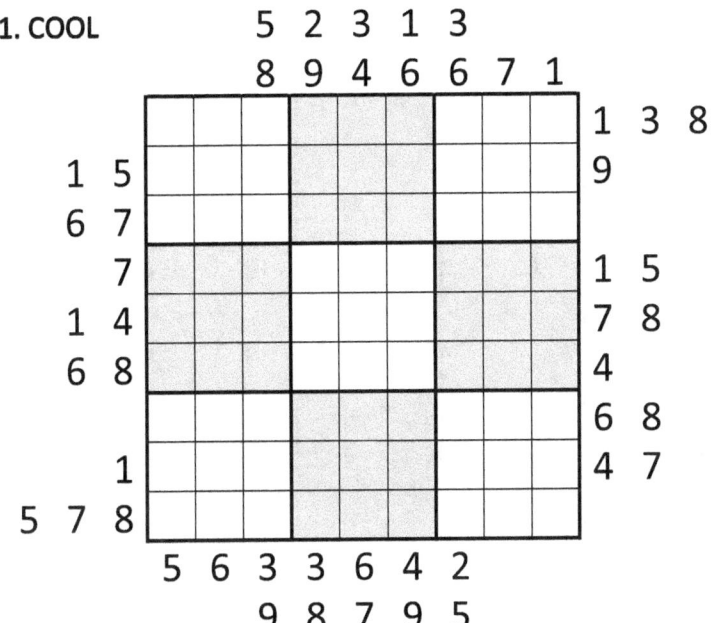

#32. COOL

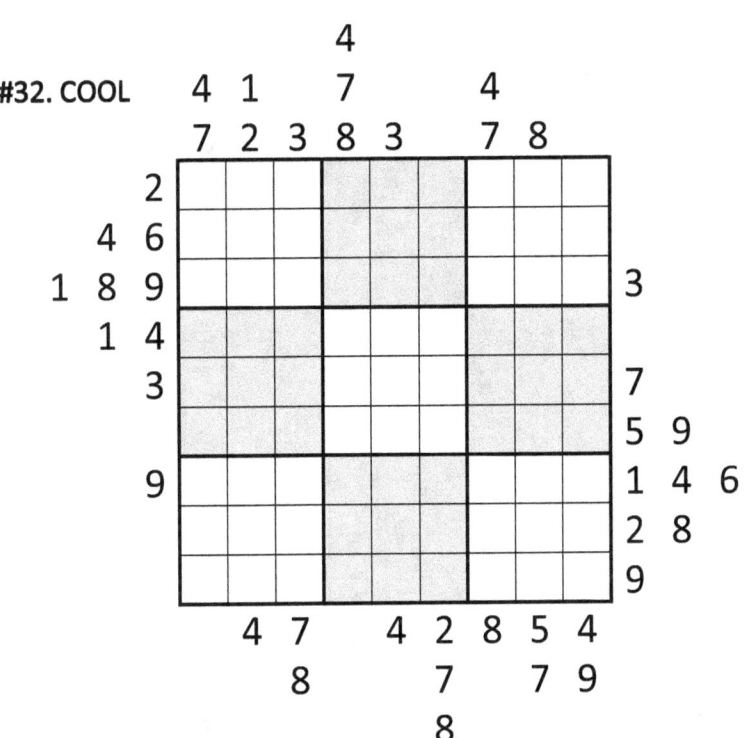

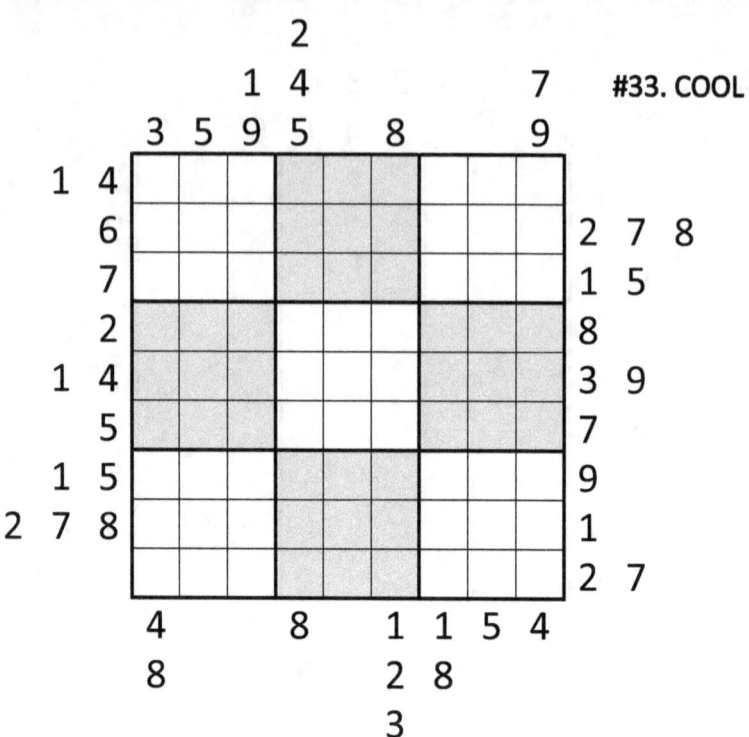

#33. COOL

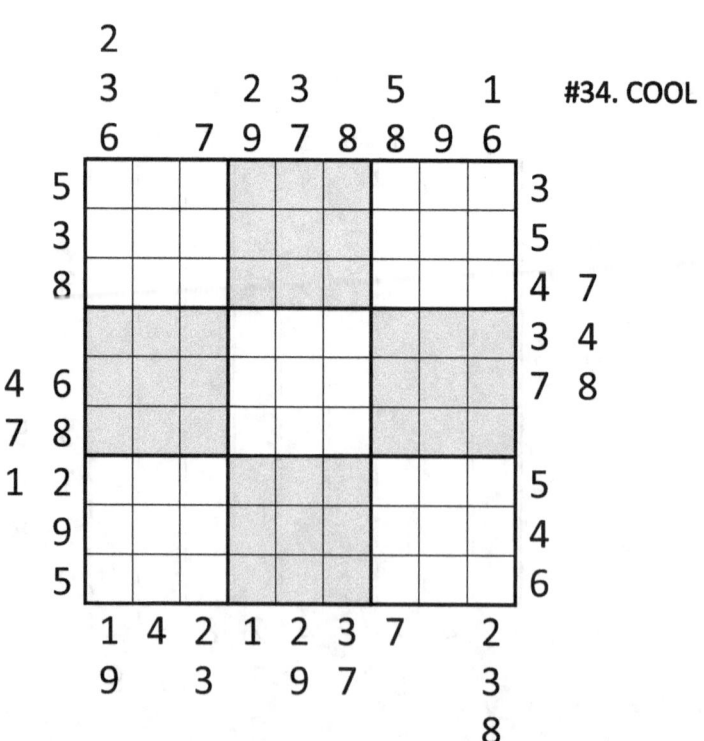

#34. COOL

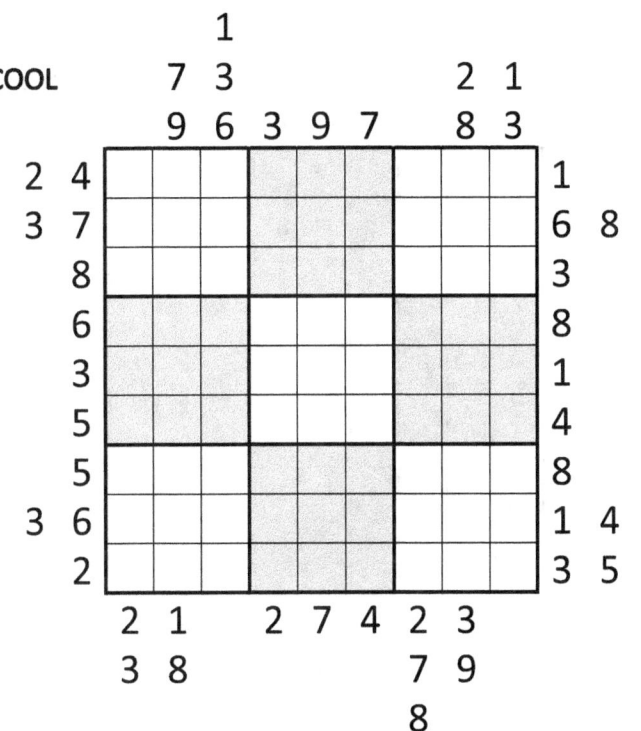

#35. COOL

#36. THINKER

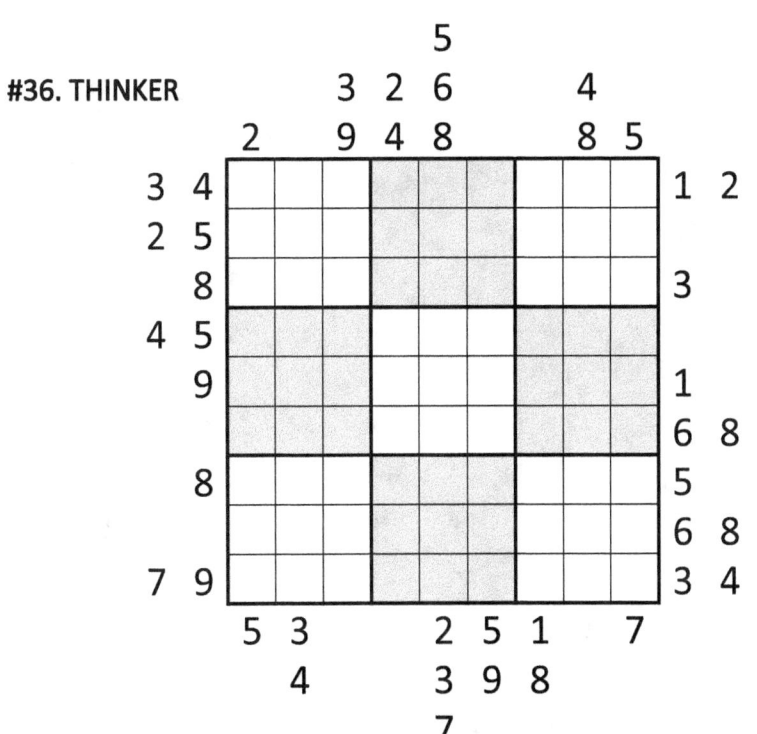

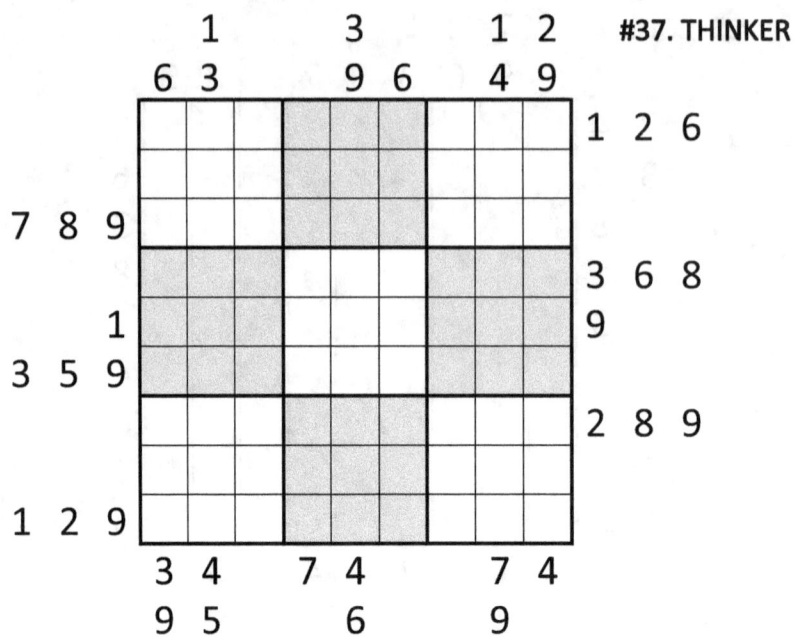

#37. THINKER

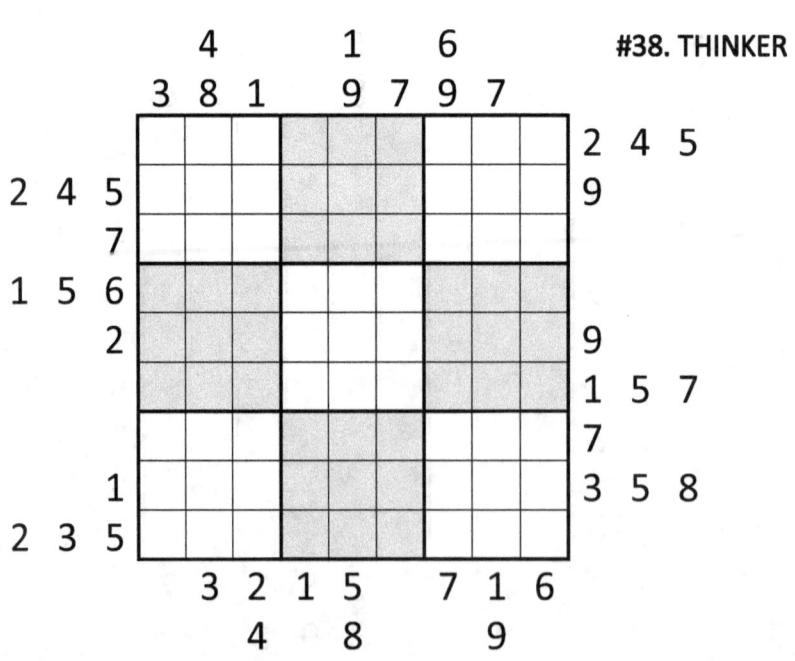

#38. THINKER

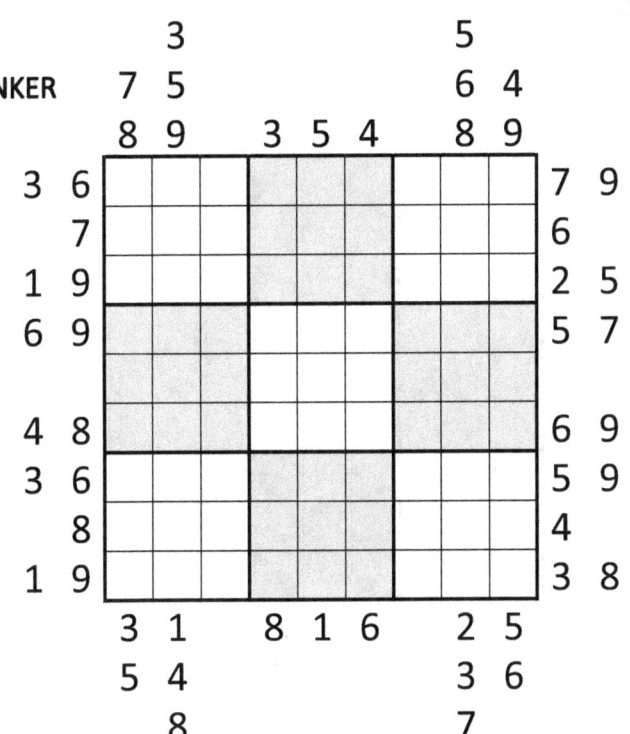

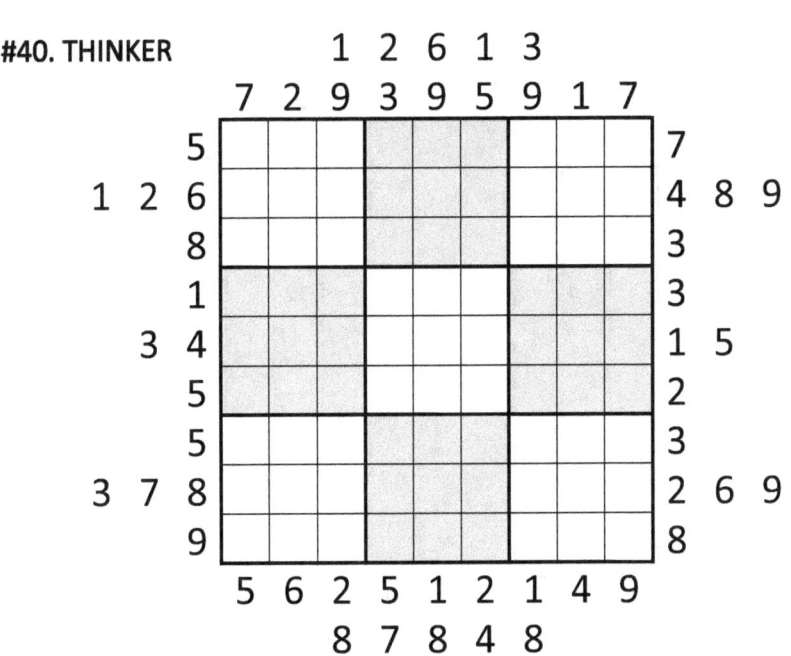

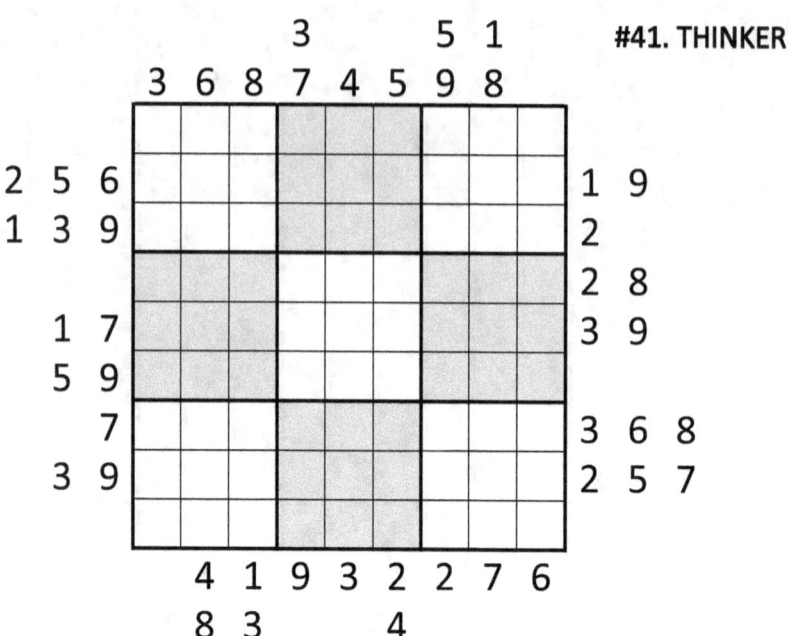

#41. THINKER

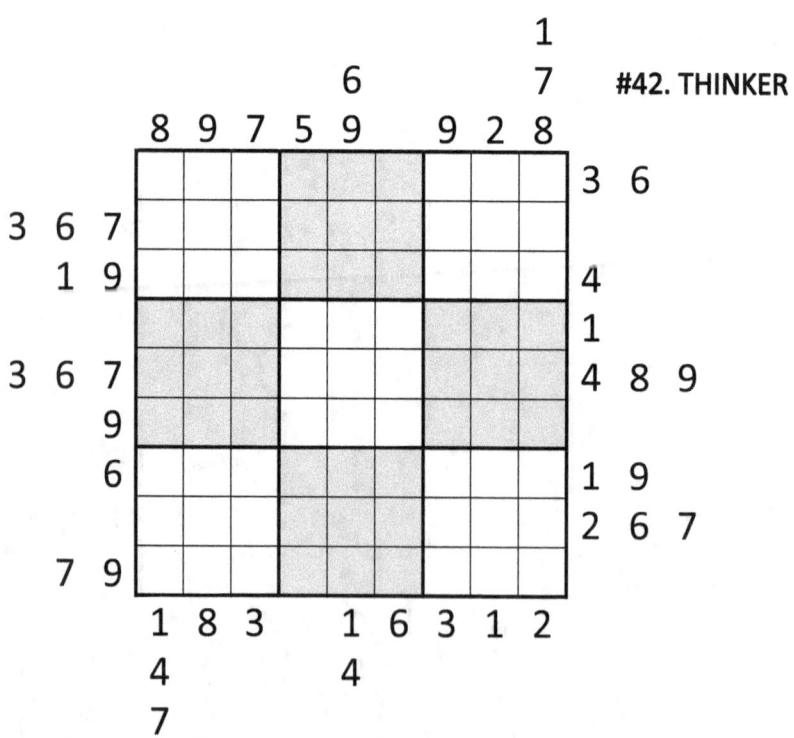

#42. THINKER

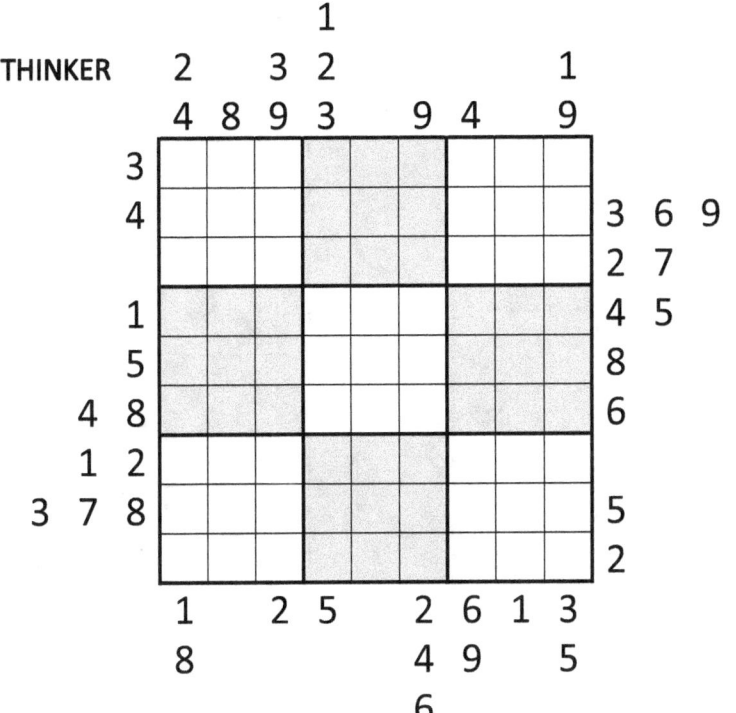

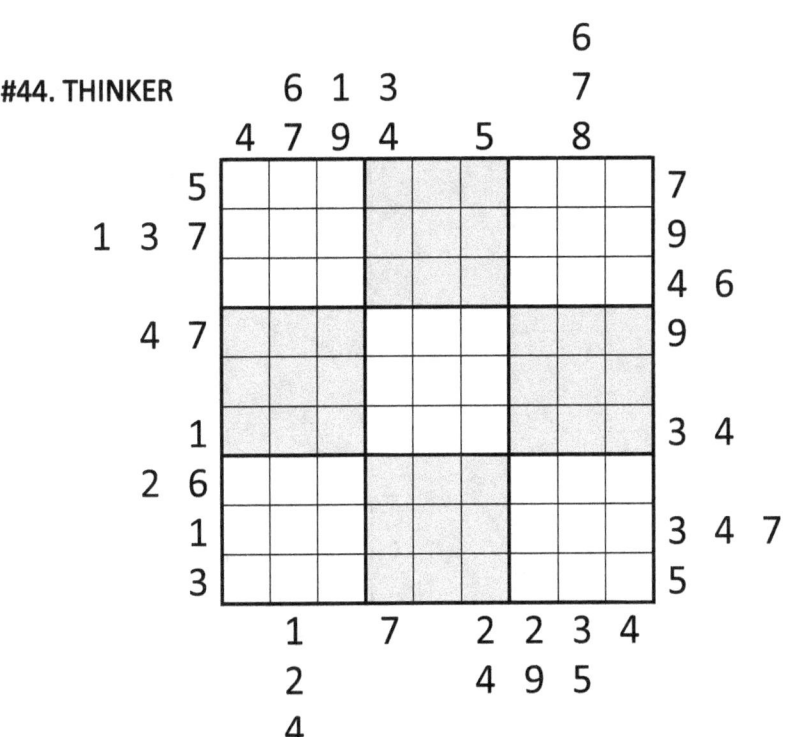

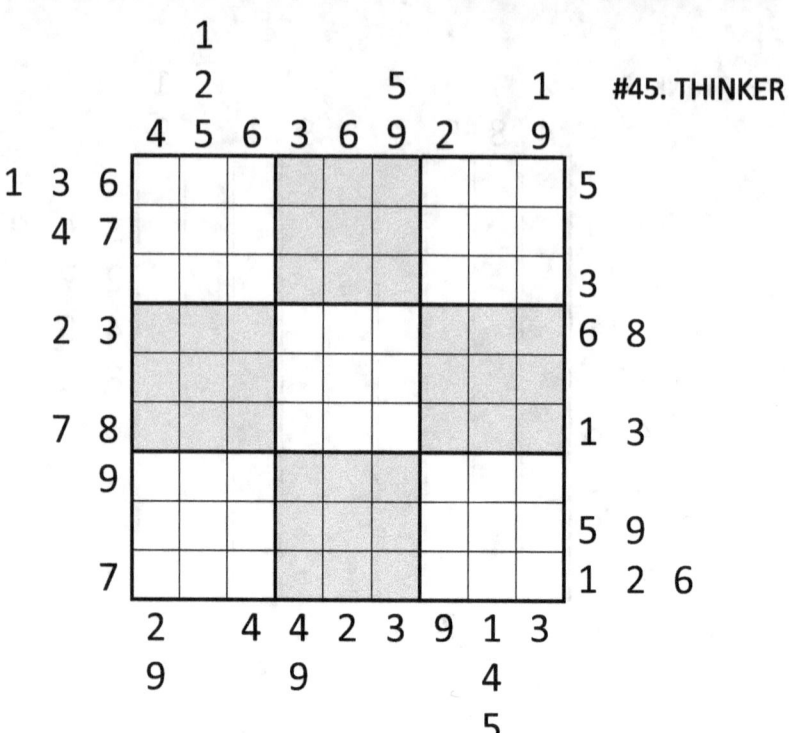

#45. THINKER

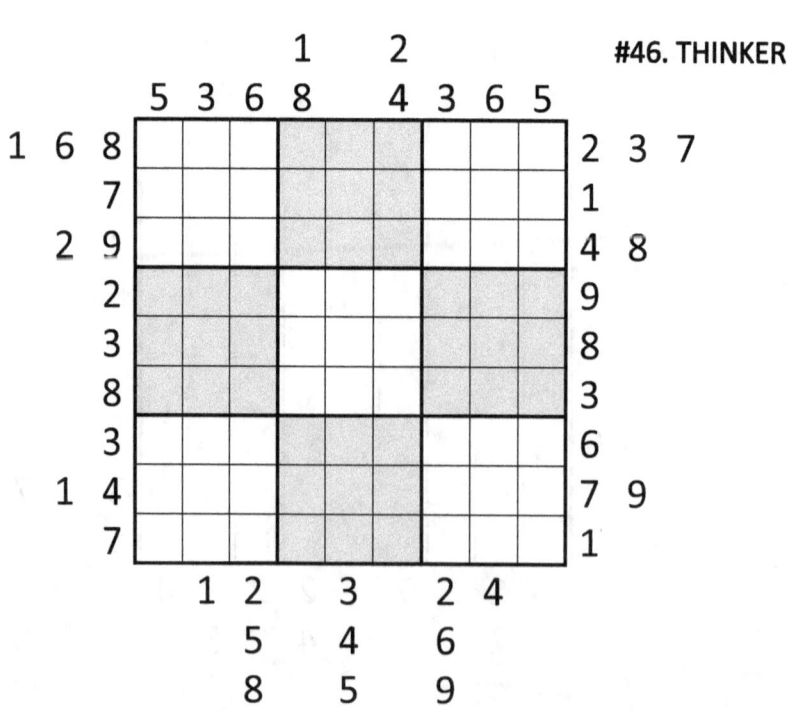

#46. THINKER

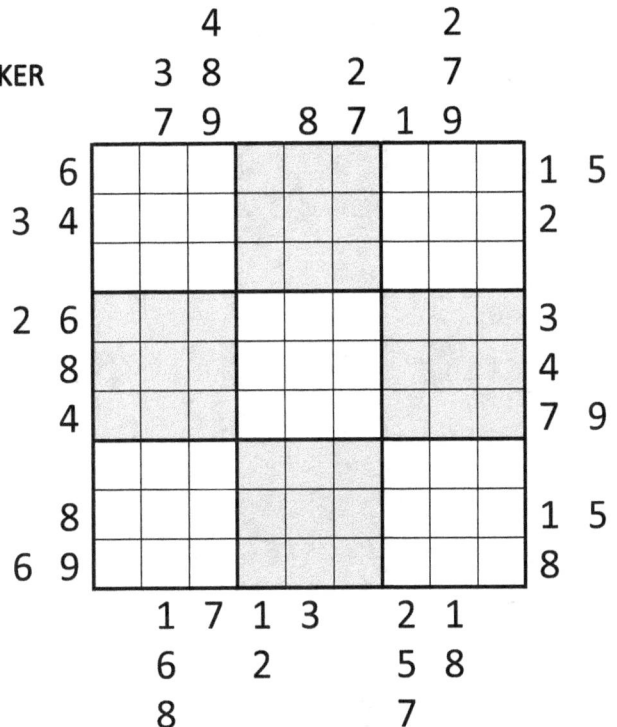

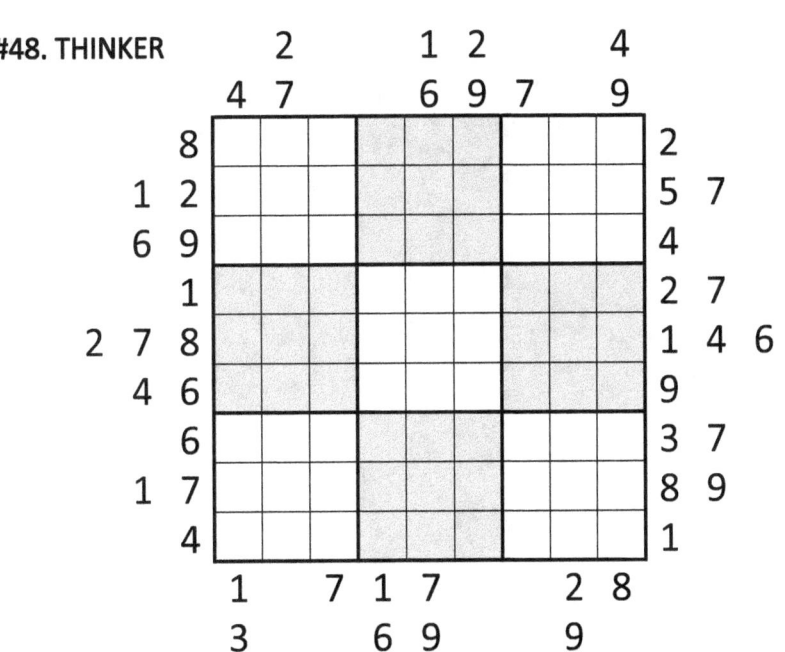

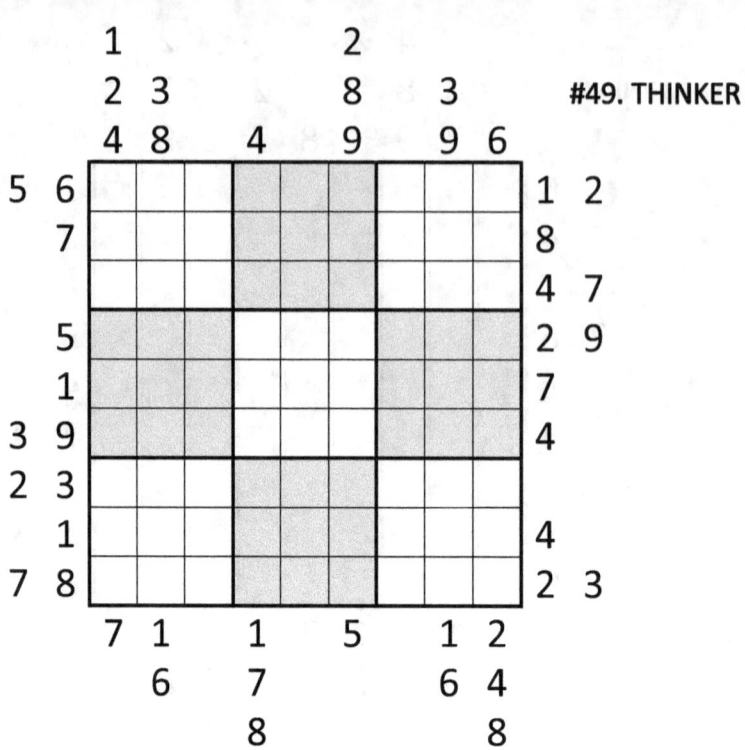

#49. THINKER

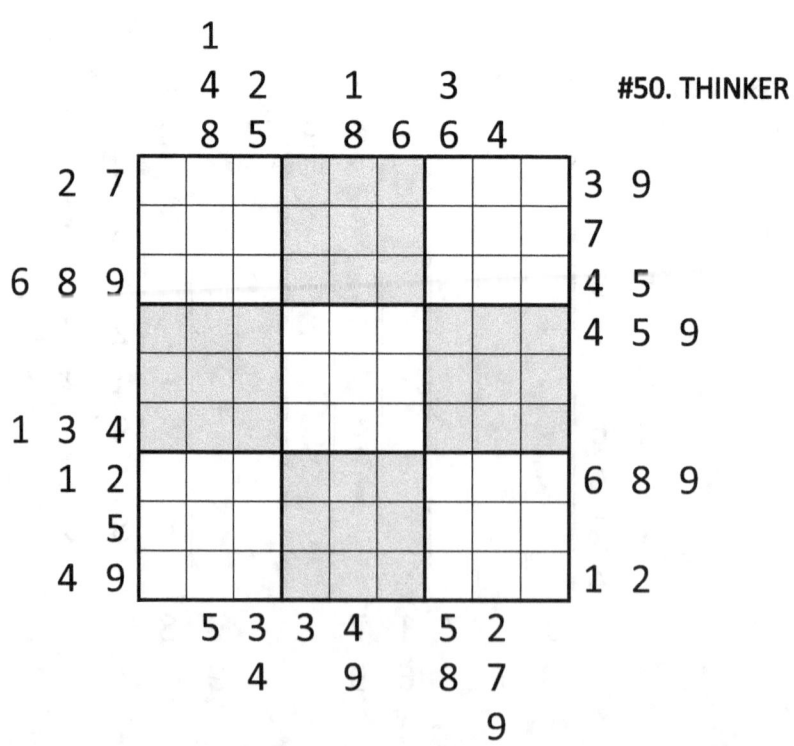

#50. THINKER

#51. THINKER

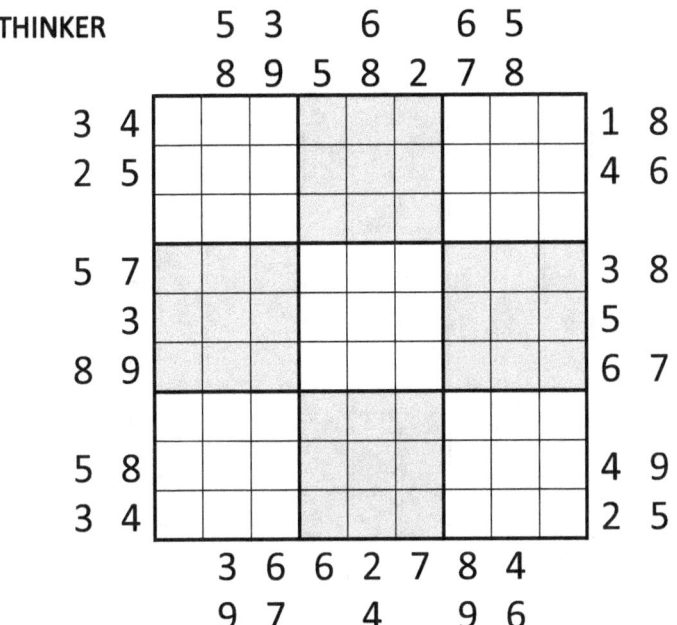

#52. THINKER

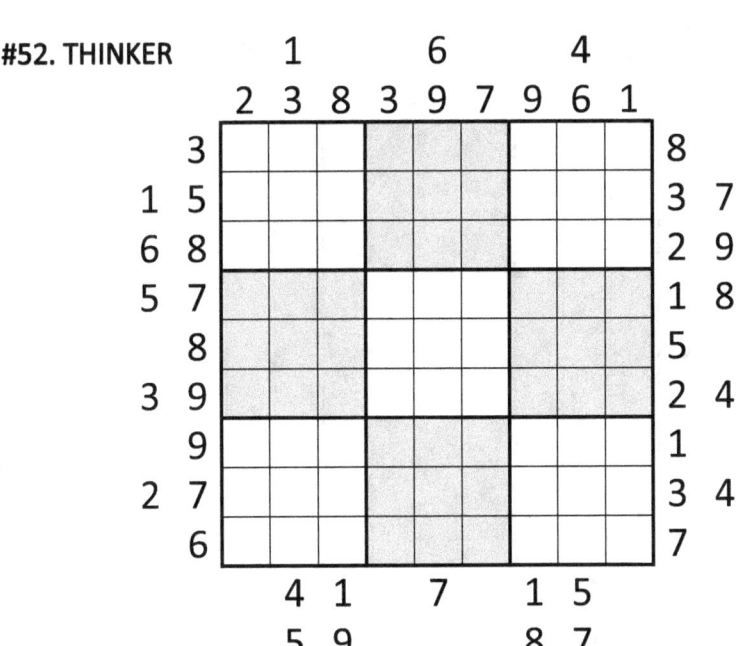

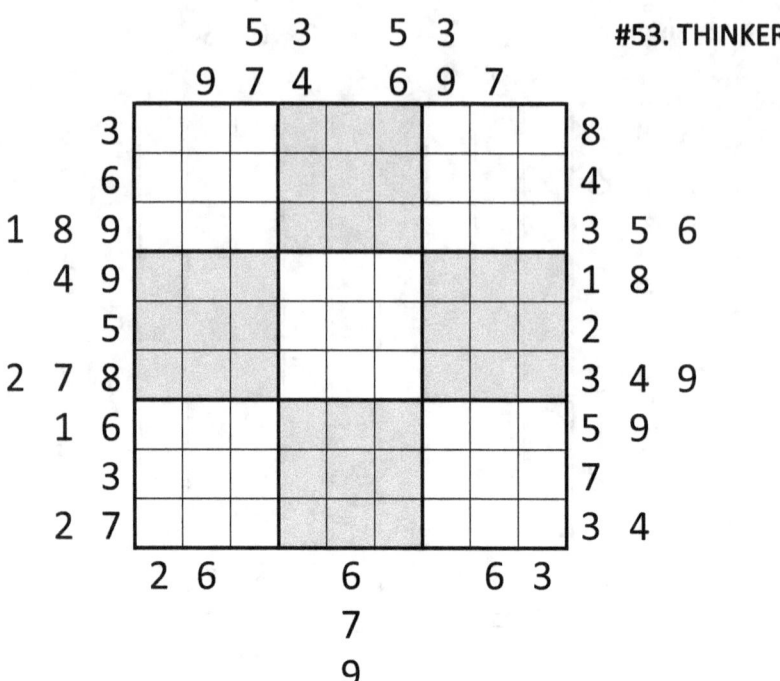

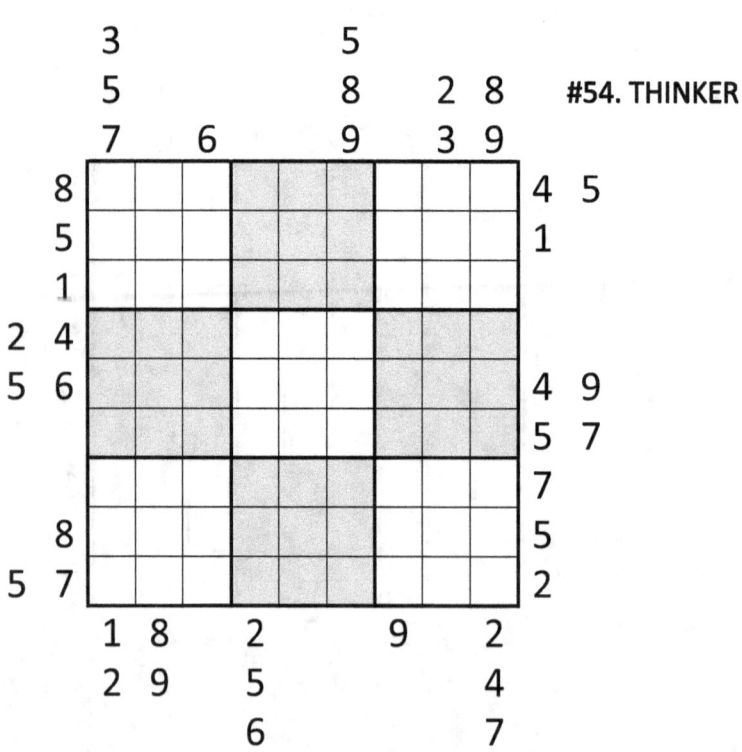

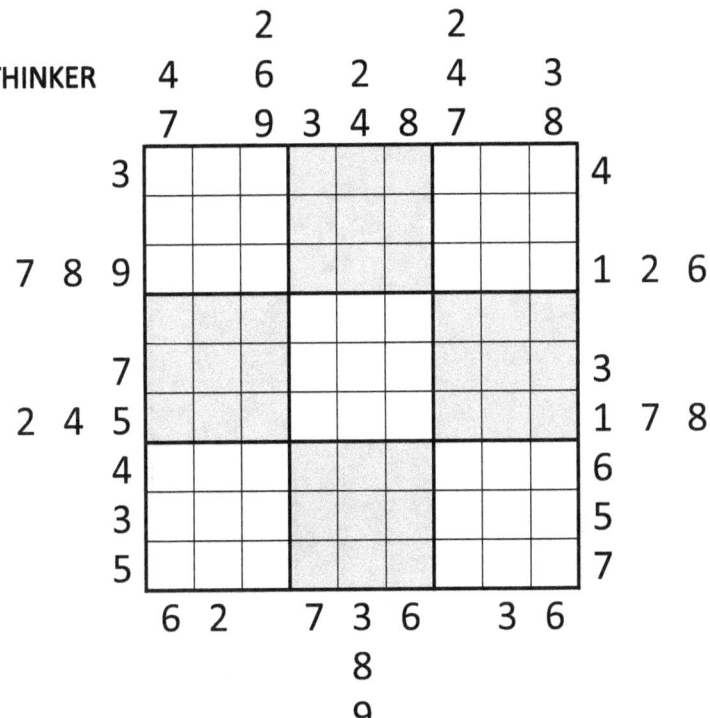

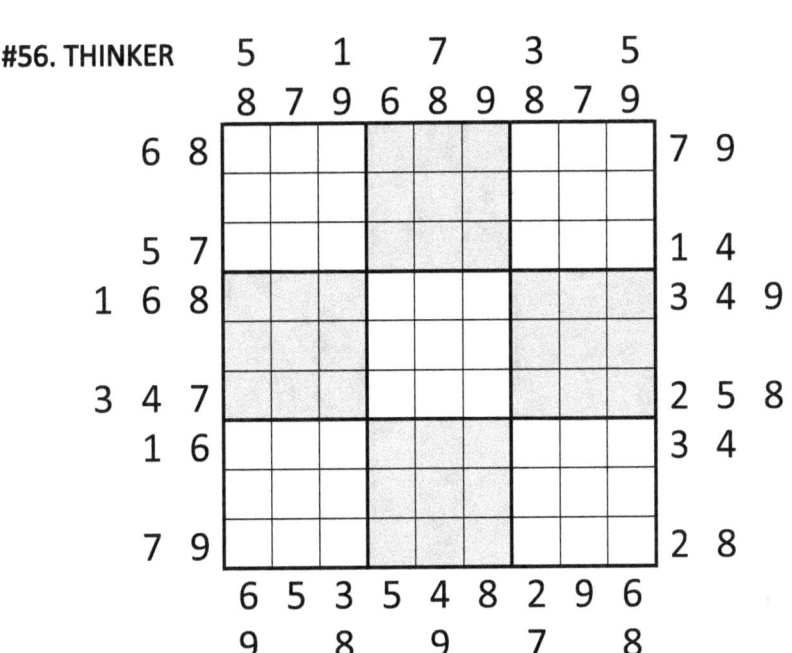

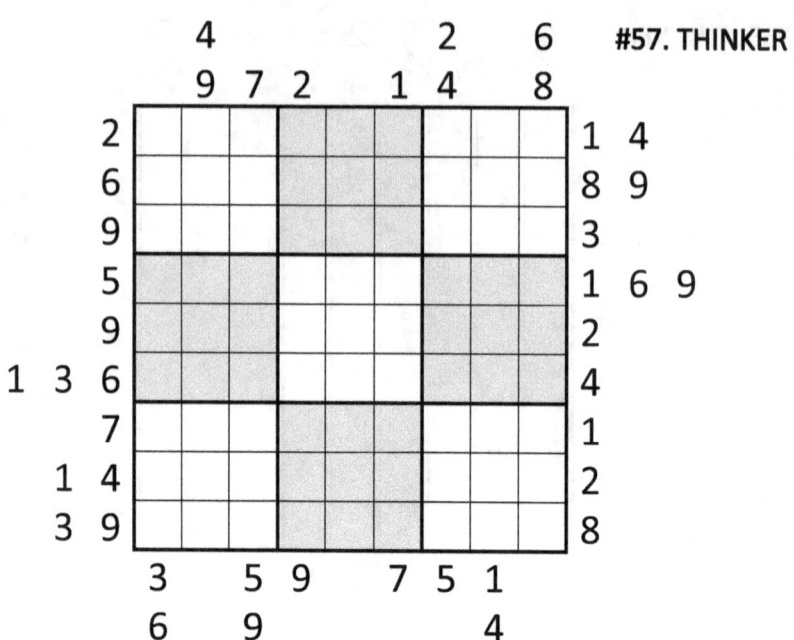

#57. THINKER

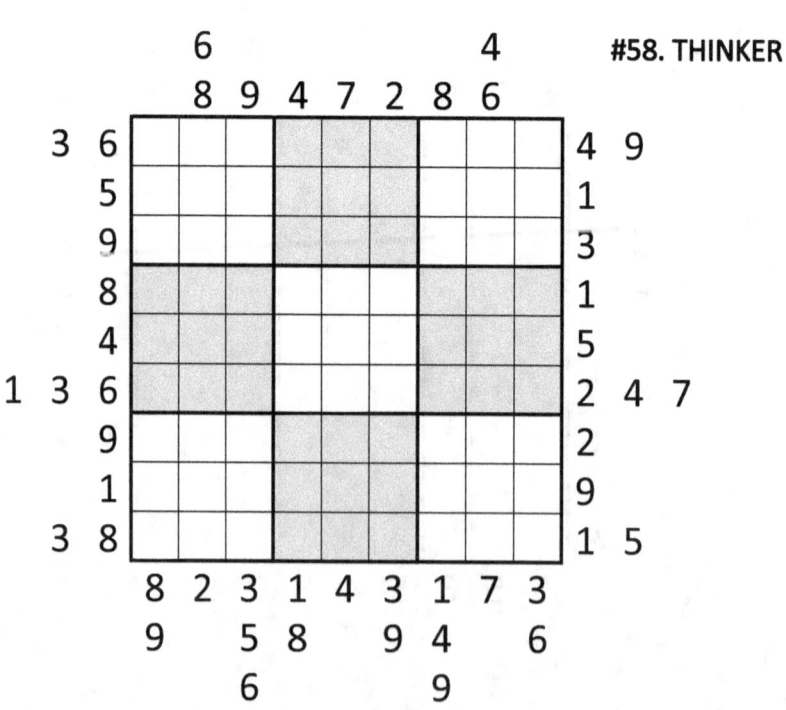

#58. THINKER

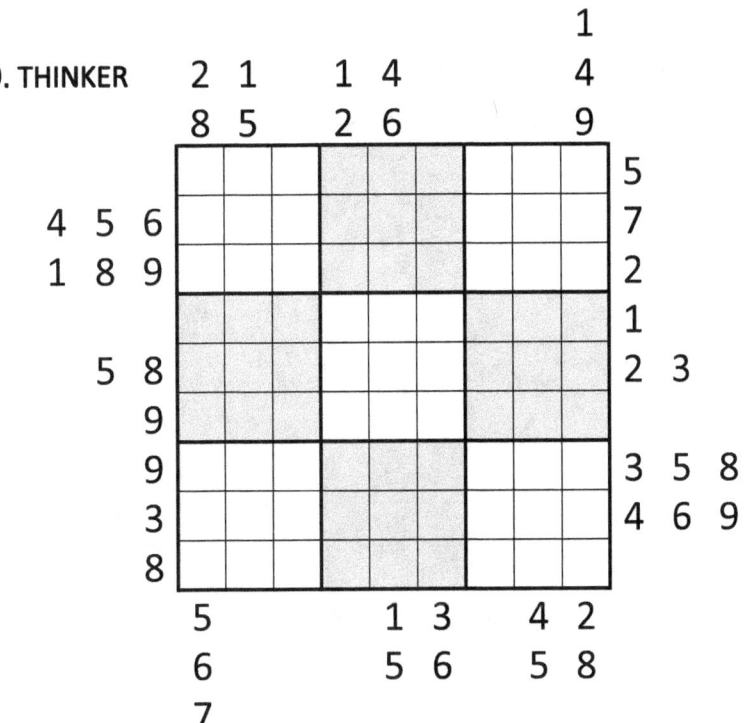

#59. THINKER

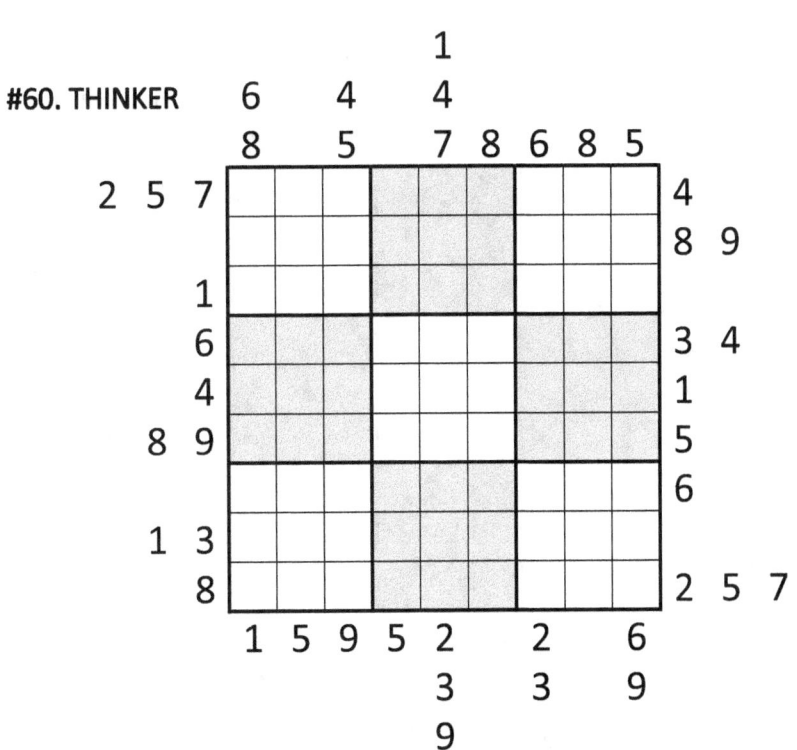

#60. THINKER

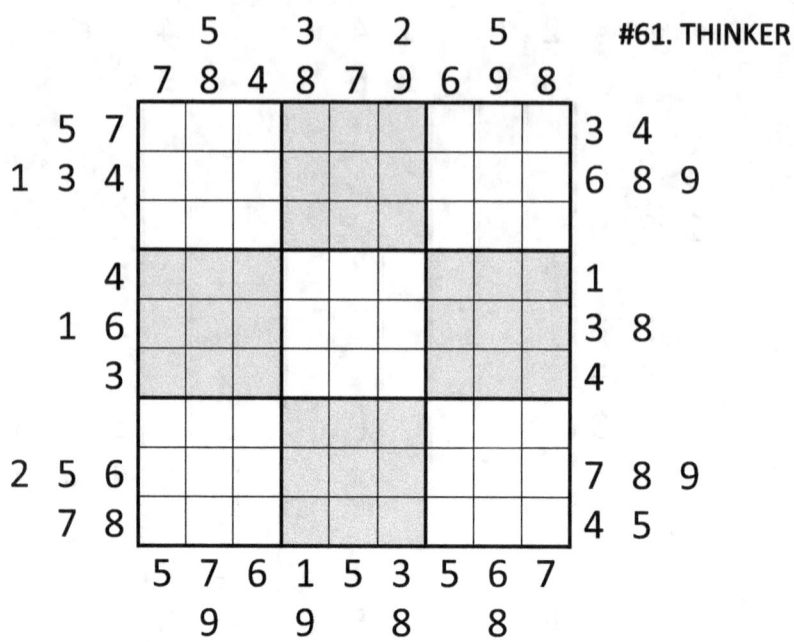

#61. THINKER

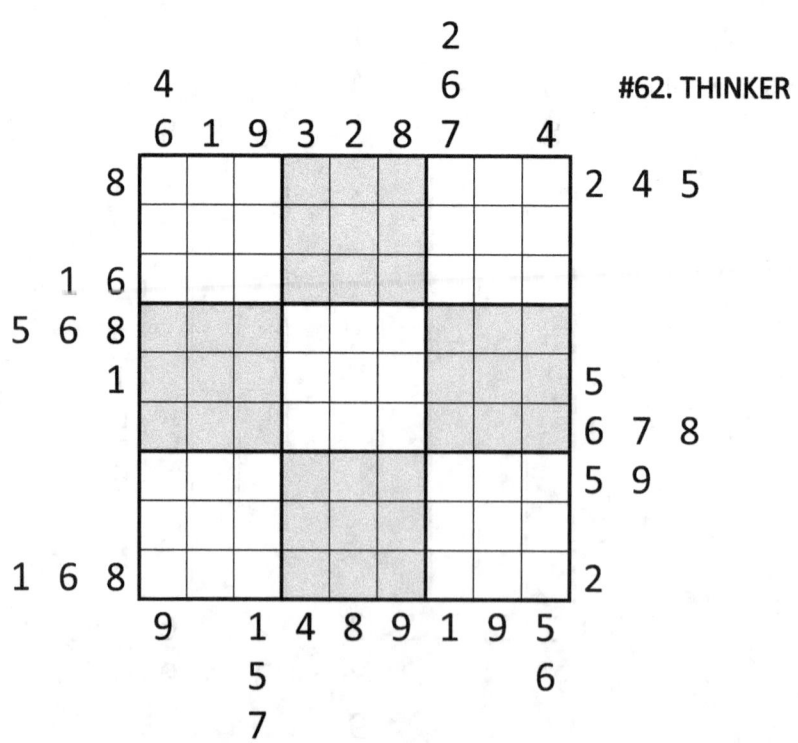

#62. THINKER

#63. THINKER

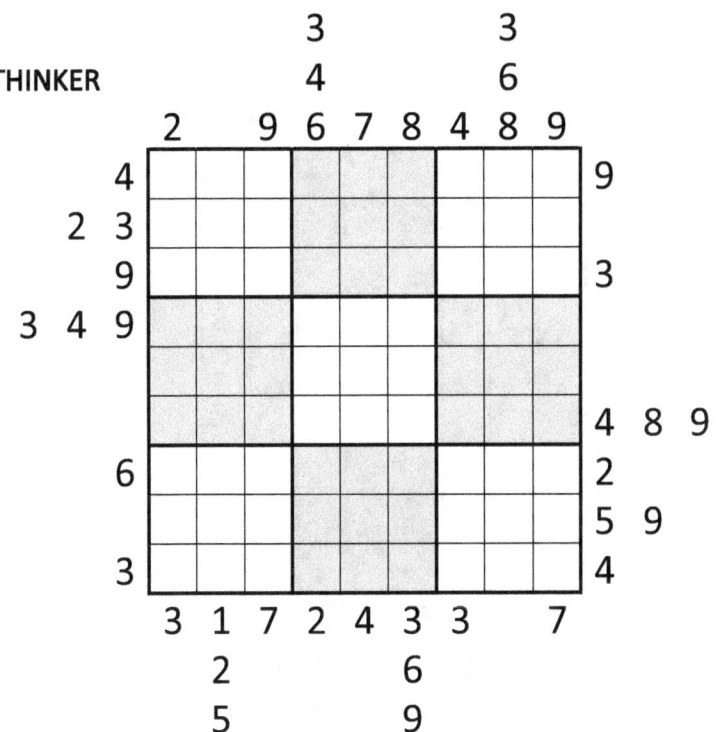

#64. THINKER

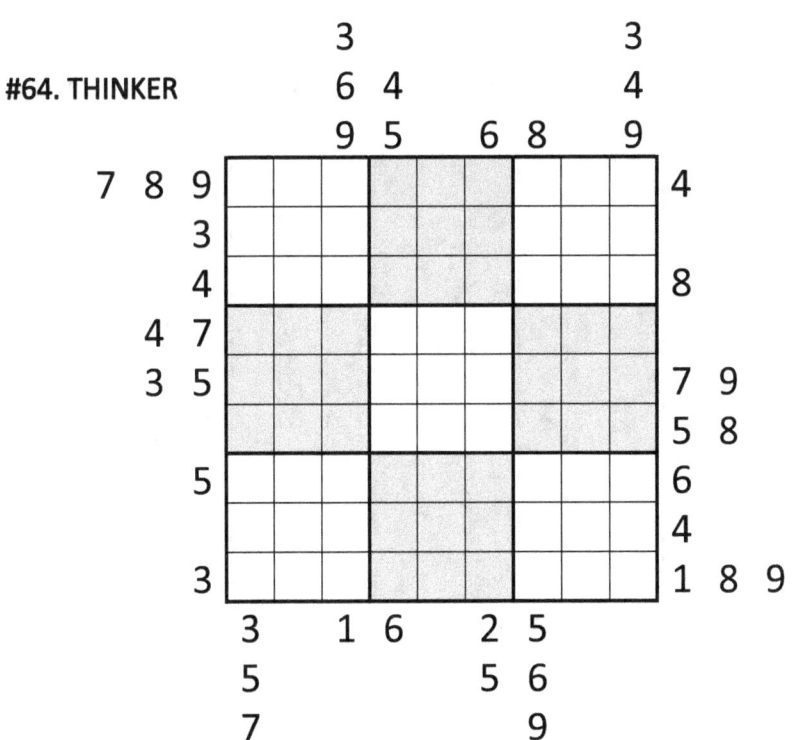

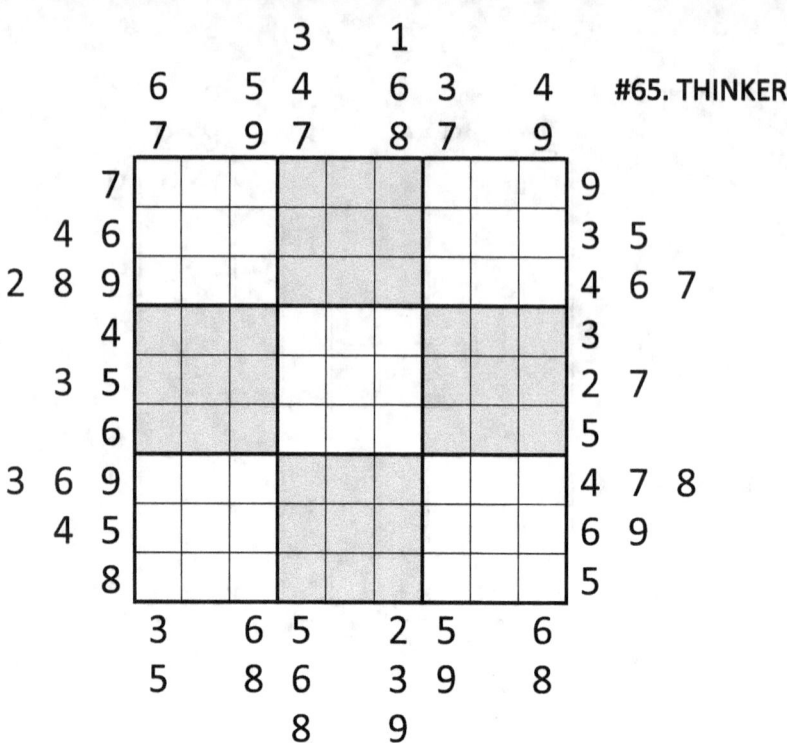

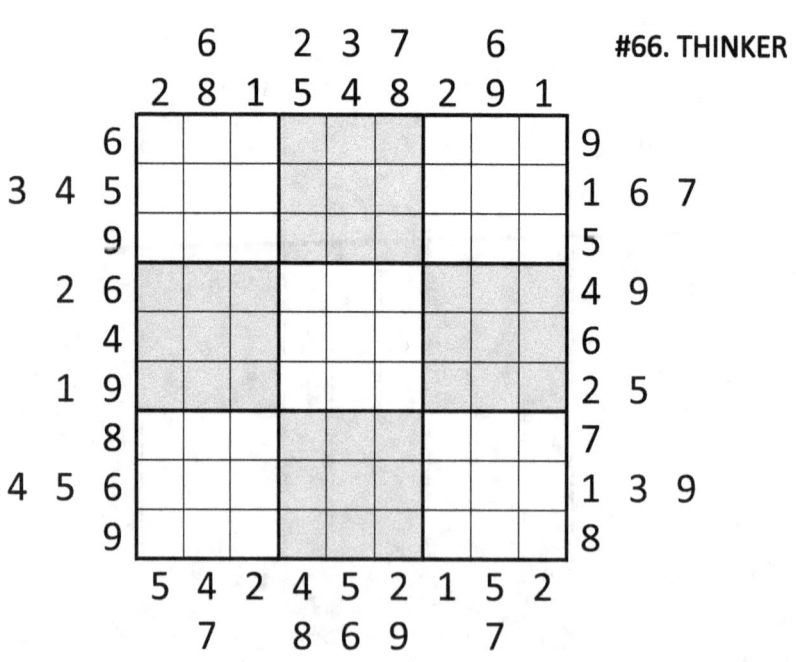

#67. THINKER

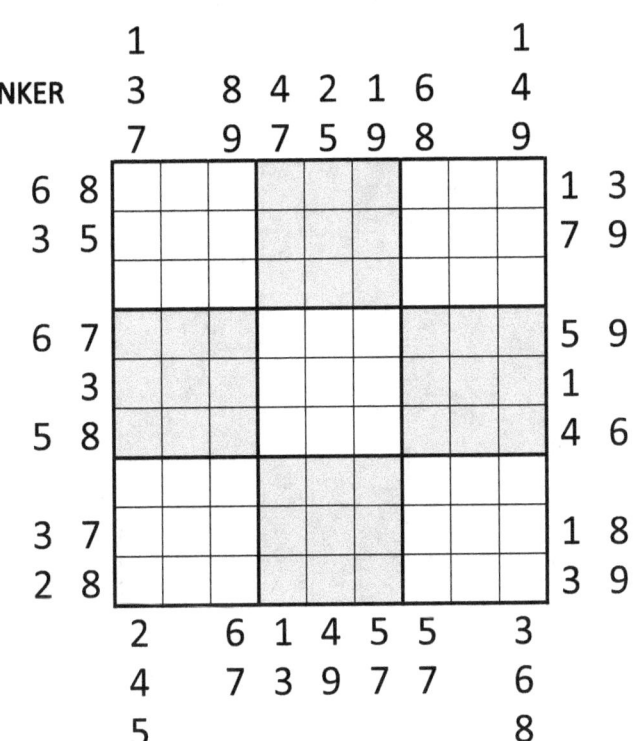

#68. BRAIN

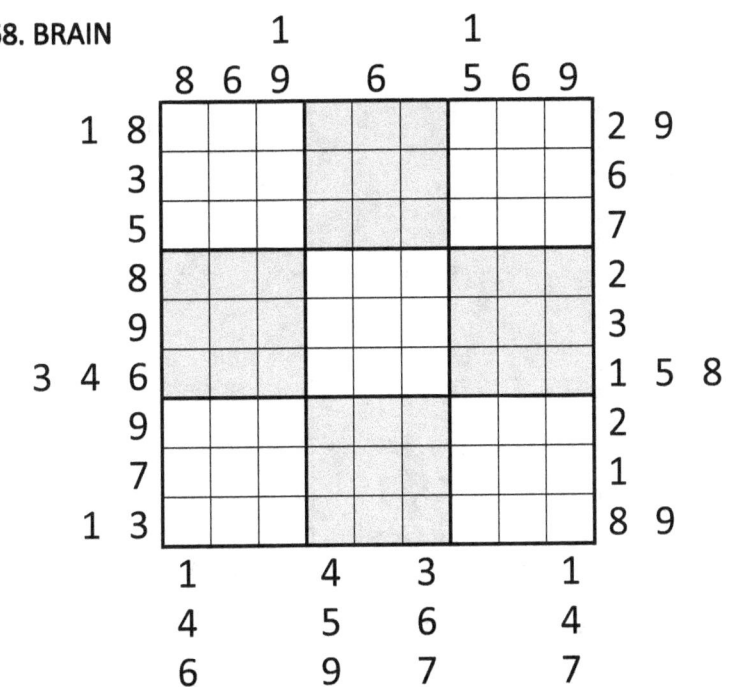

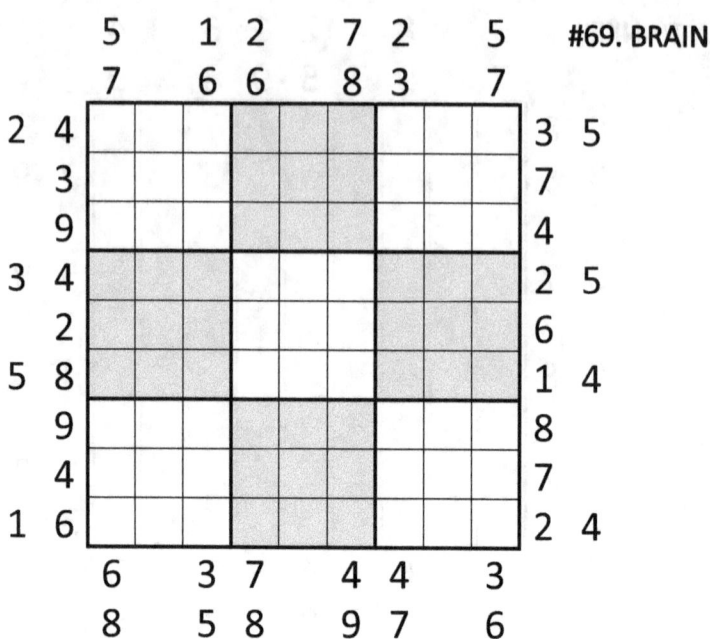

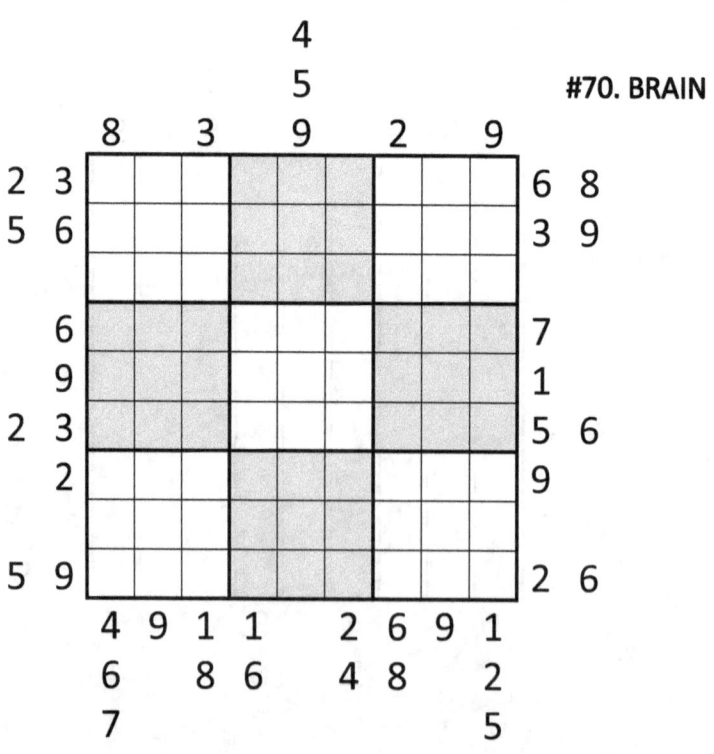

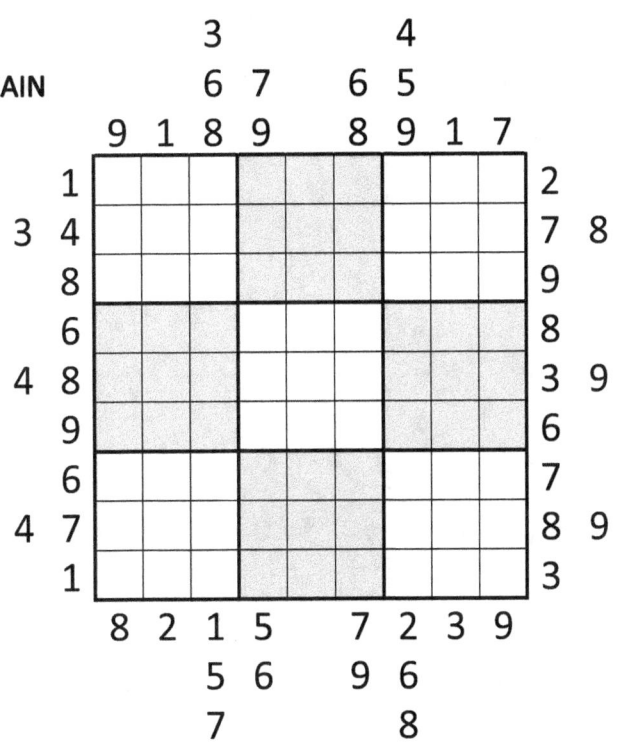

#71. BRAIN

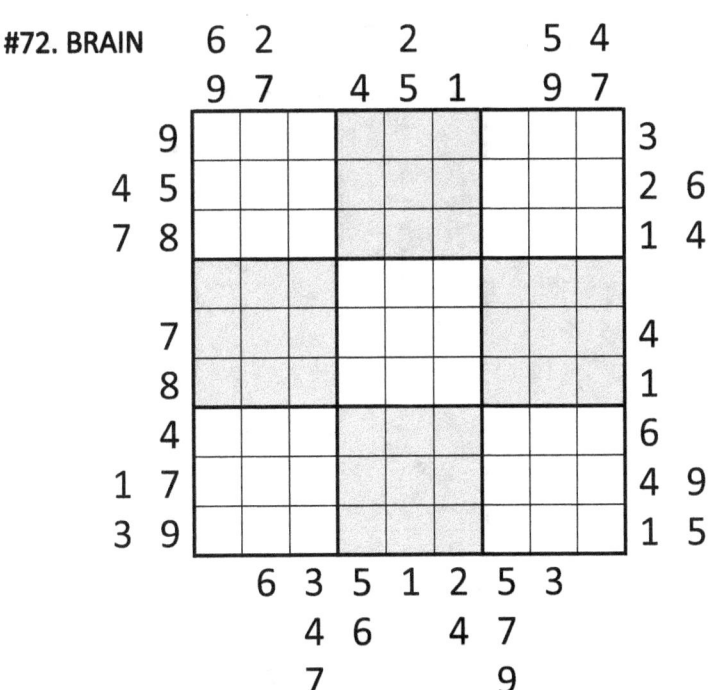

#72. BRAIN

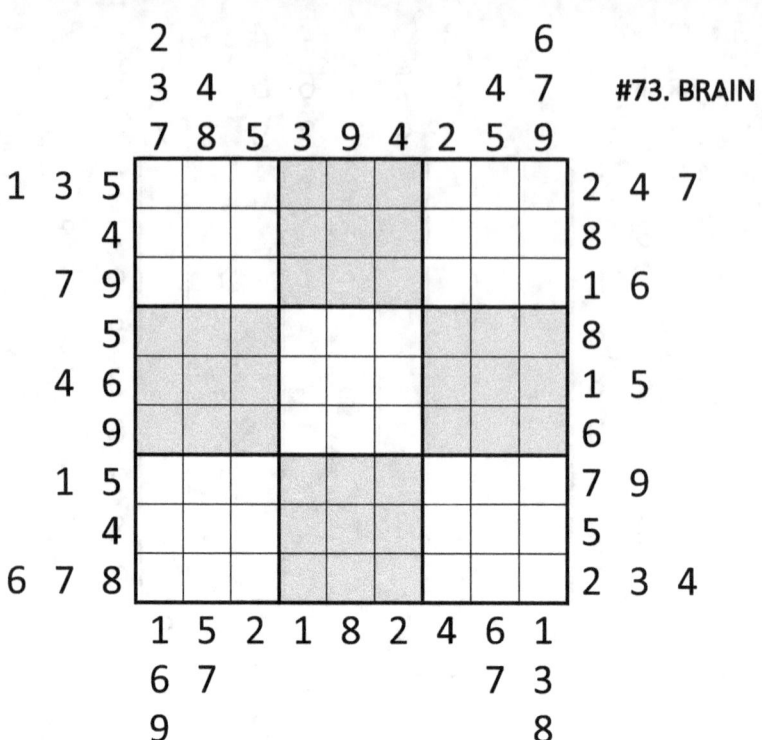

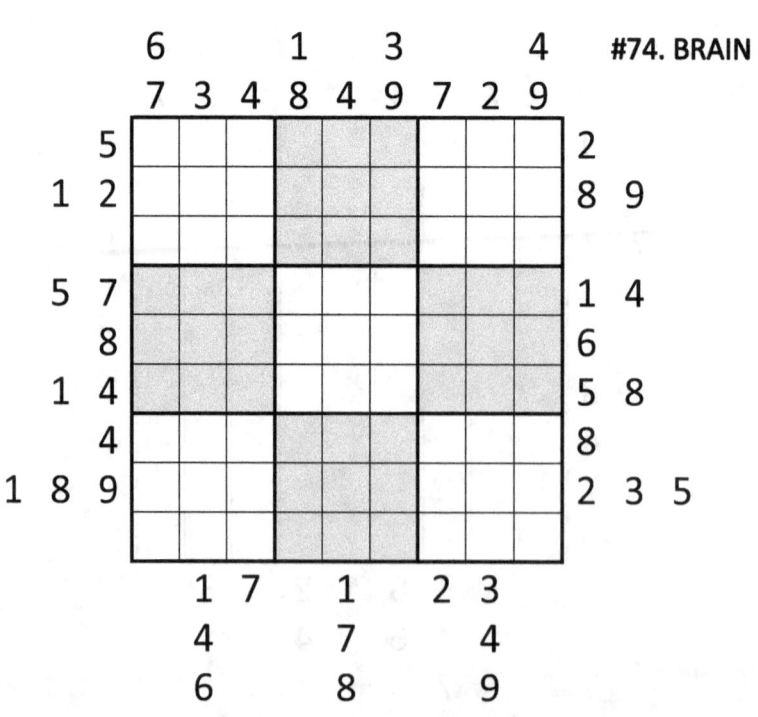

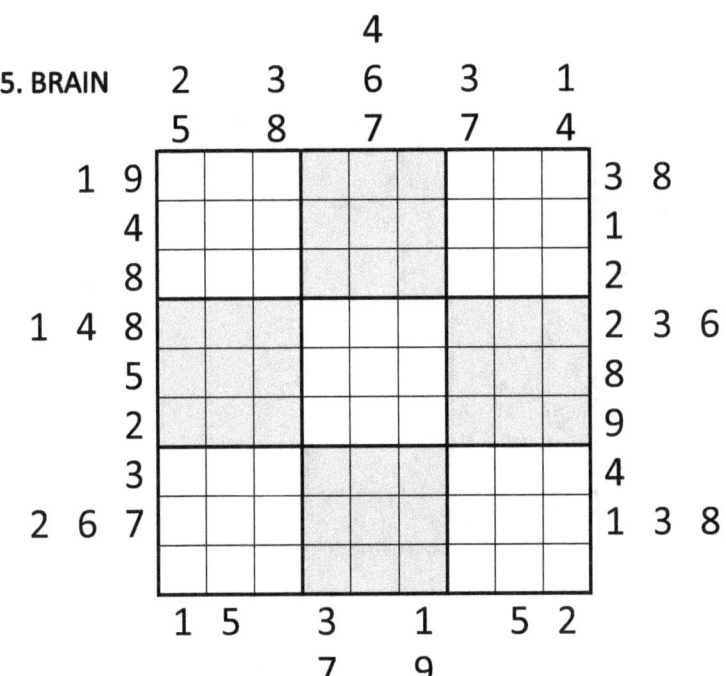

#75. BRAIN

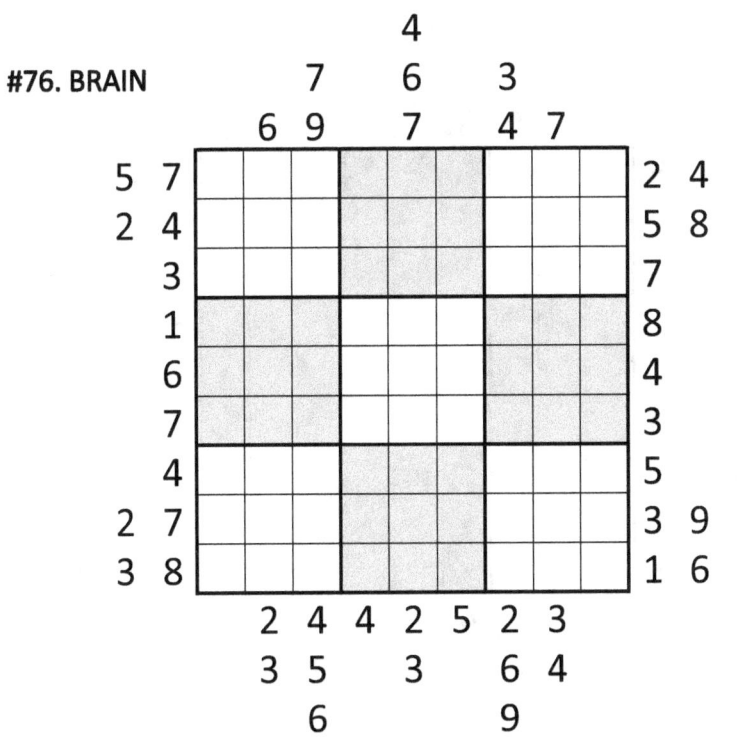

#76. BRAIN

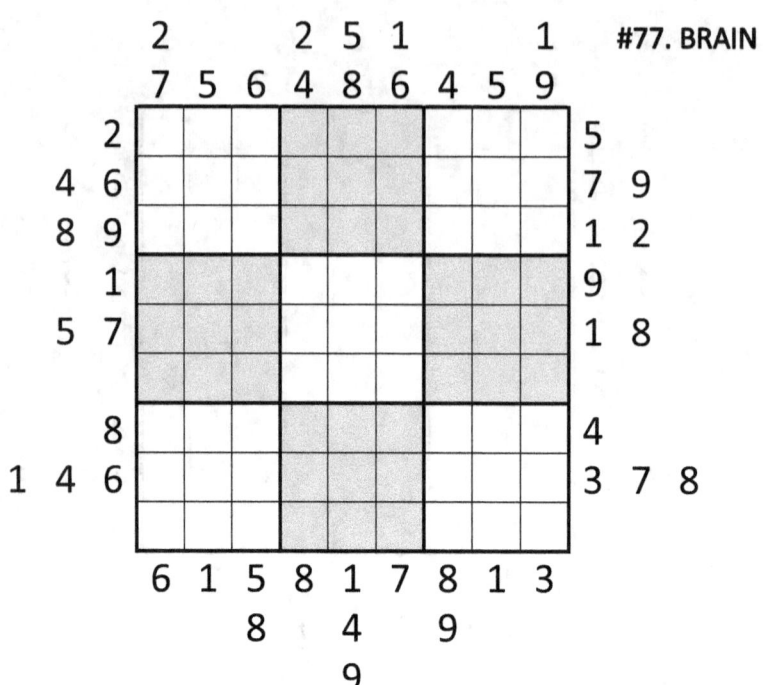

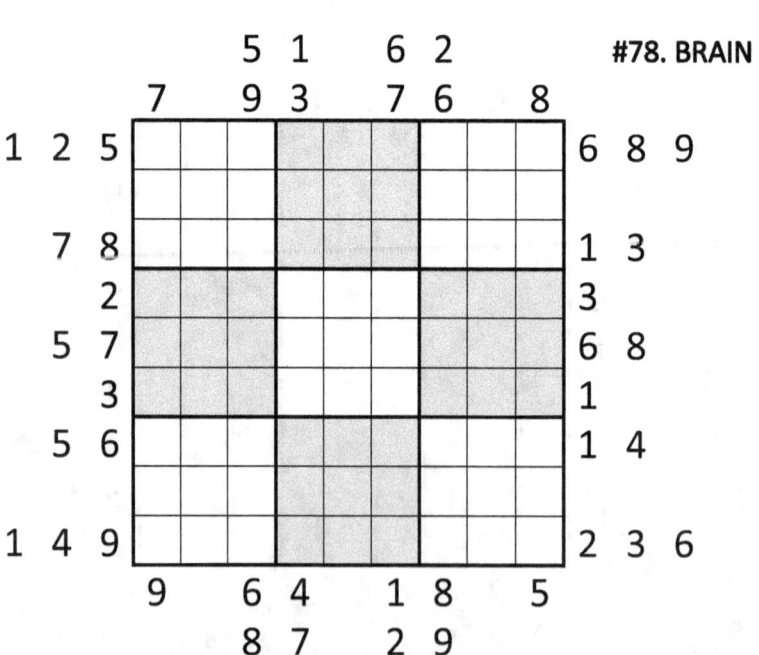

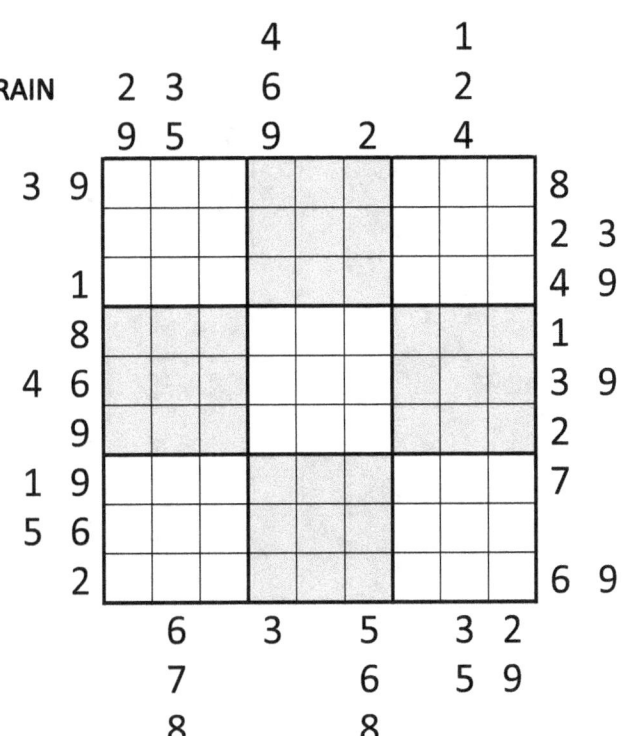

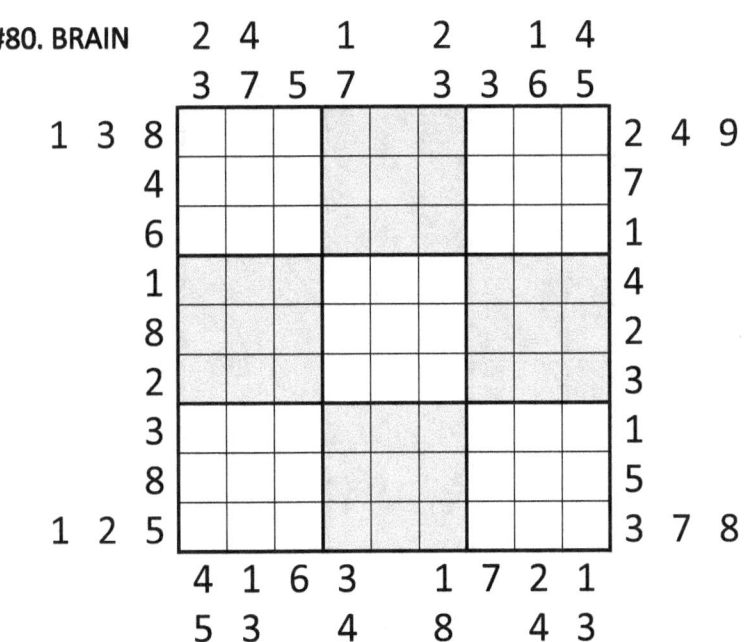

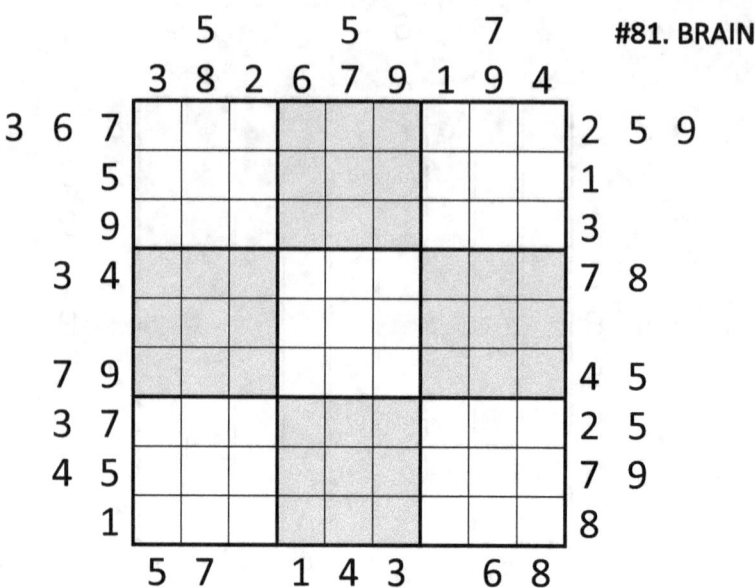

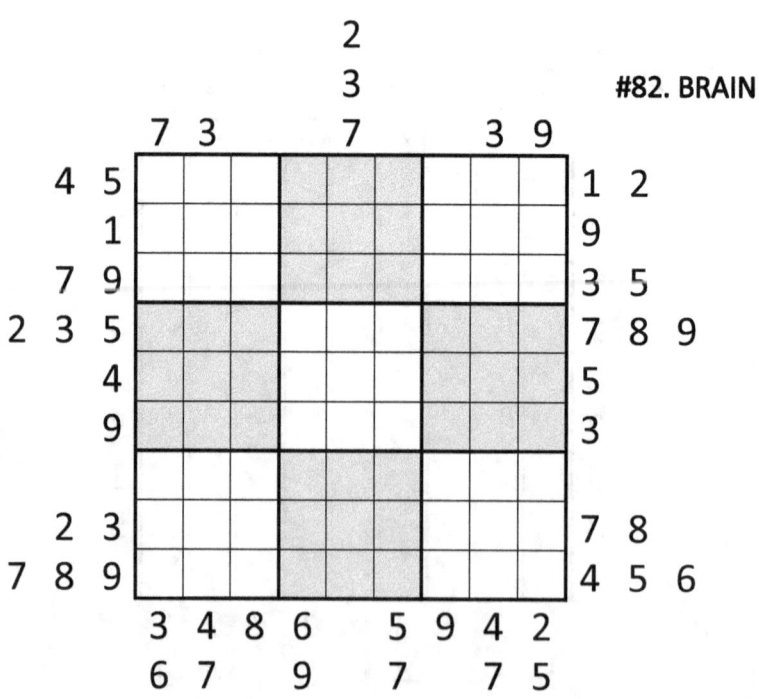

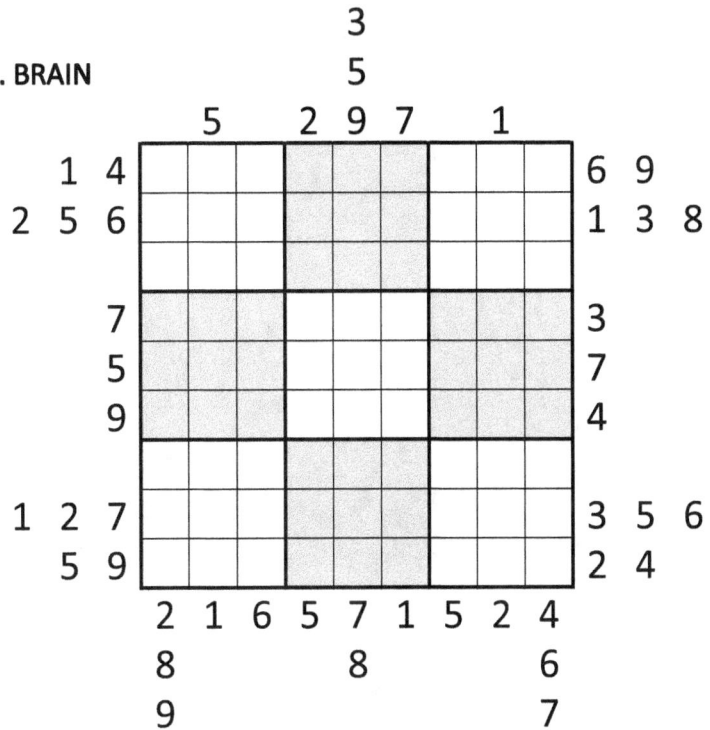

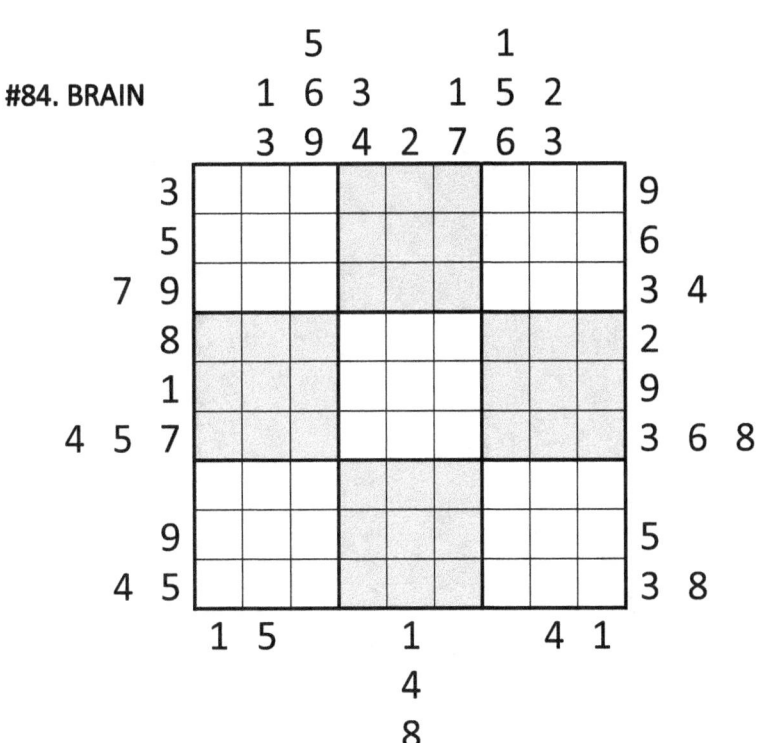

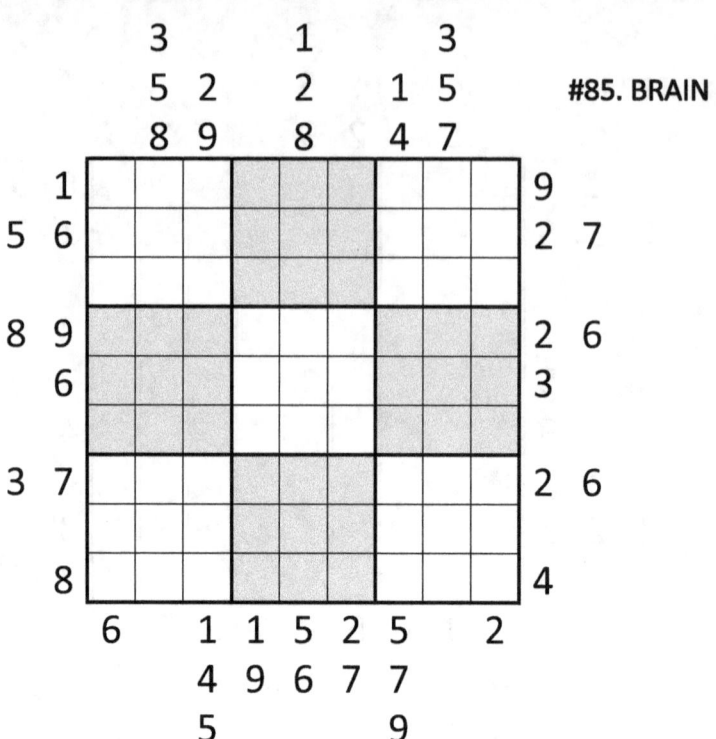

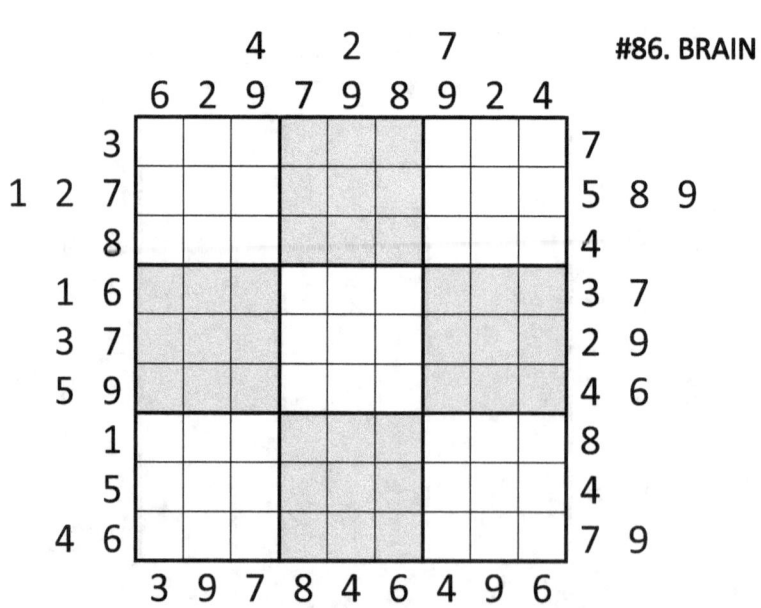

#87. BRAIN

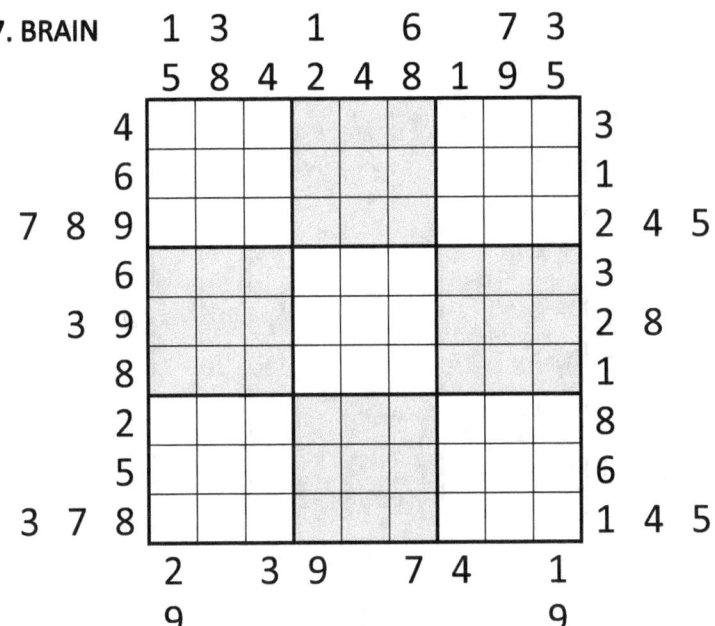

#88. BRAIN

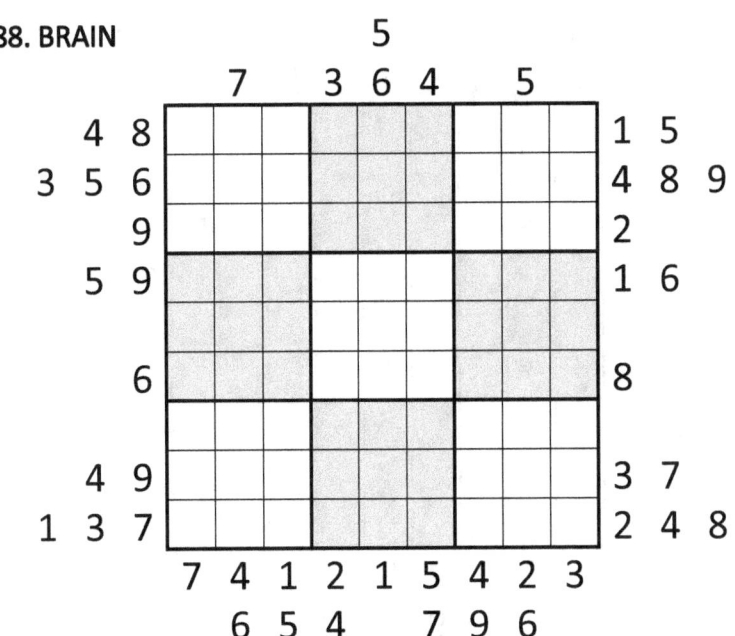

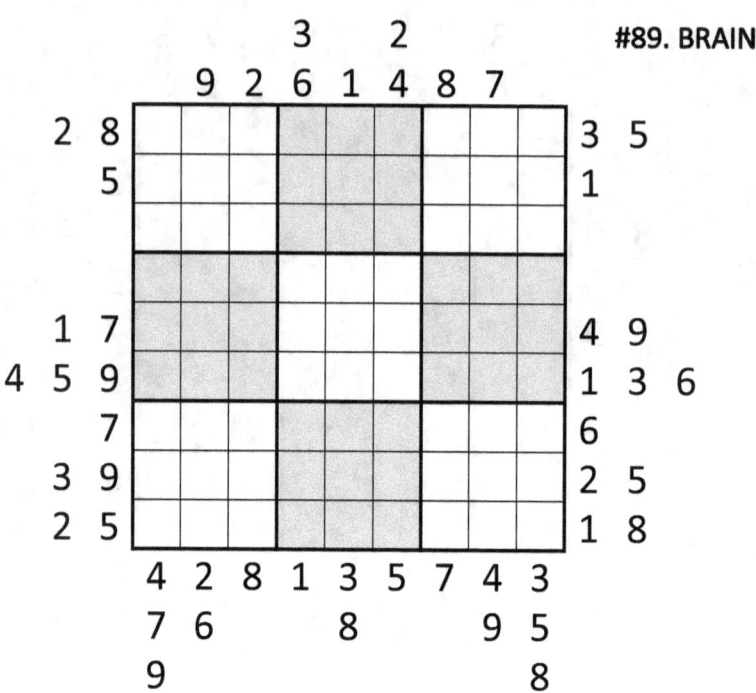

#89. BRAIN

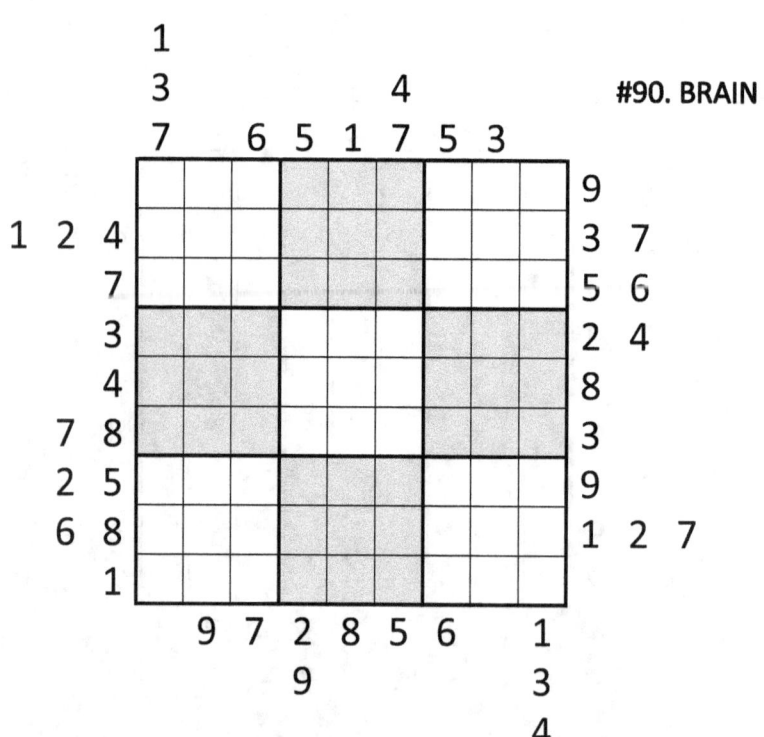

#90. BRAIN

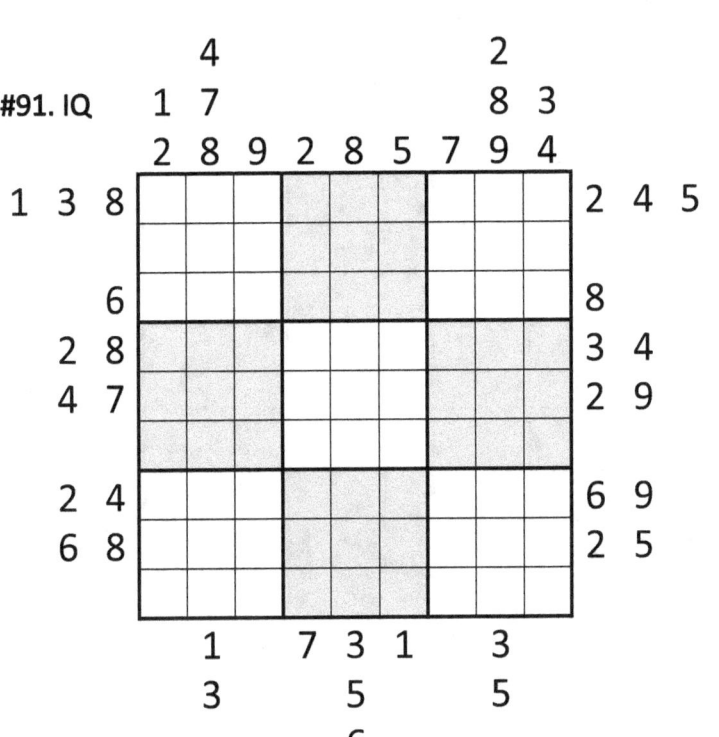

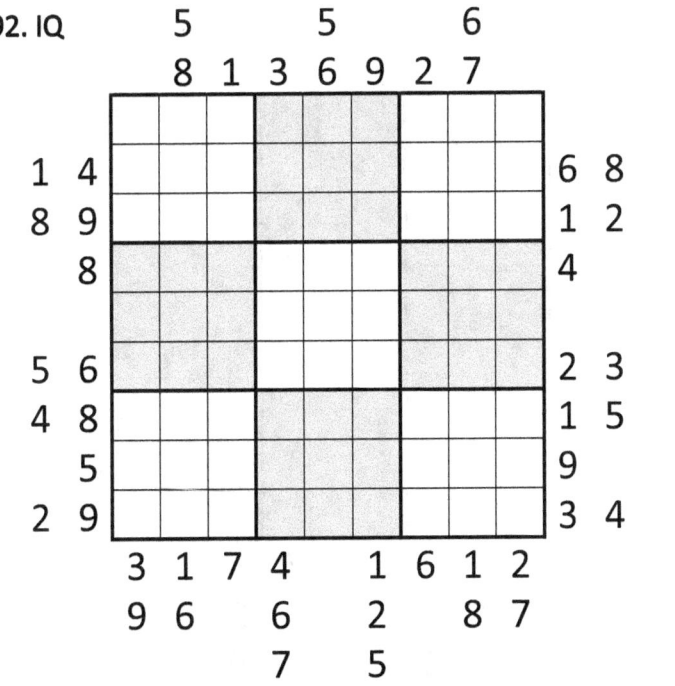

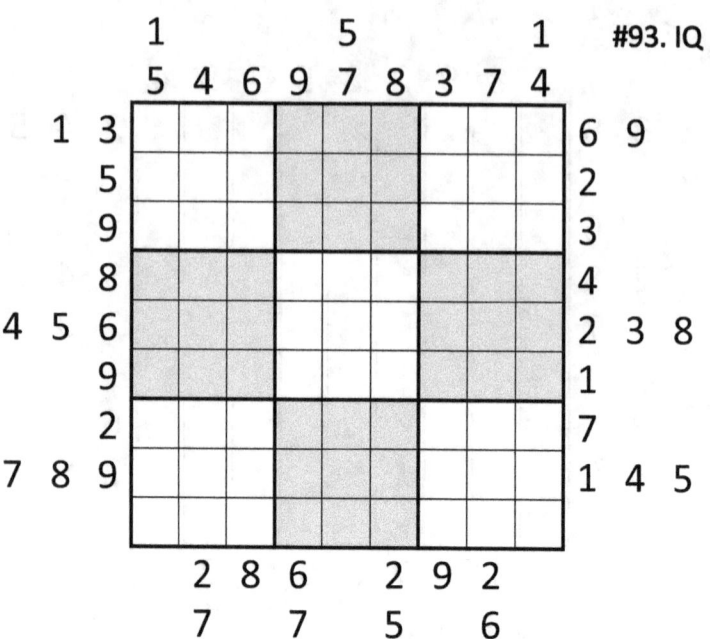

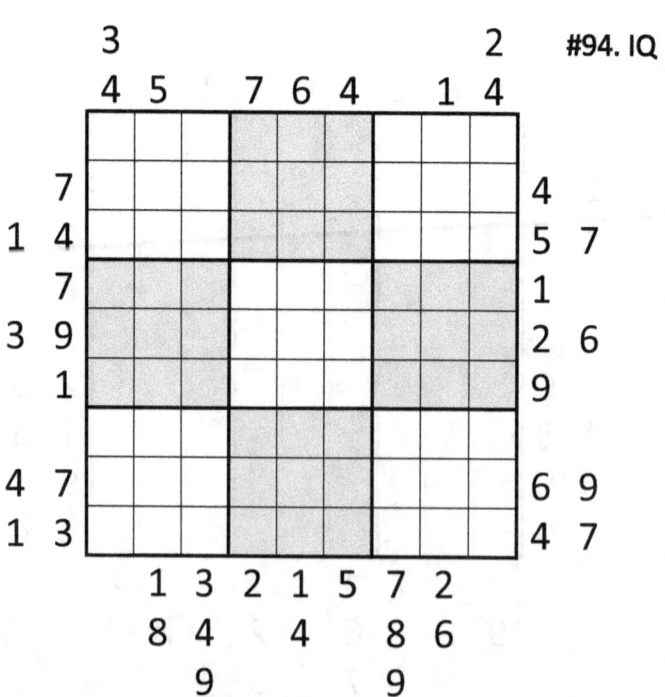

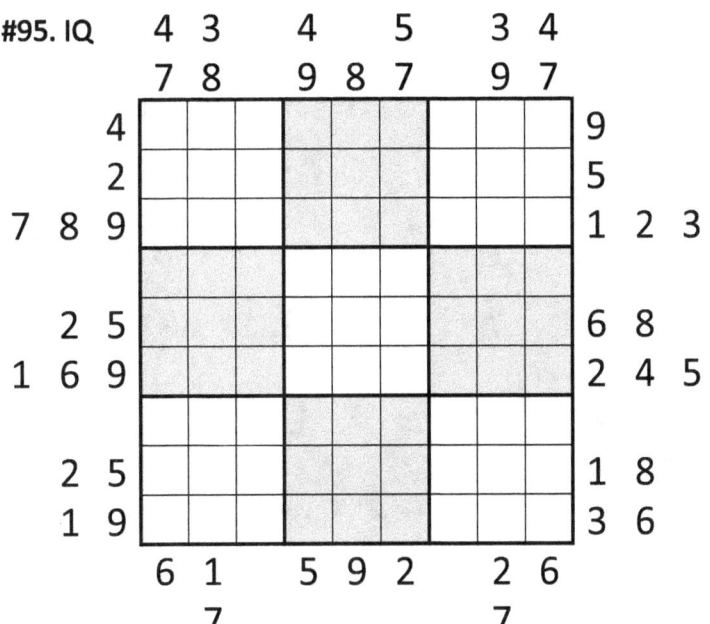

#95. IQ

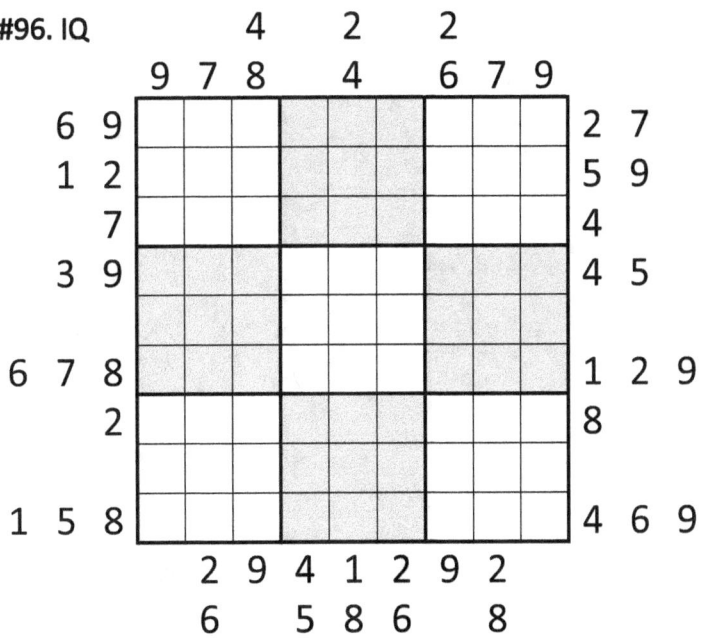

#96. IQ

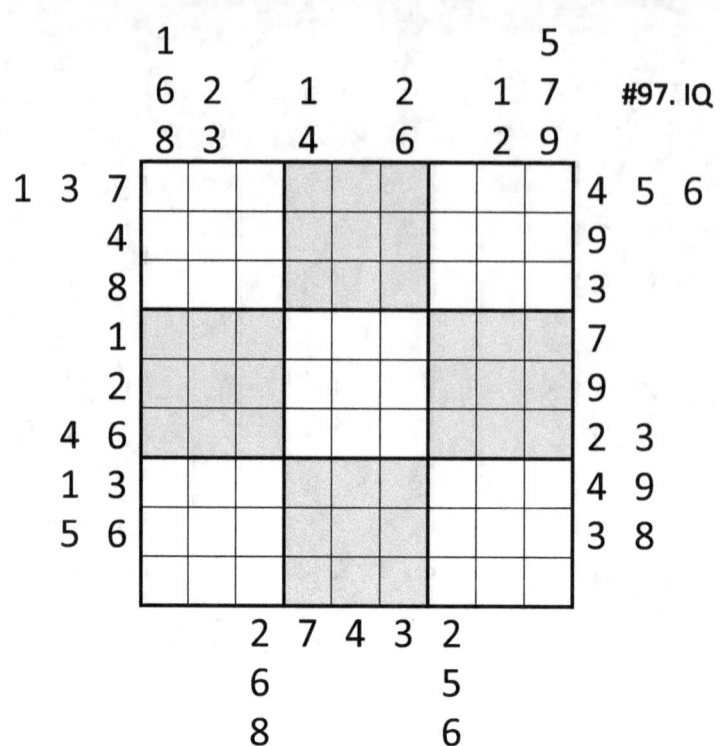

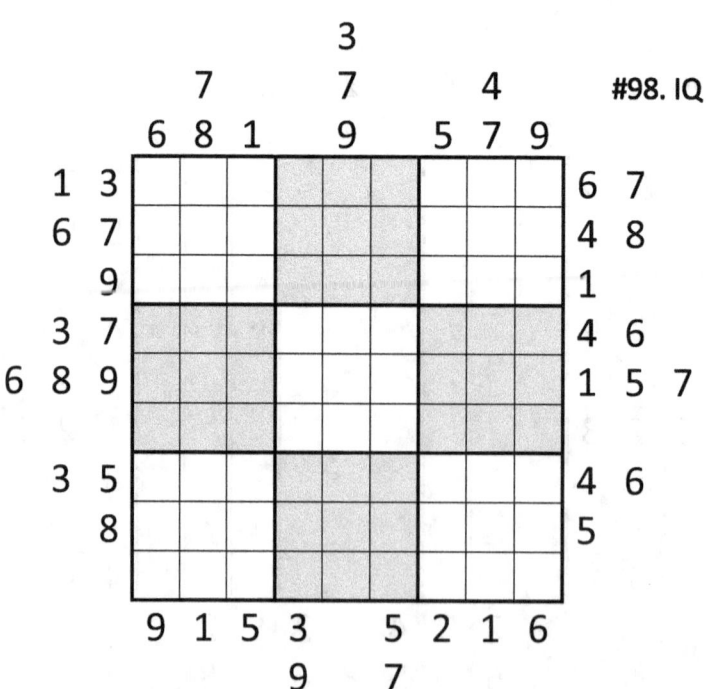

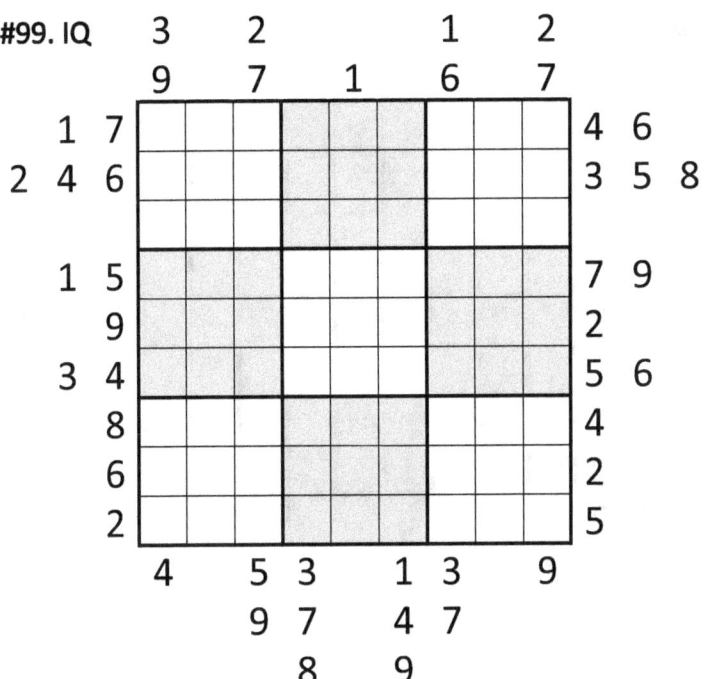

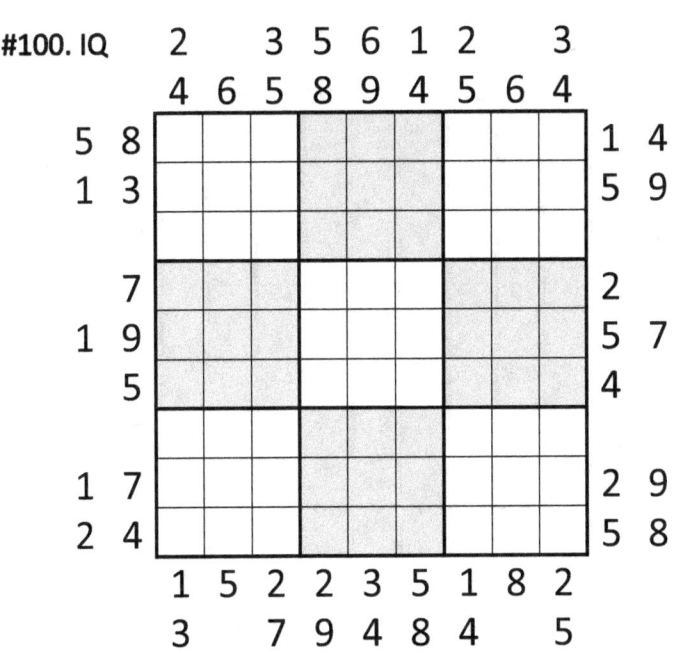

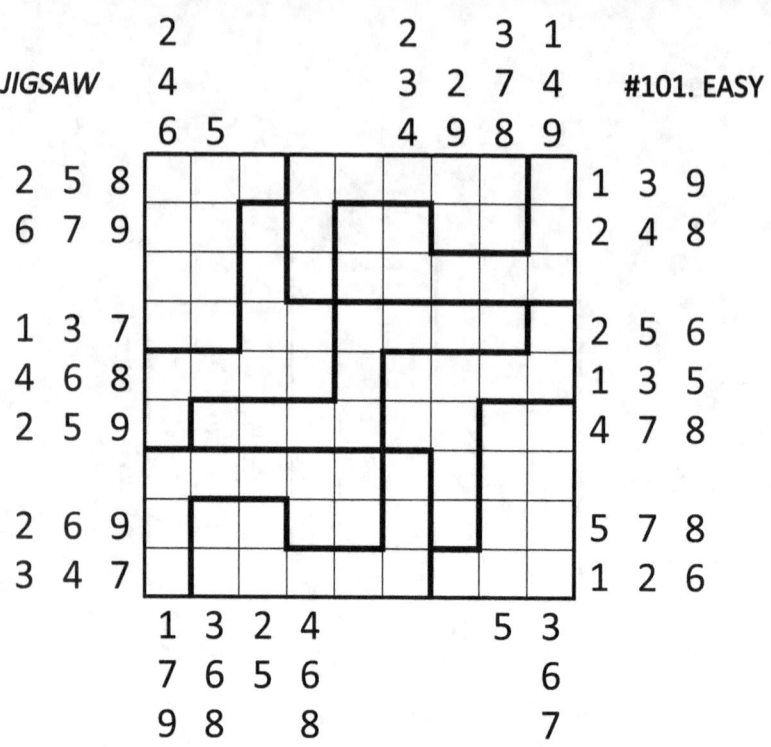

JIGSAW   #101. EASY

Outside Sudoku by amazon.com/djape, page 56

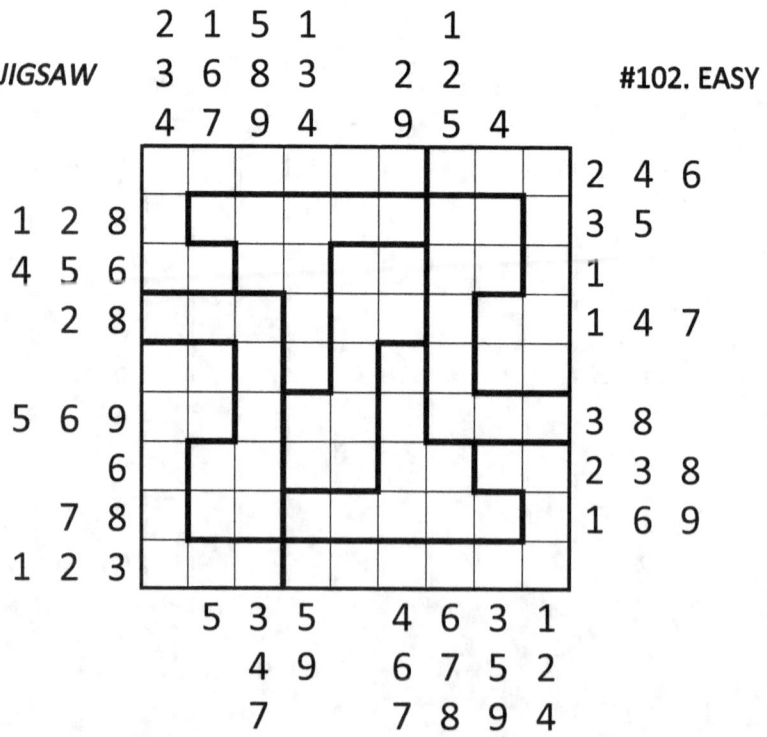

JIGSAW   #102. EASY

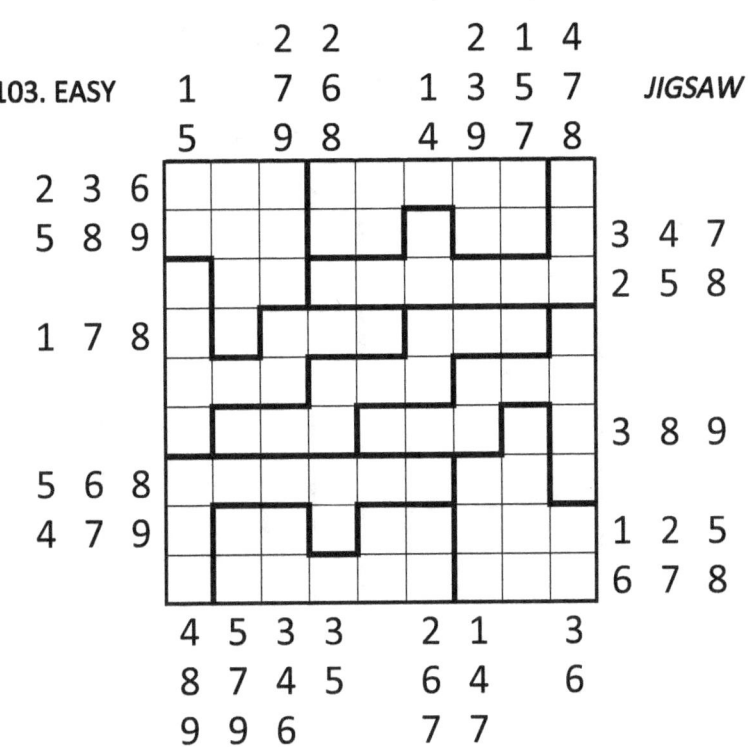

#103. EASY   JIGSAW

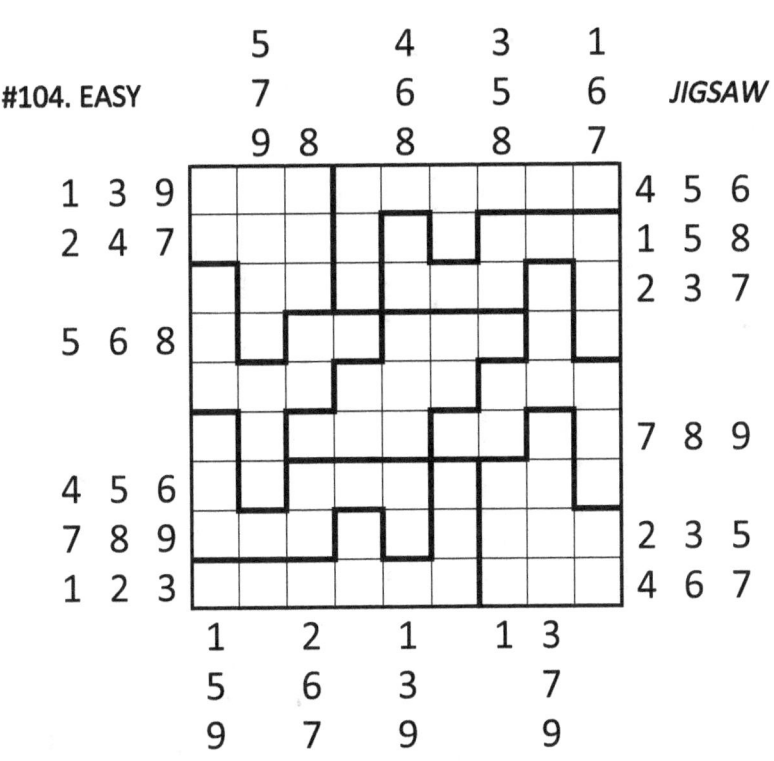

#104. EASY   JIGSAW

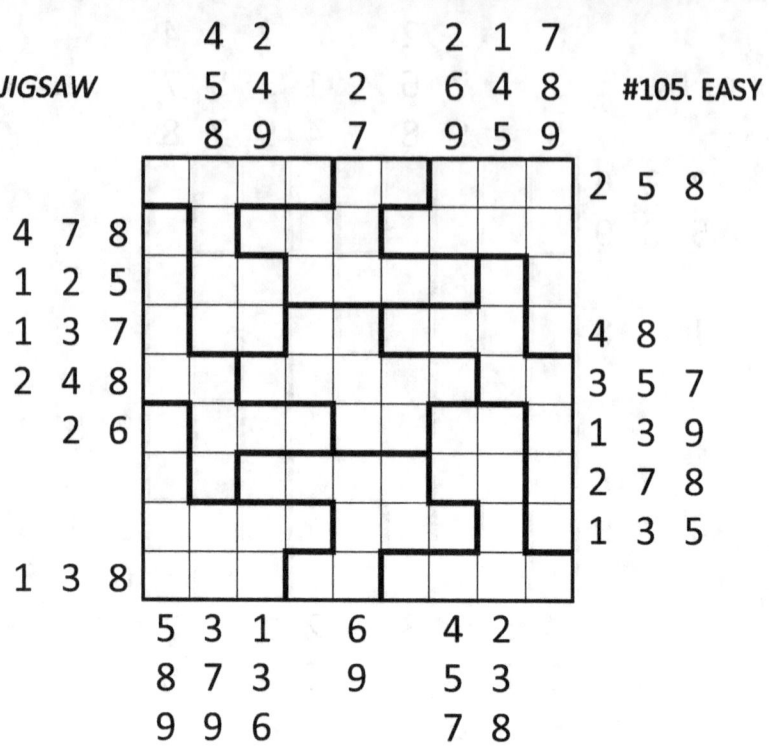

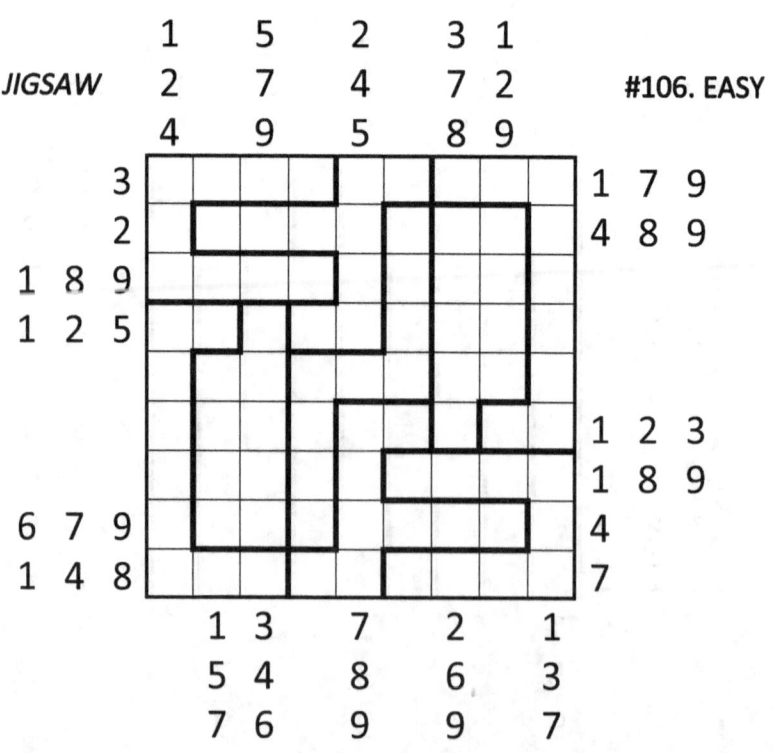

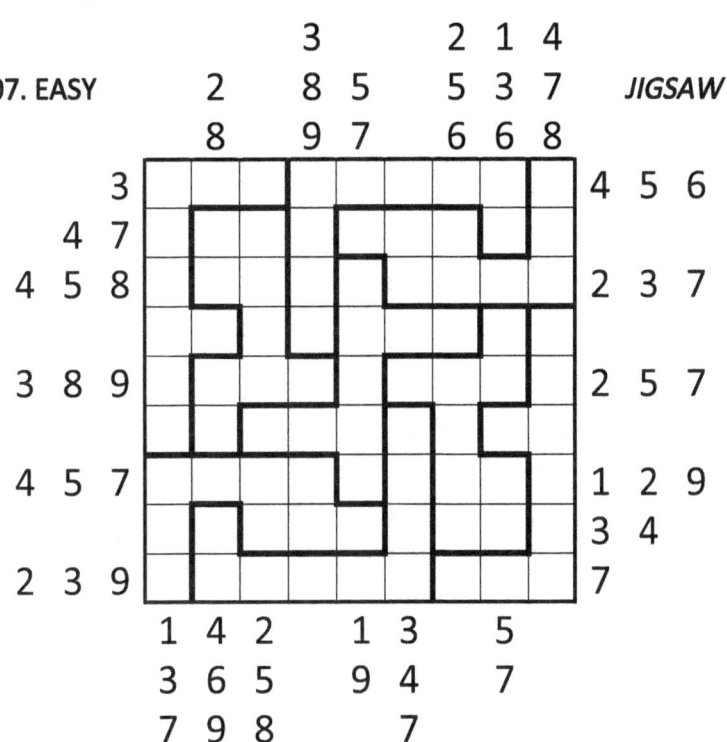

Outside Sudoku by amazon.com/djape, page 59

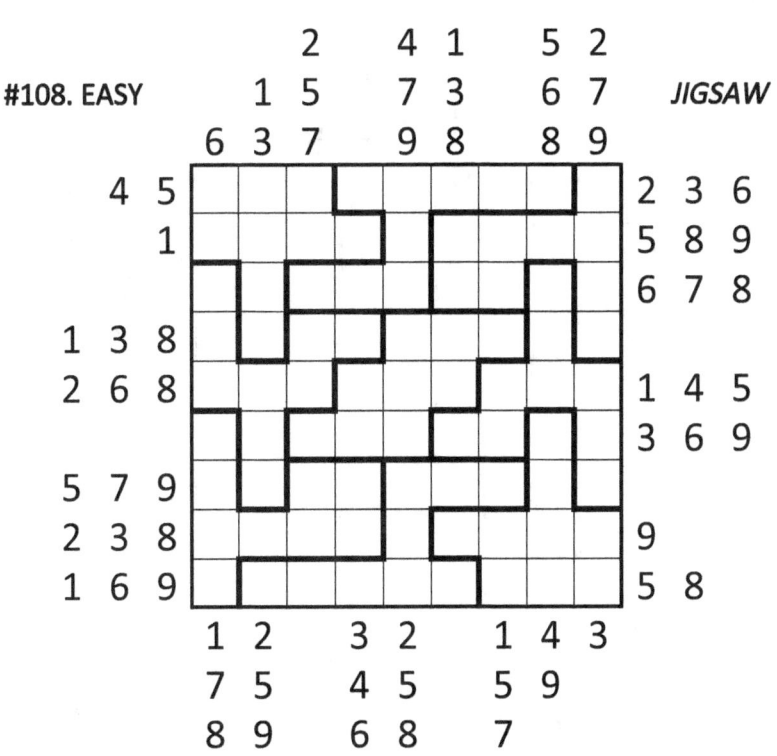

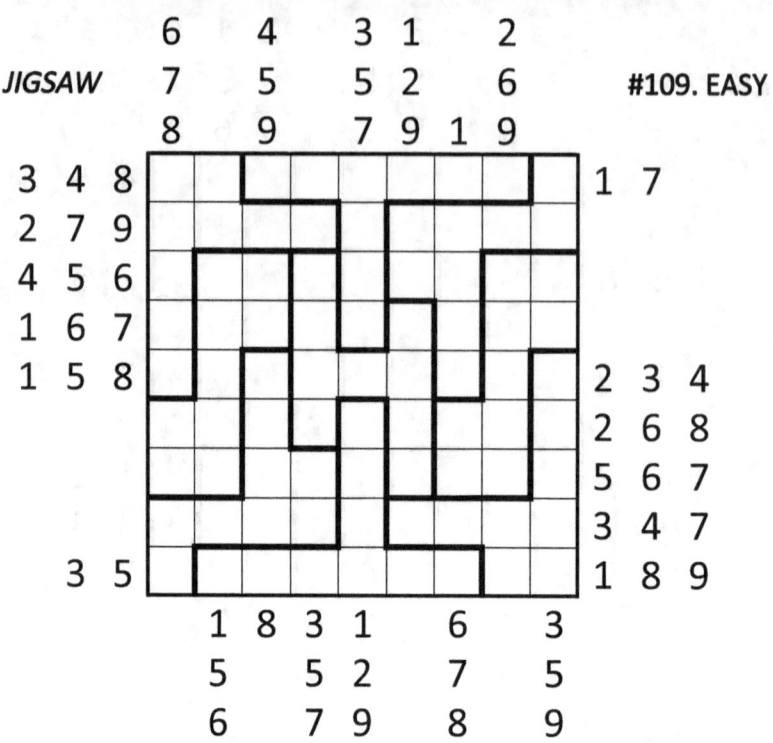

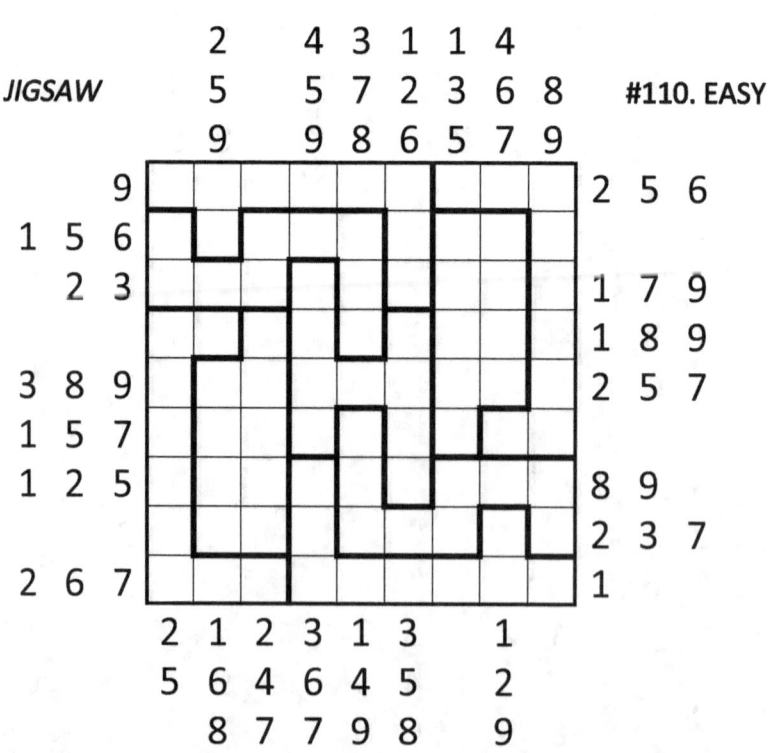

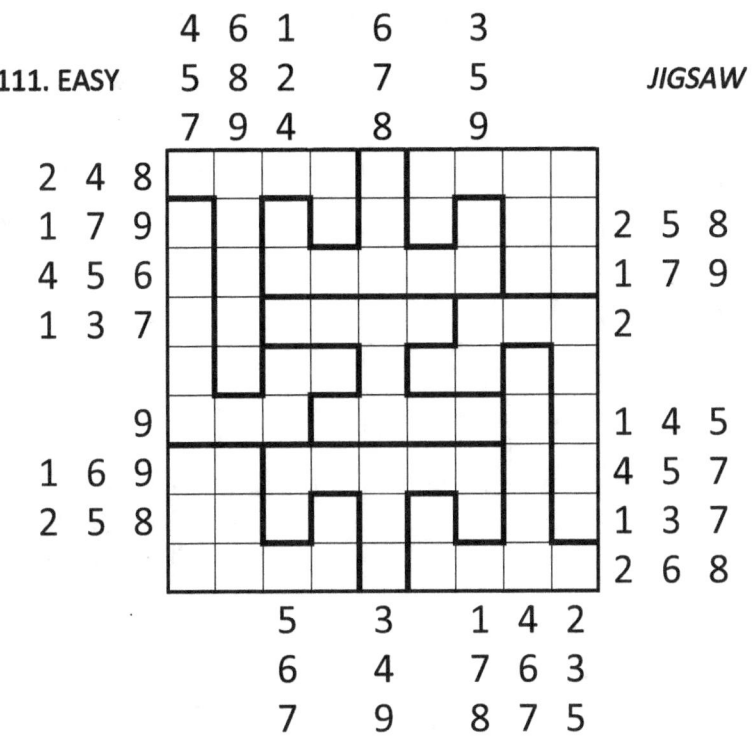

#111. EASY — JIGSAW

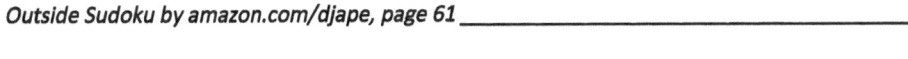

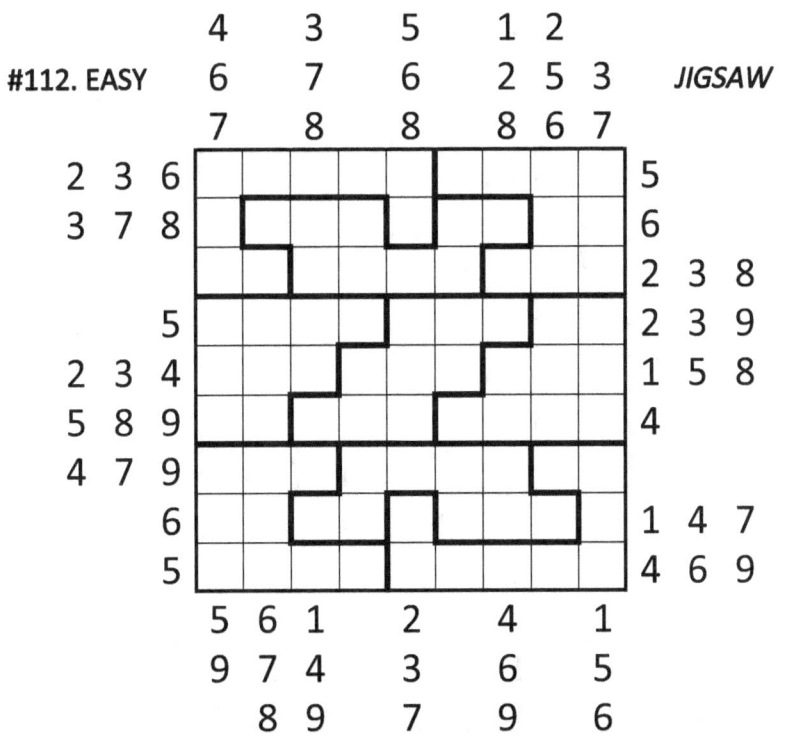

#112. EASY — JIGSAW

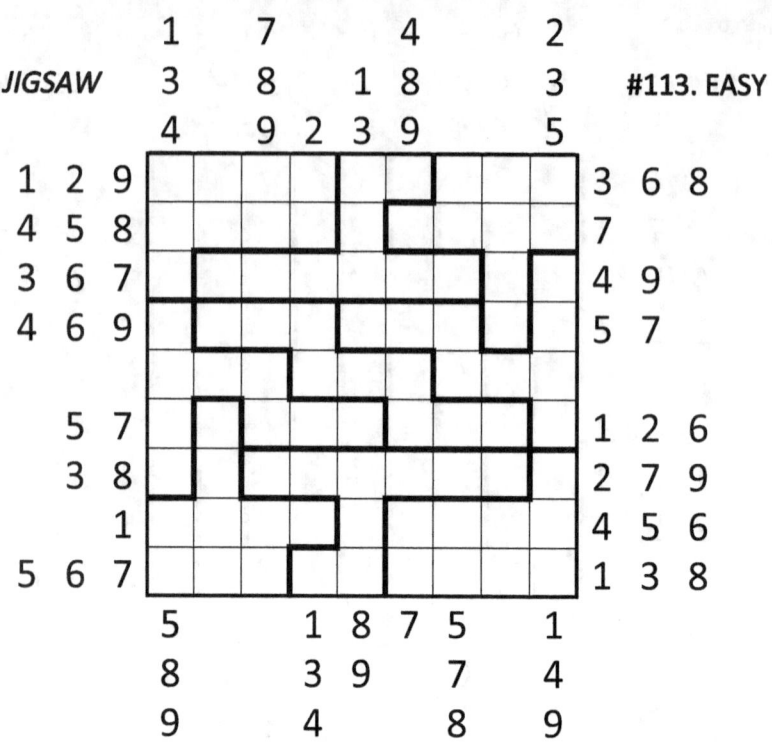

JIGSAW   #113. EASY

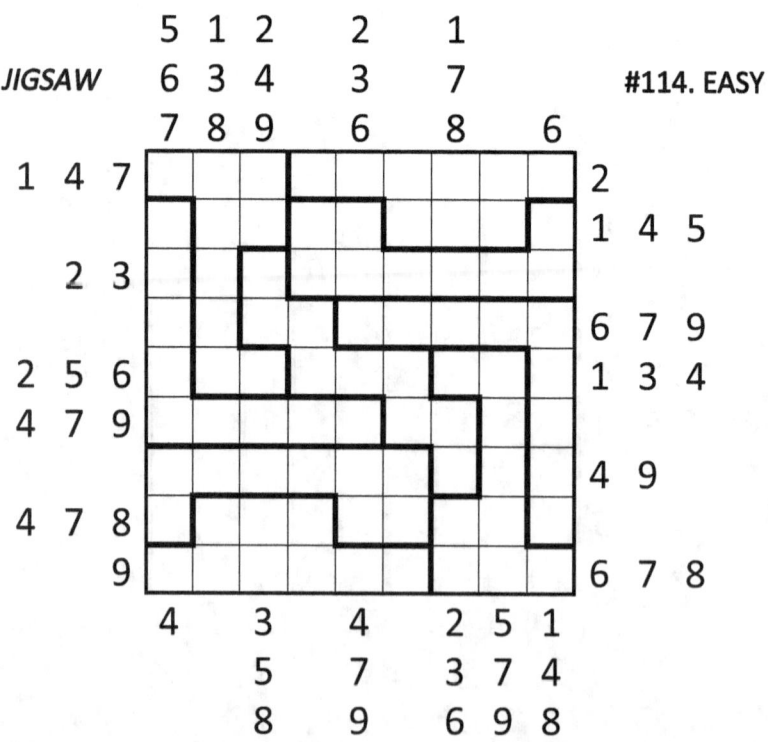

JIGSAW   #114. EASY

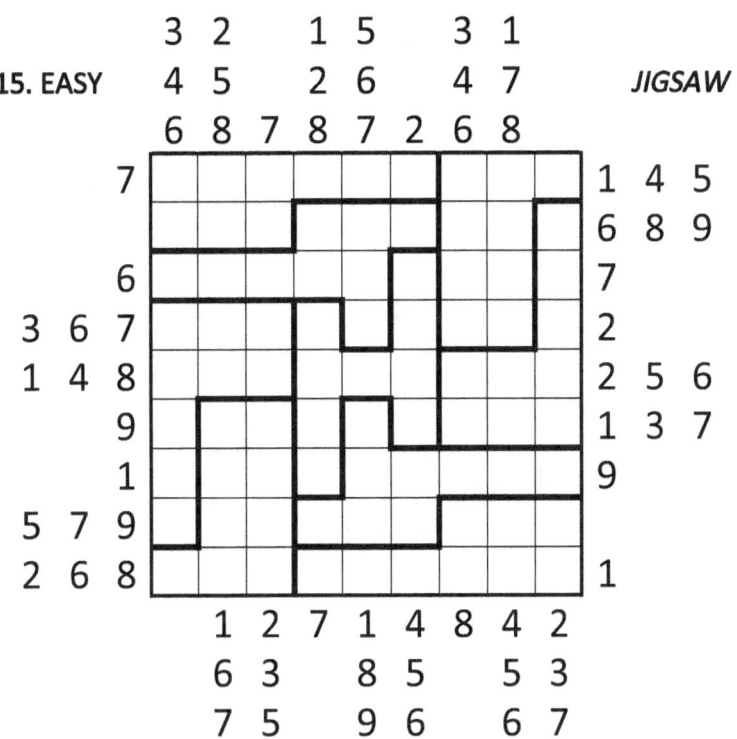

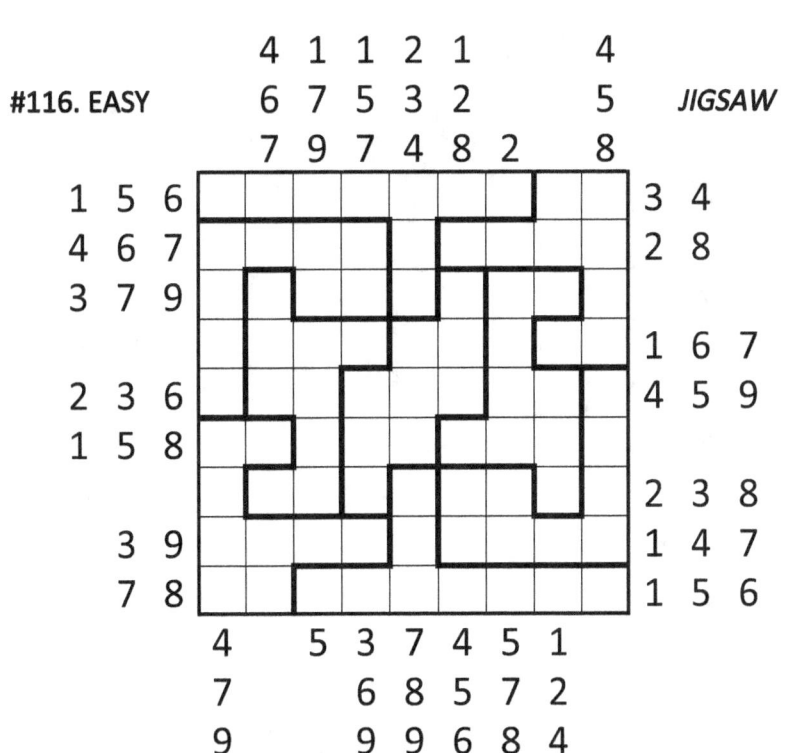

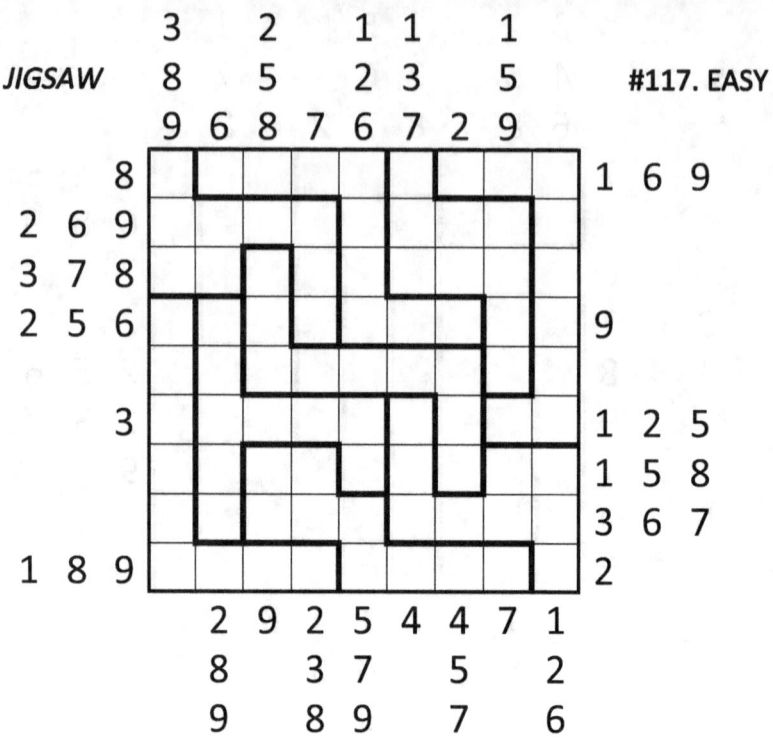

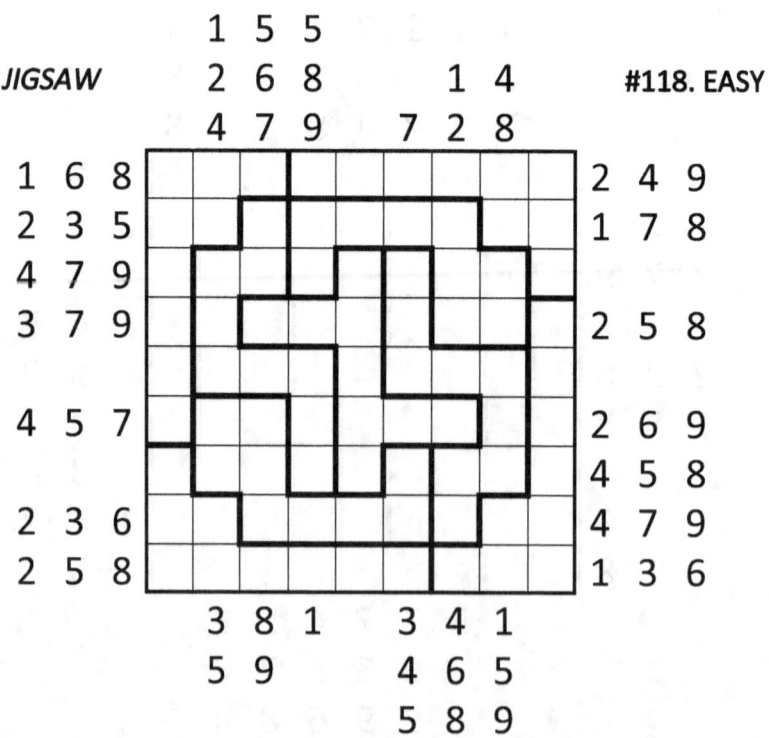

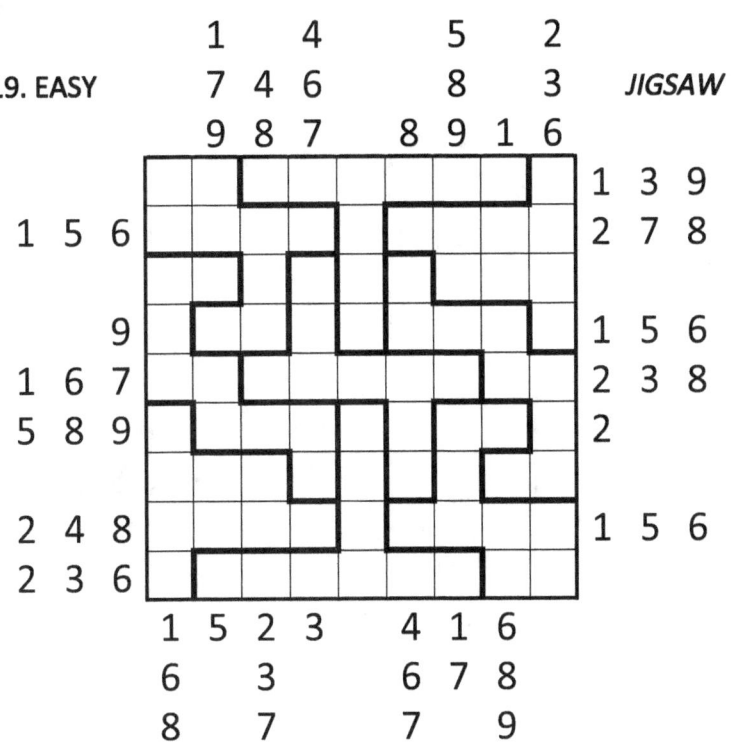

#119. EASY   *JIGSAW*

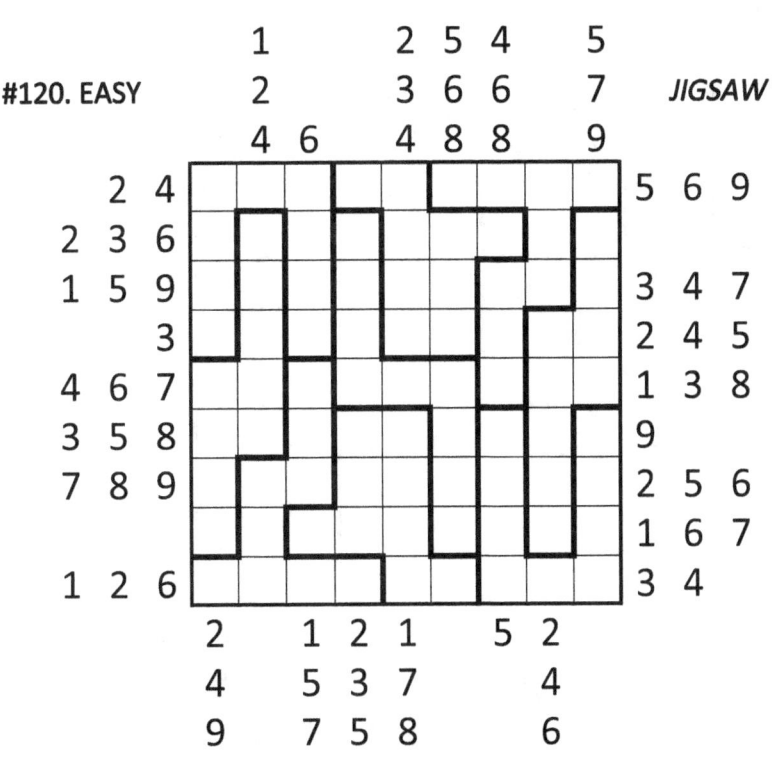

#120. EASY   *JIGSAW*

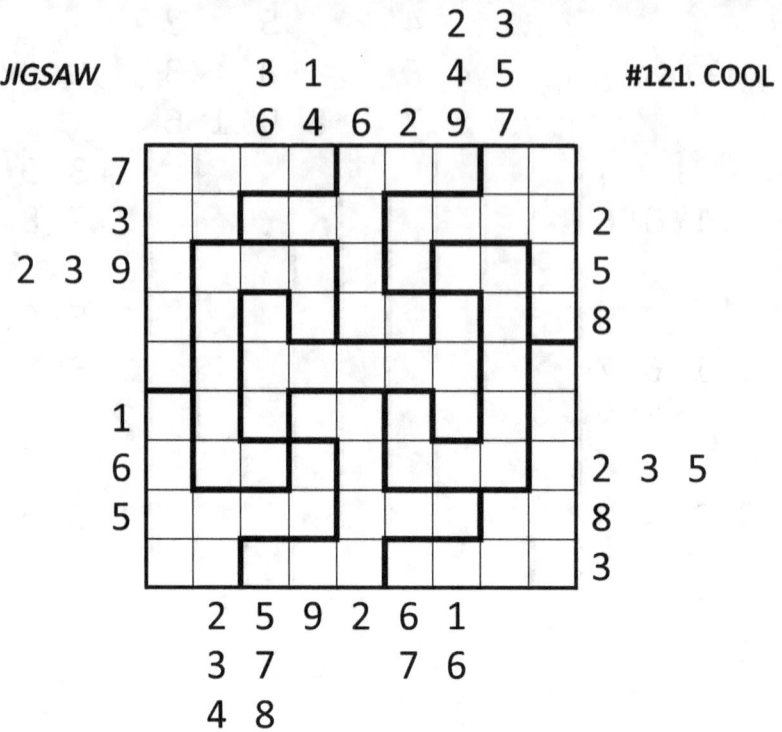

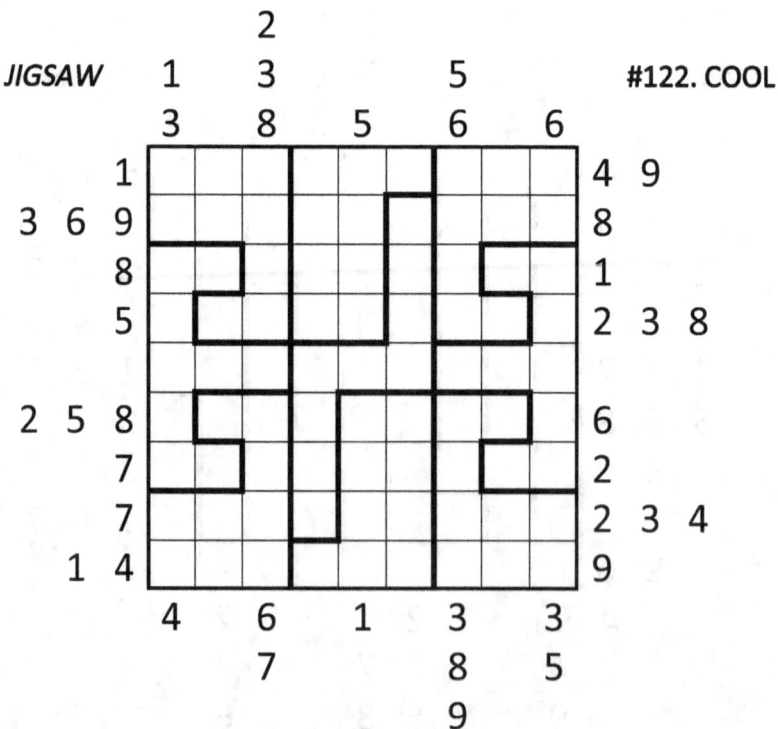

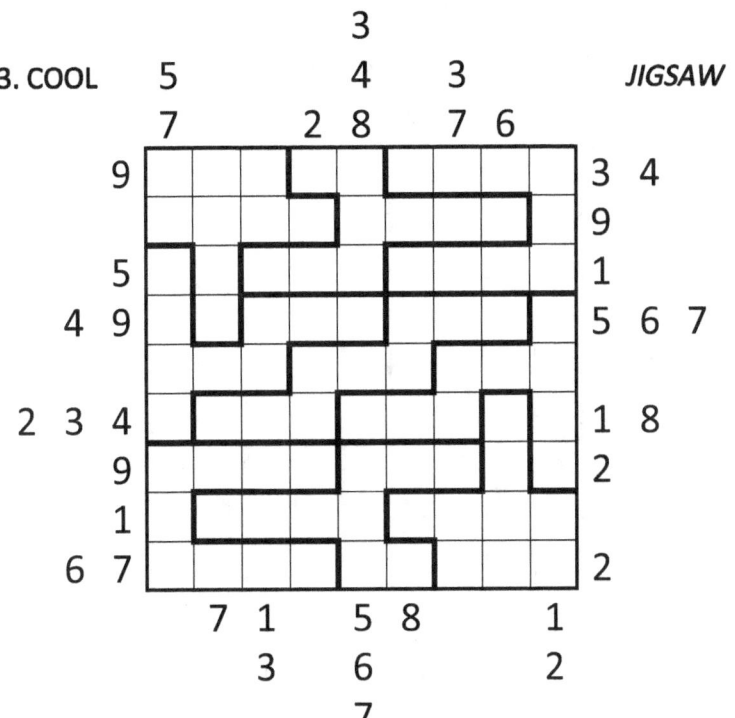

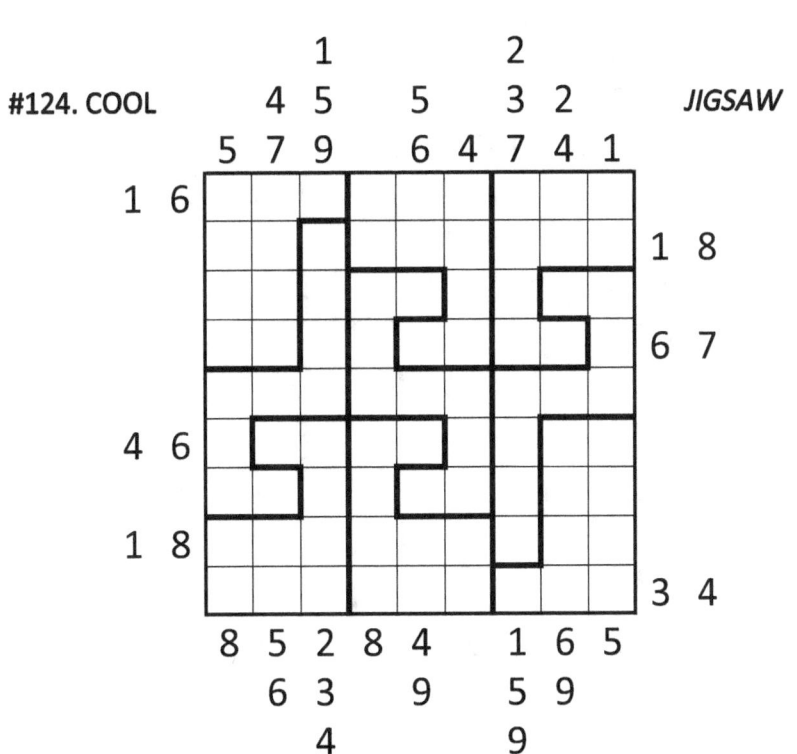

JIGSAW

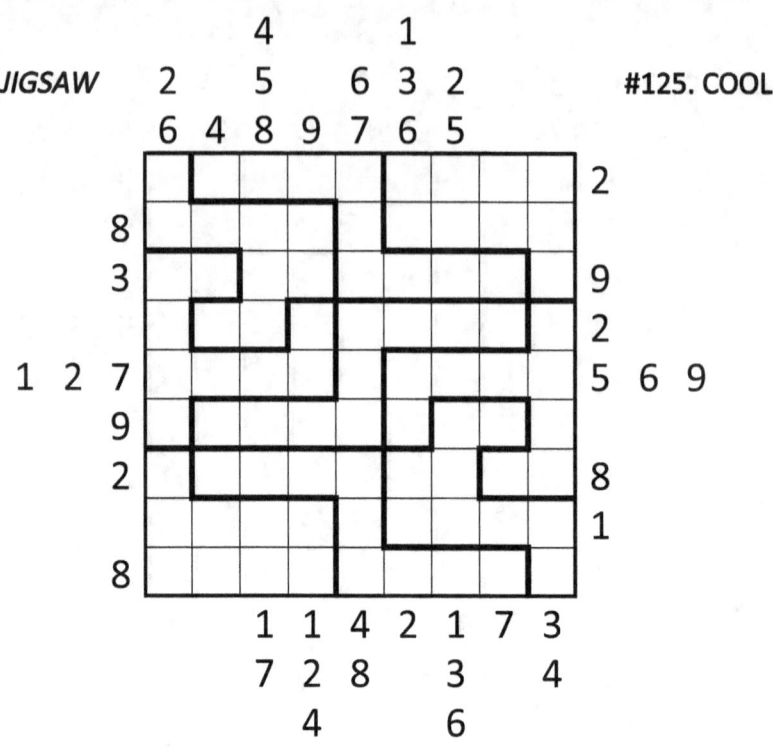

JIGSAW

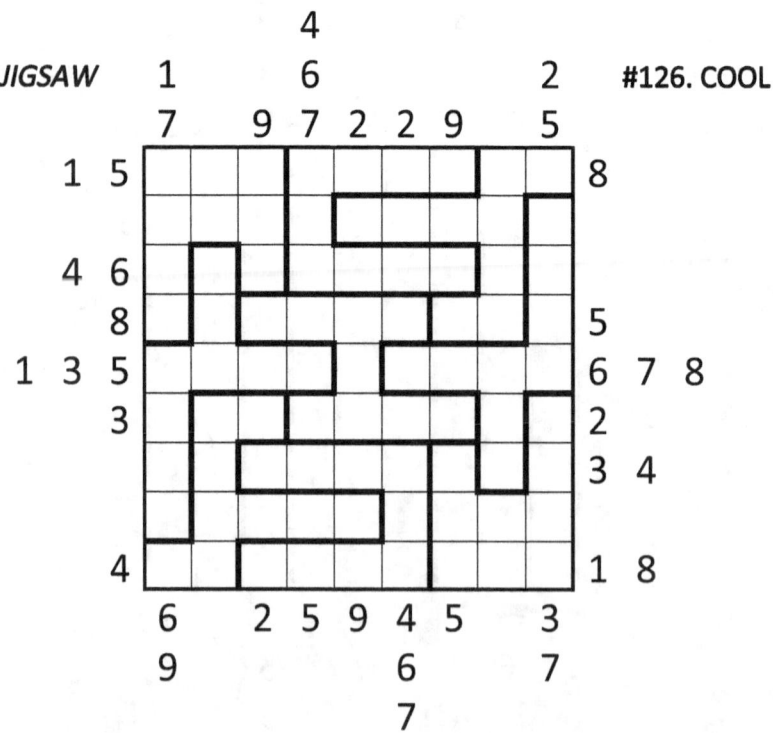

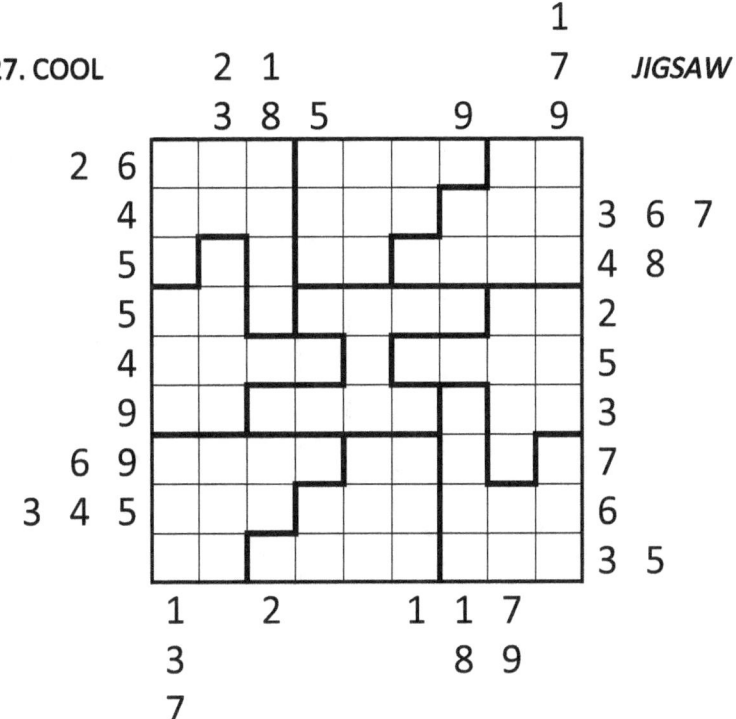

#127. COOL — JIGSAW

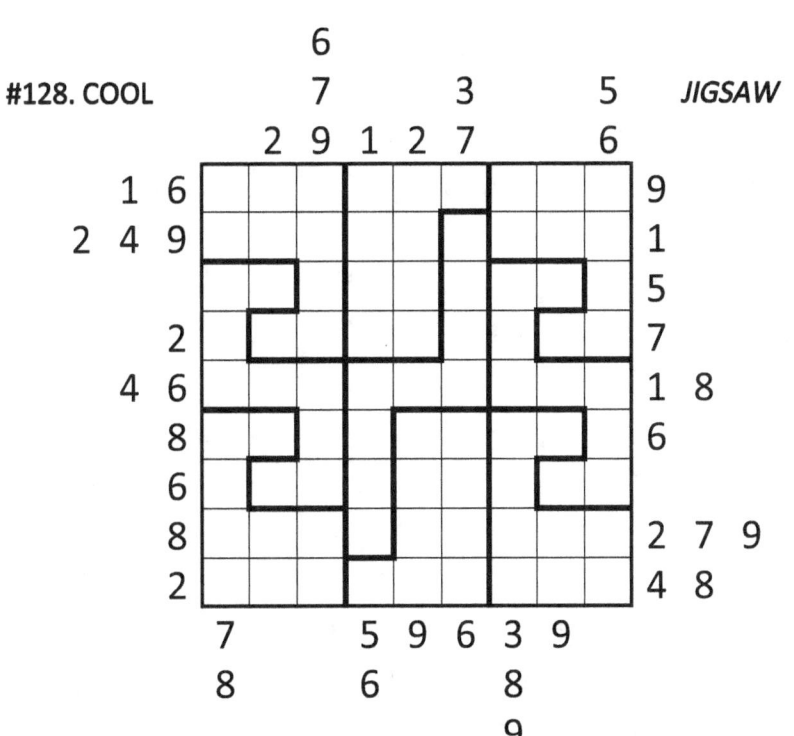

Outside Sudoku by amazon.com/djape, page 69

#128. COOL — JIGSAW

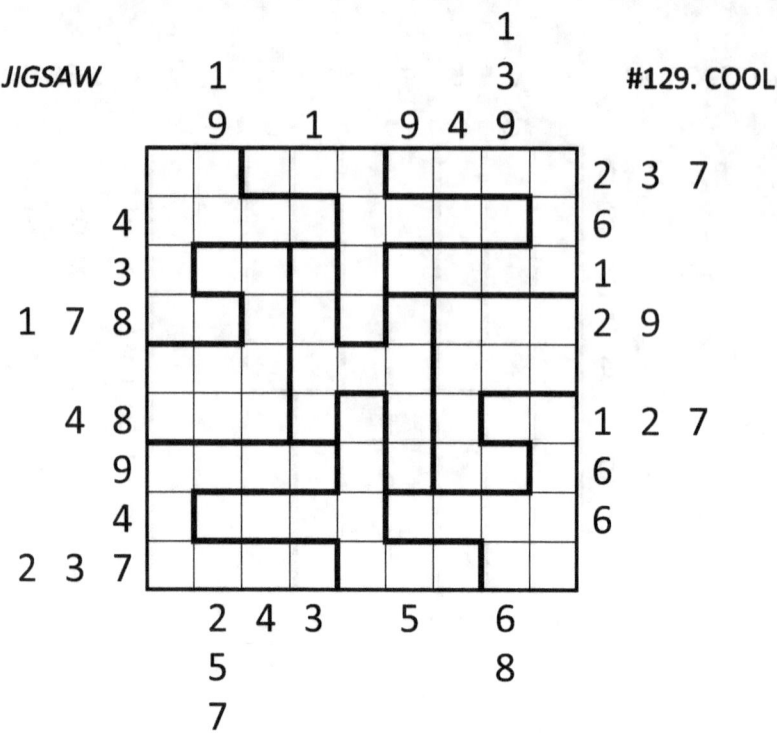

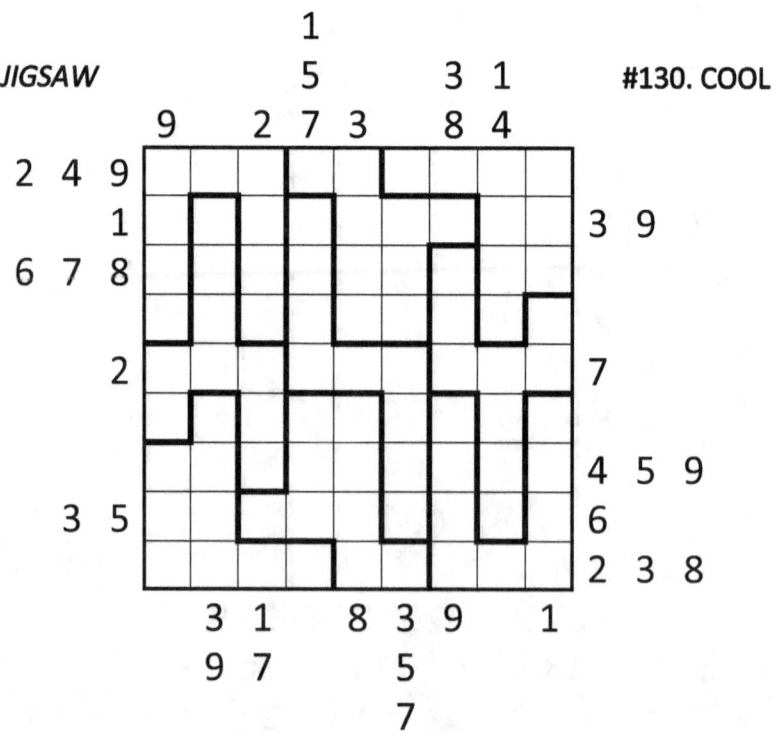

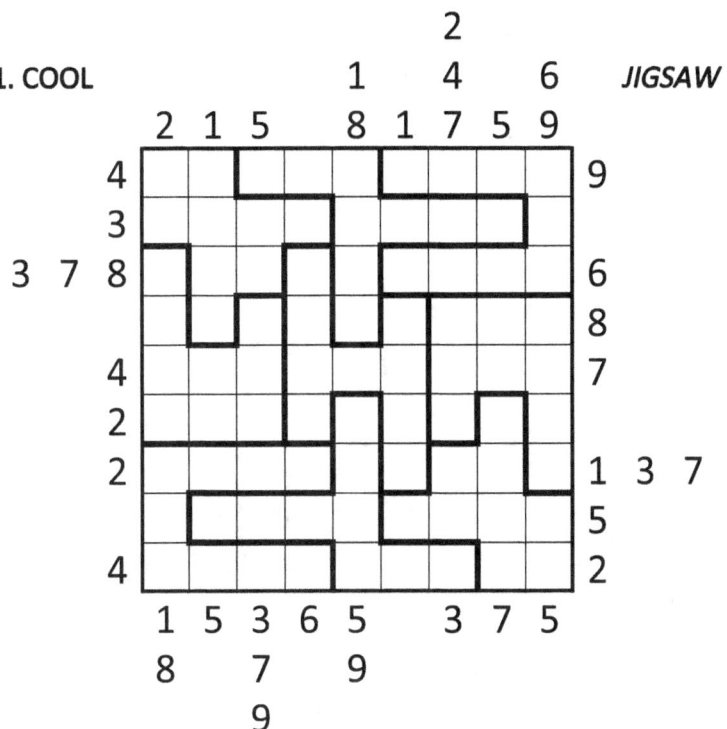

#131. COOL   JIGSAW

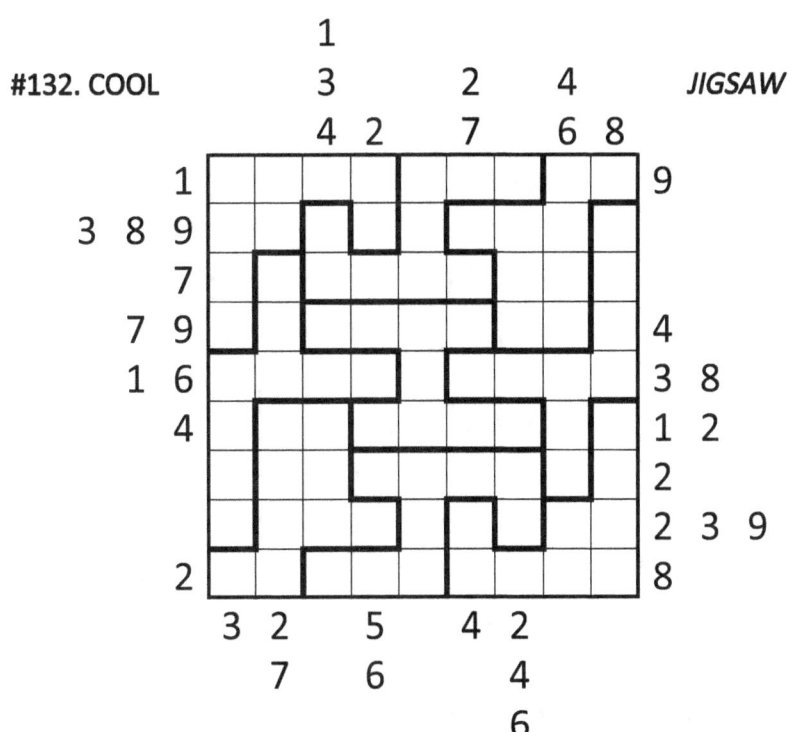

#132. COOL   JIGSAW

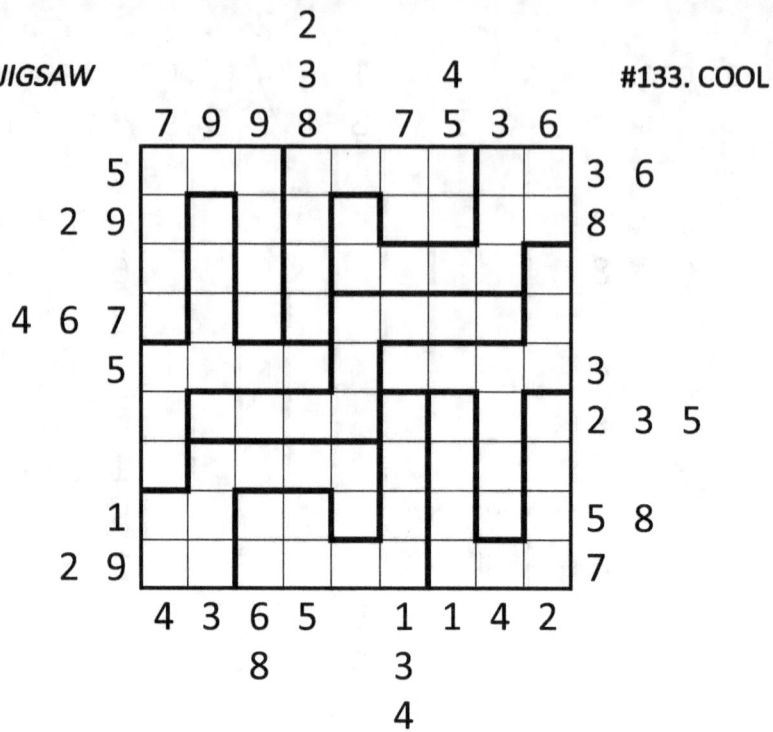

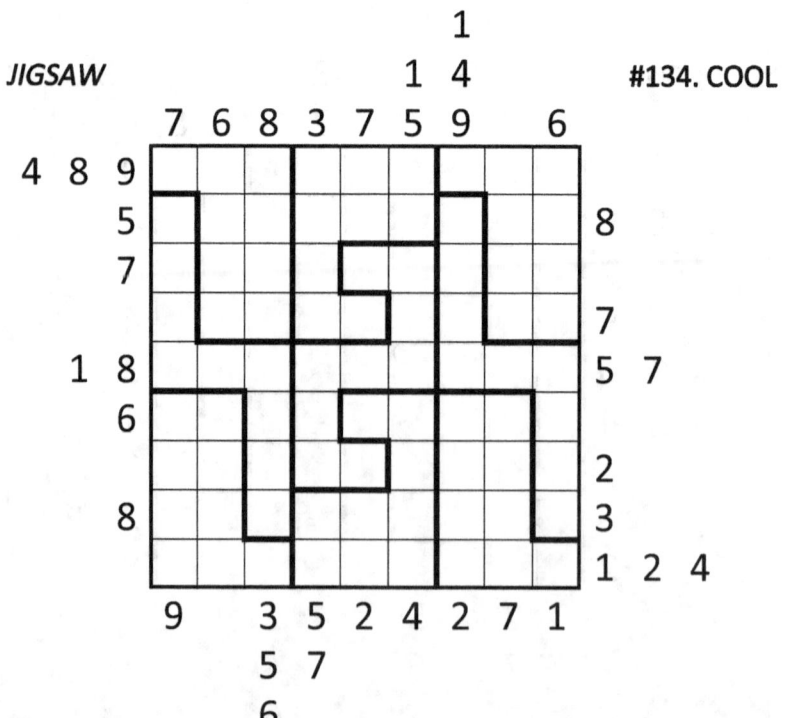

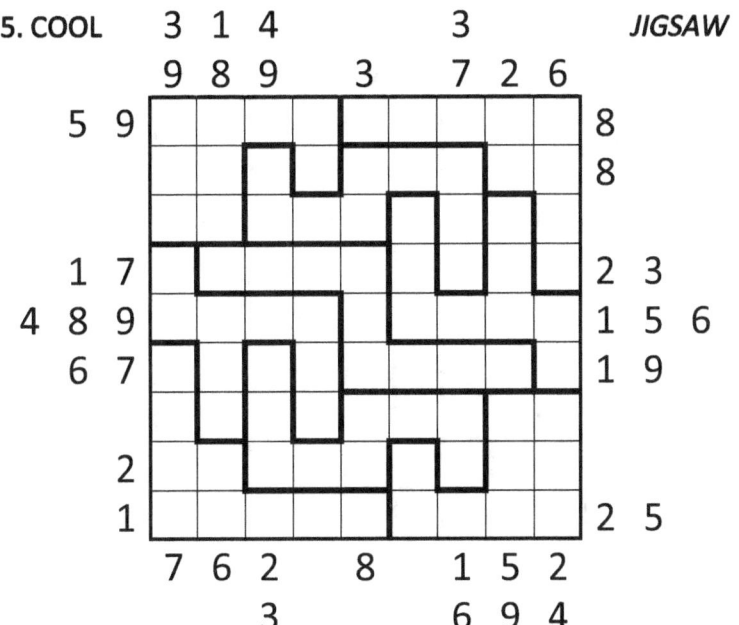

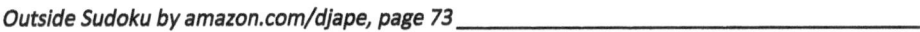

#136. COOL   *JIGSAW*

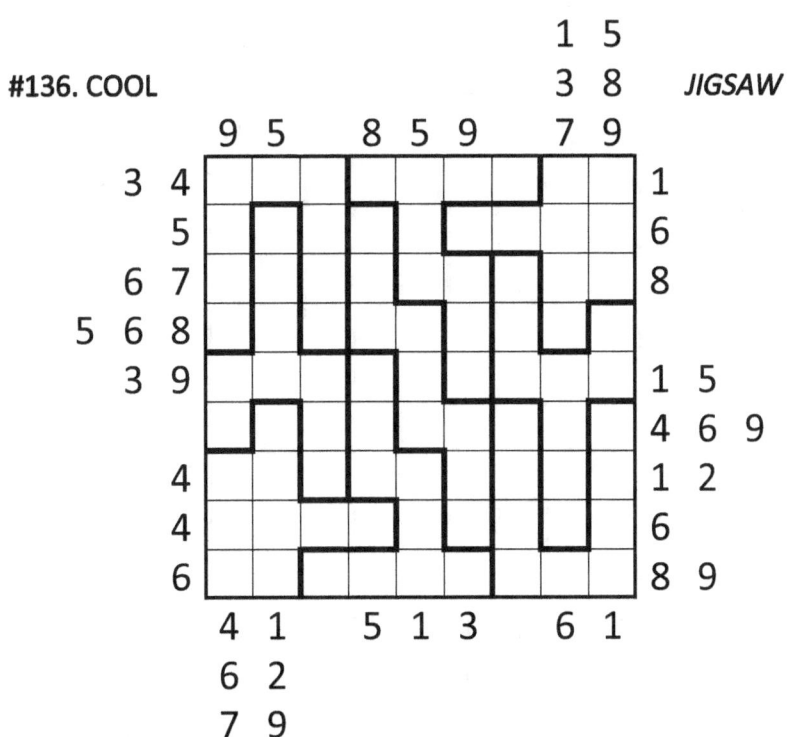

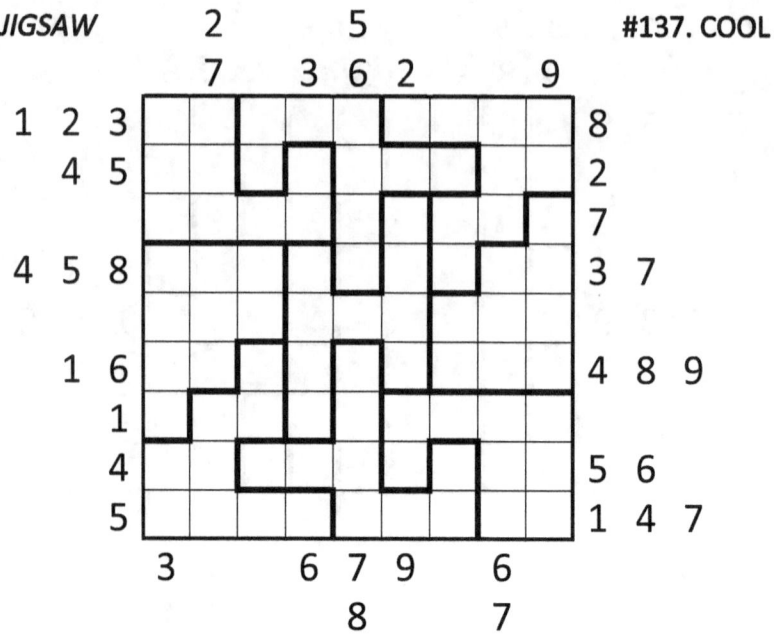

JIGSAW   #137. COOL

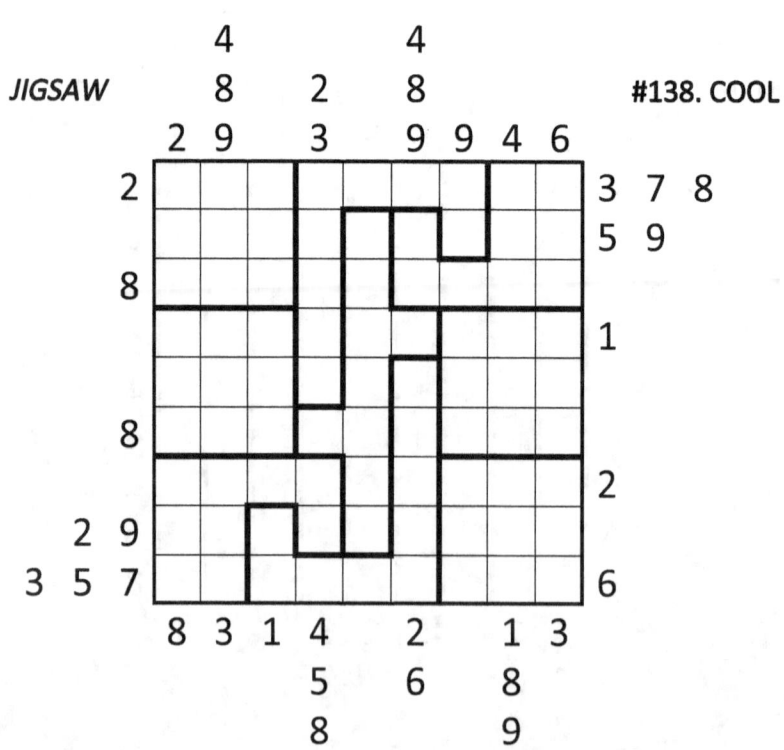

JIGSAW   #138. COOL

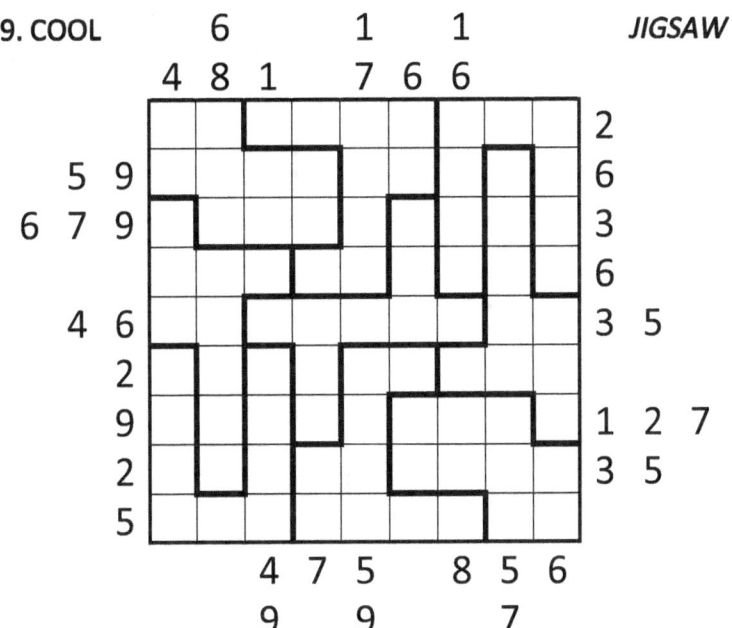

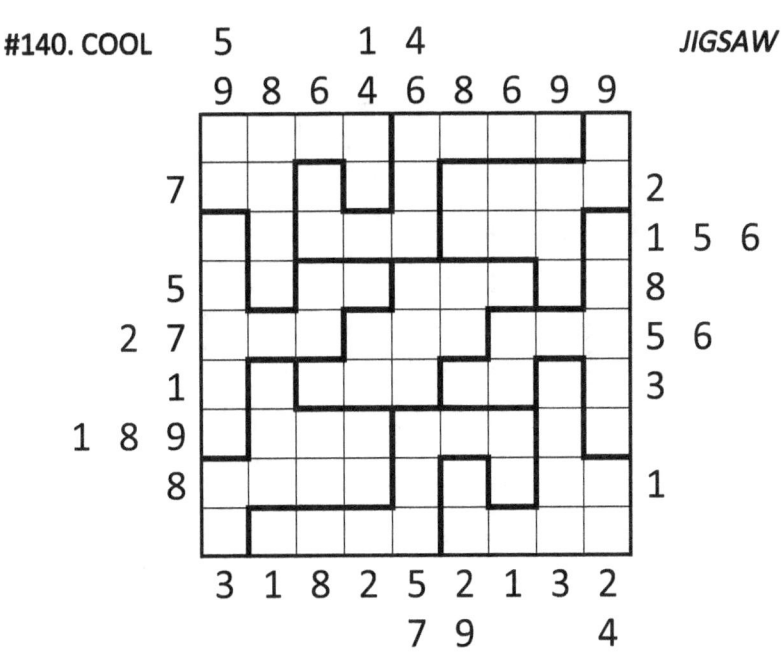

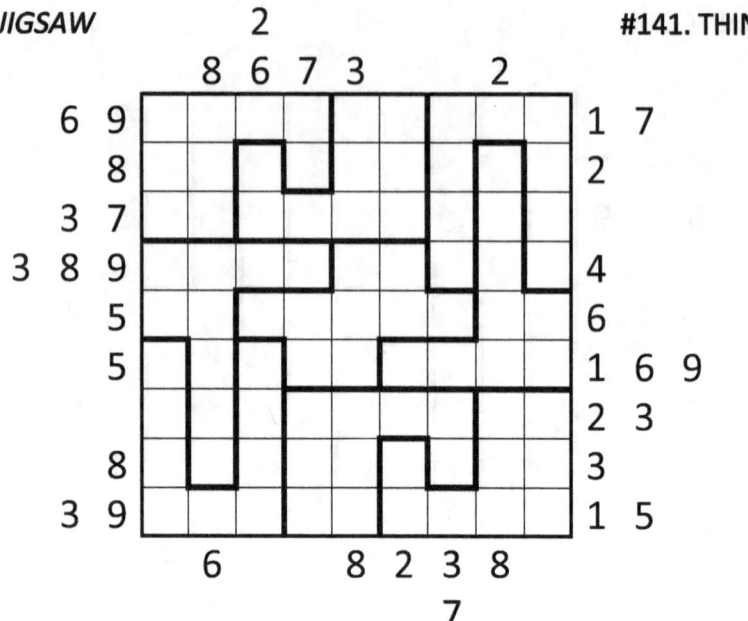

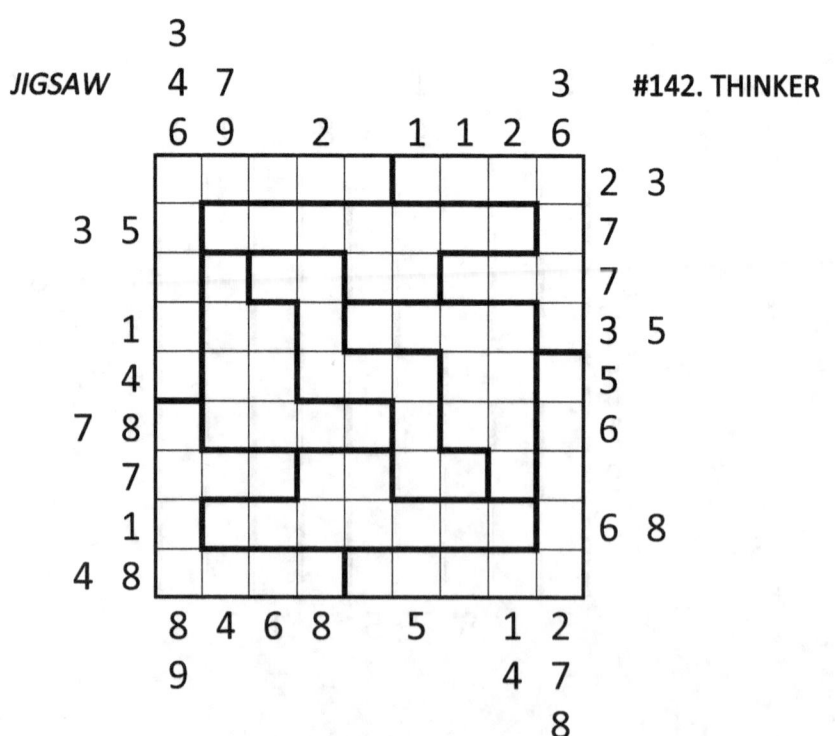

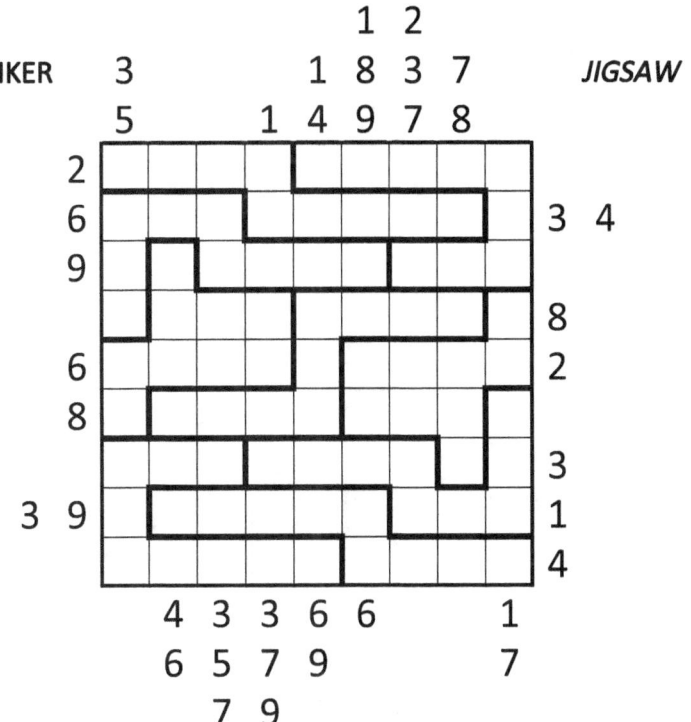

#143. THINKER   *JIGSAW*

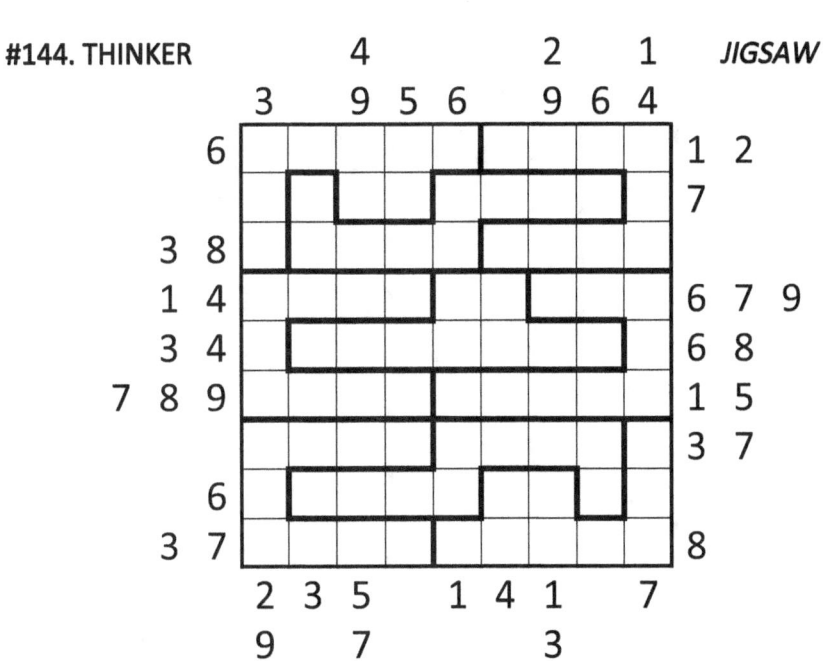

#144. THINKER   *JIGSAW*

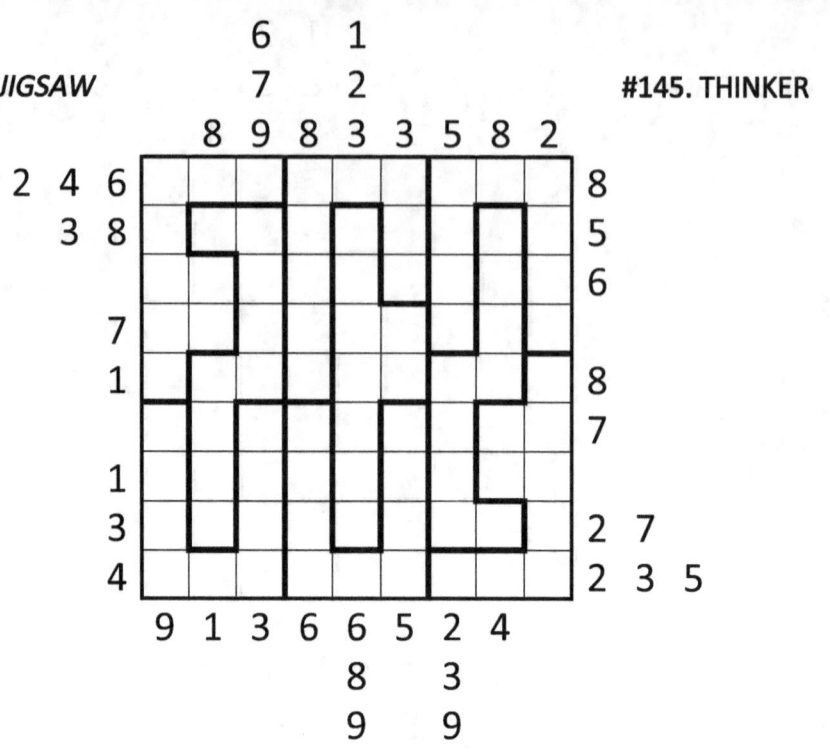

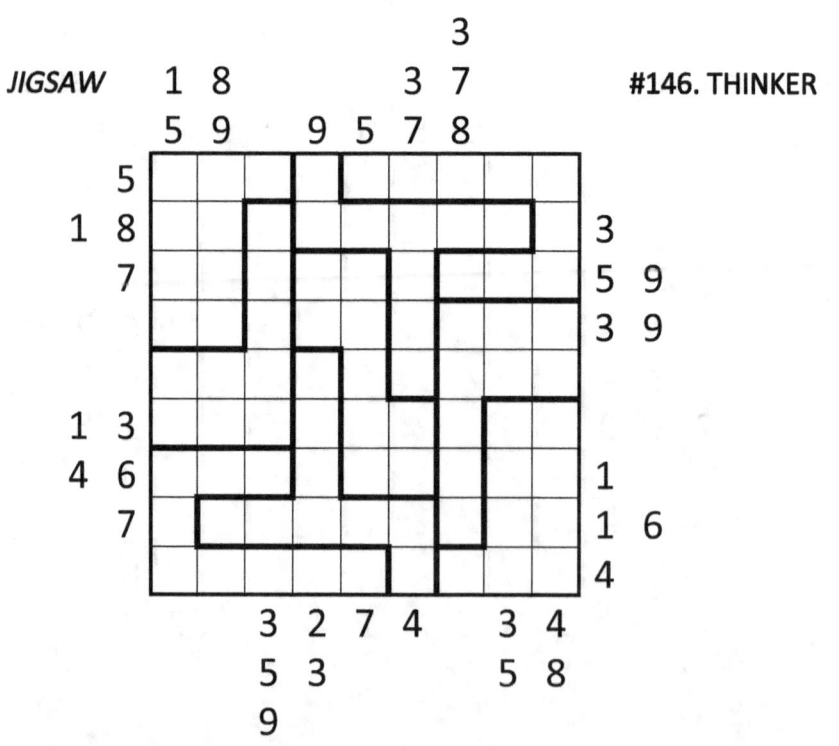

#147. THINKER

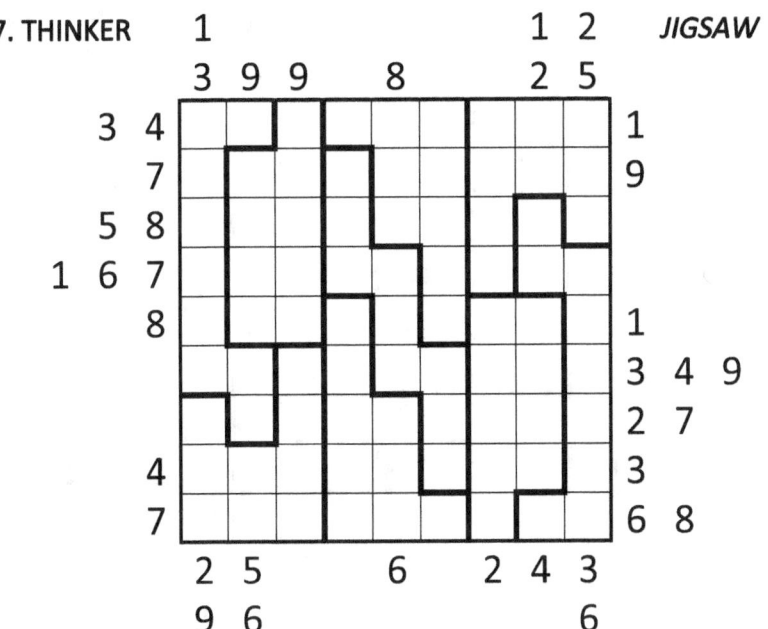

#148. THINKER

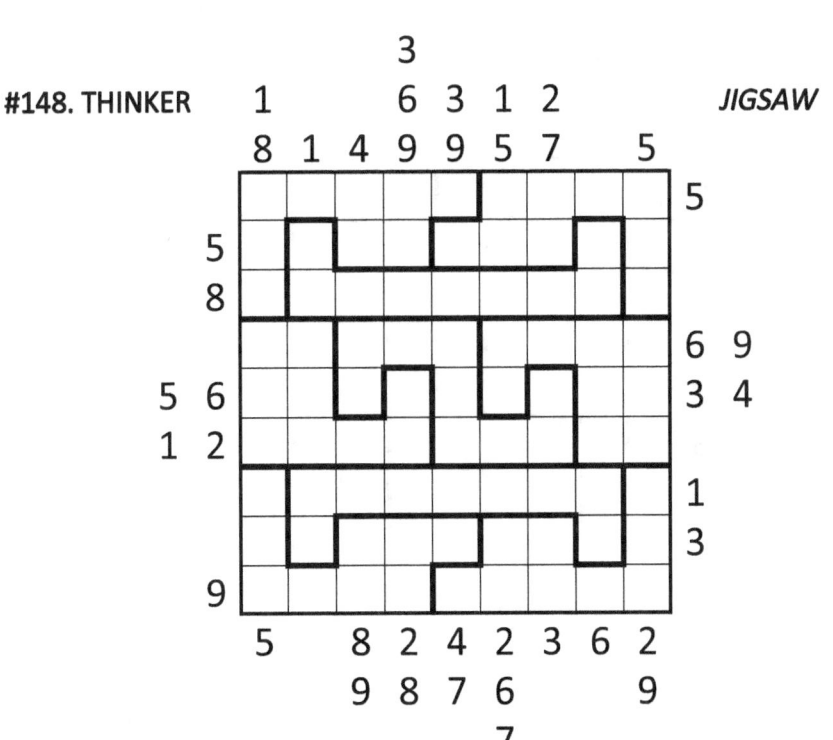

JIGSAW

#149. THINKER

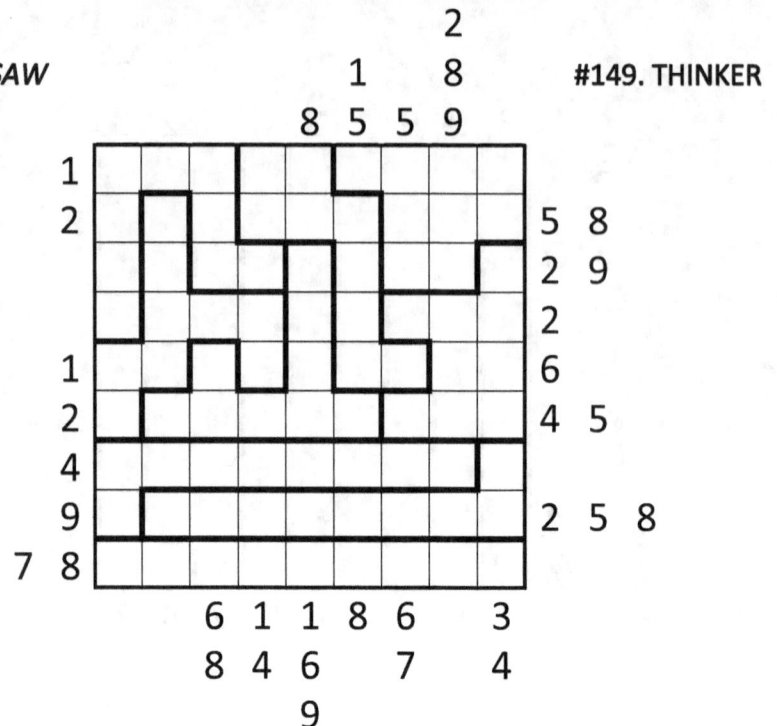

JIGSAW

#150. THINKER

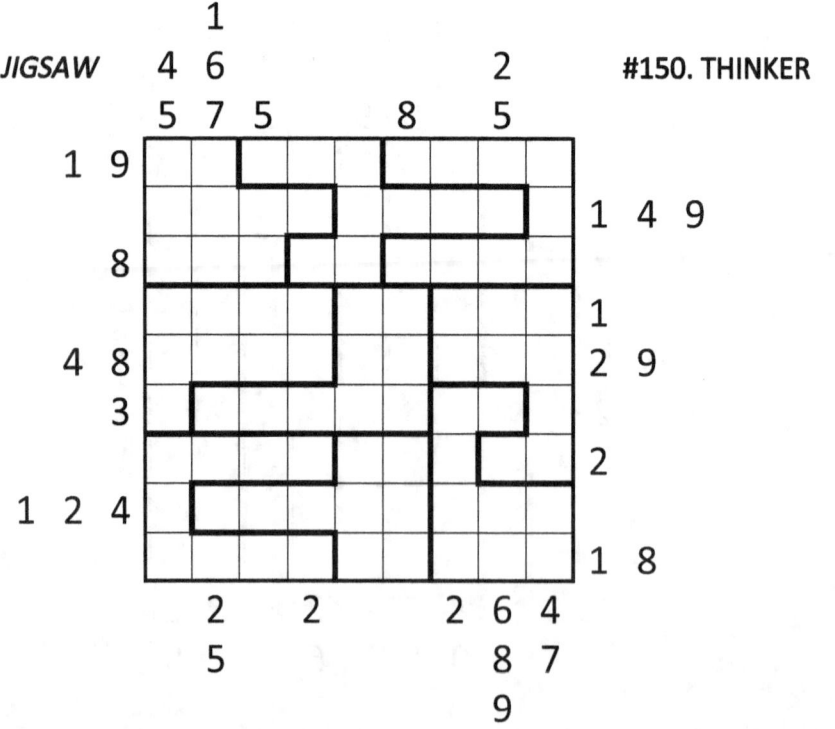

#151. THINKER   *JIGSAW*

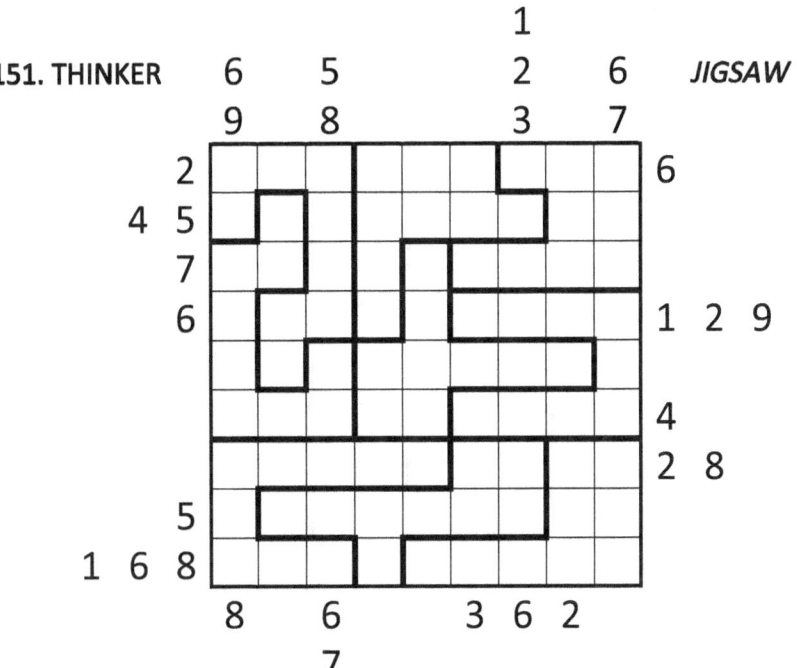

#152. THINKER   *JIGSAW*

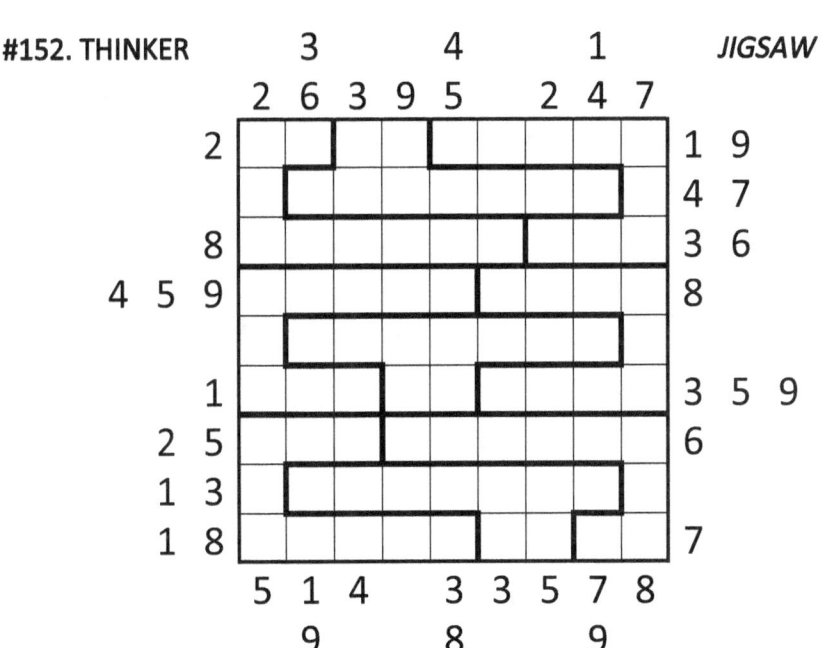

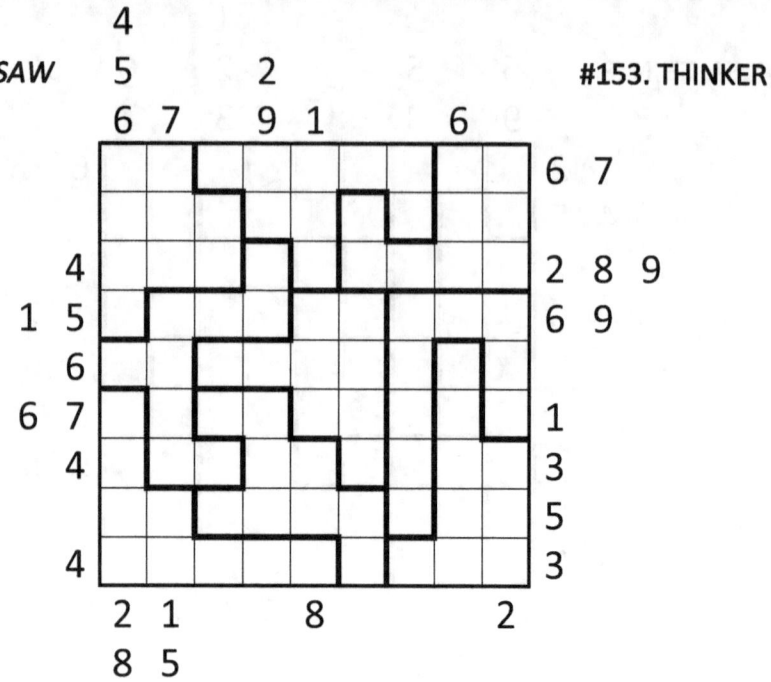

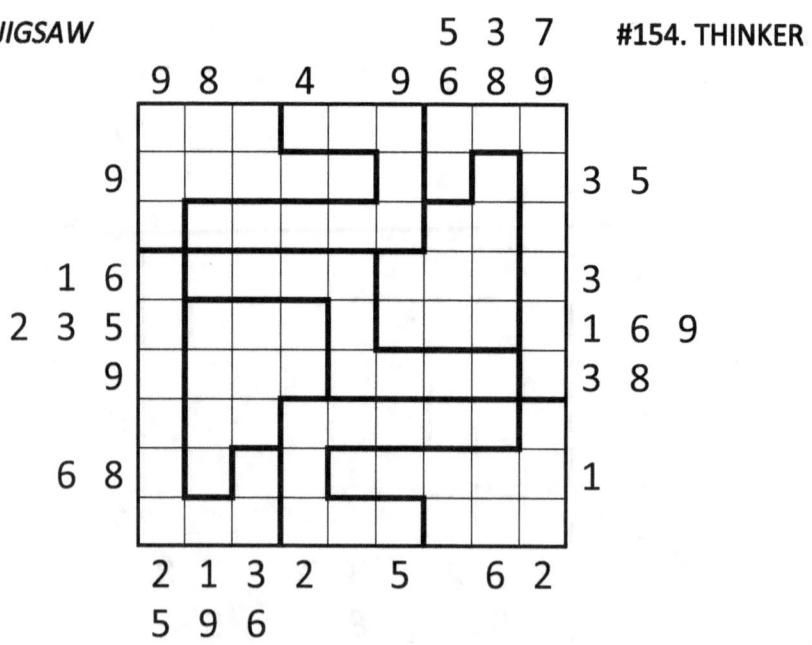

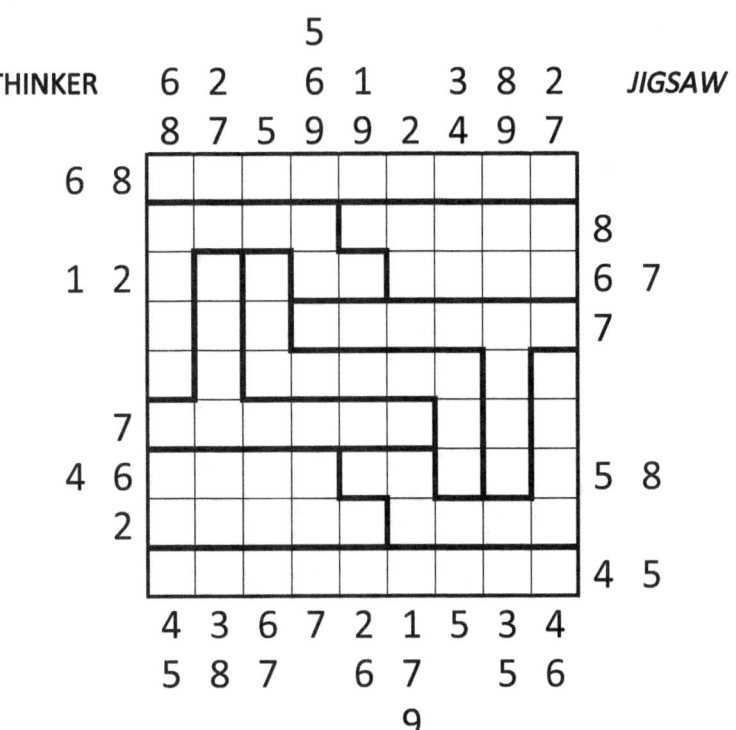

#155. THINKER *JIGSAW*

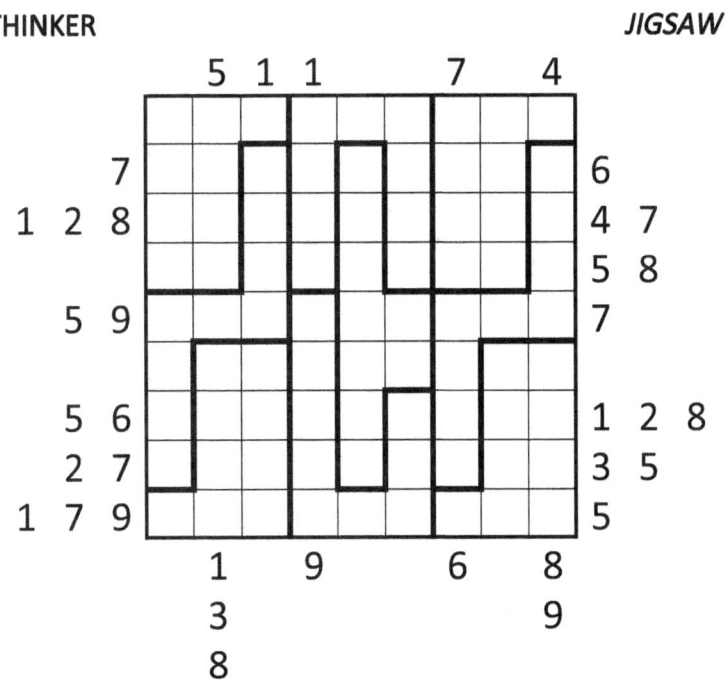

#156. THINKER *JIGSAW*

JIGSAW

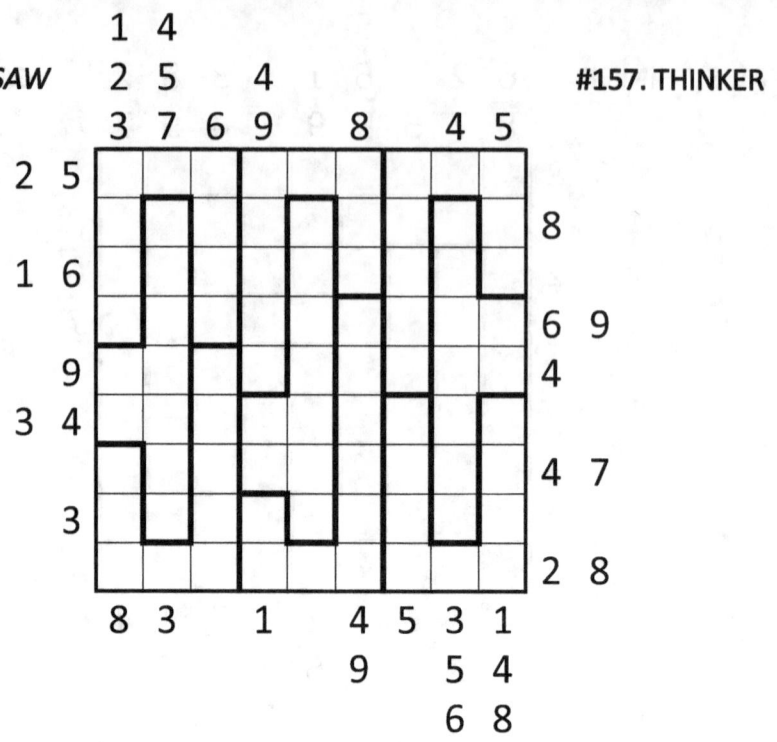

JIGSAW

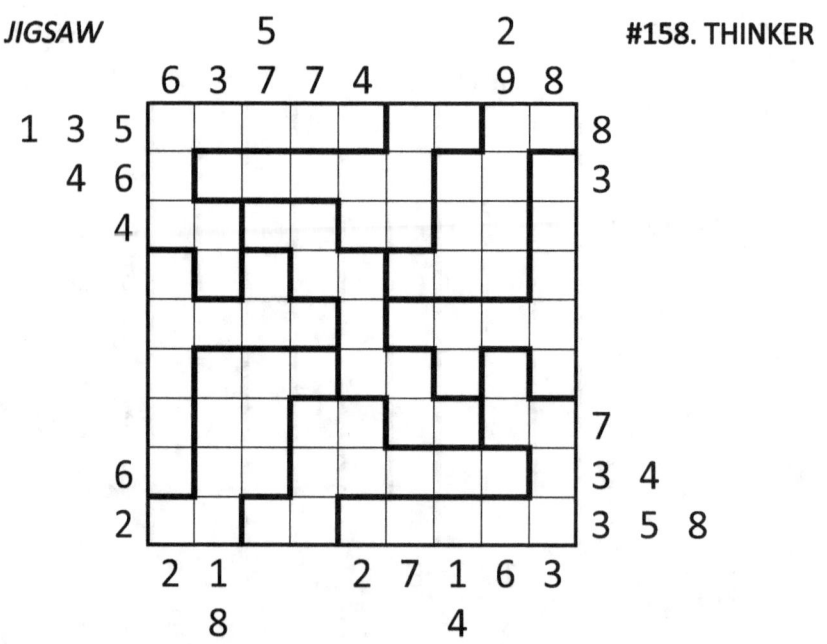

JIGSAW

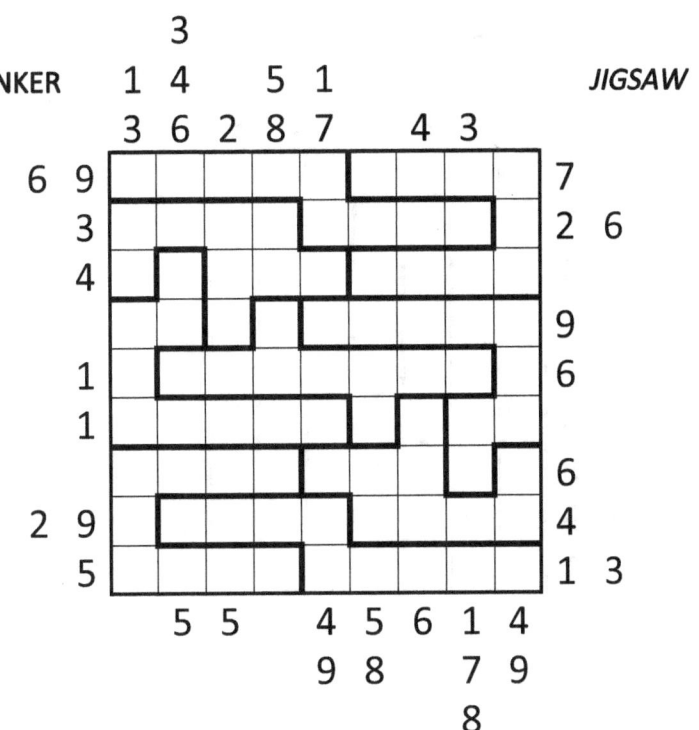

JIGSAW

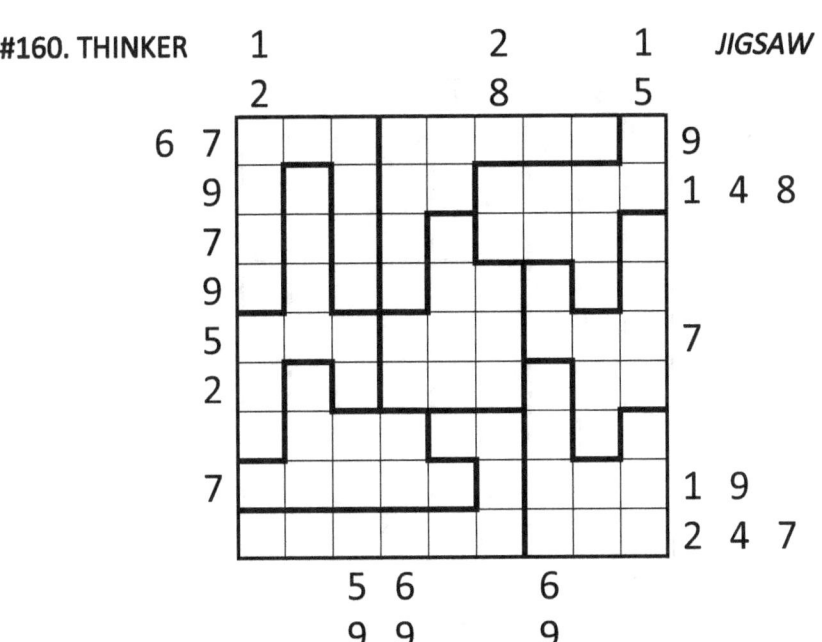

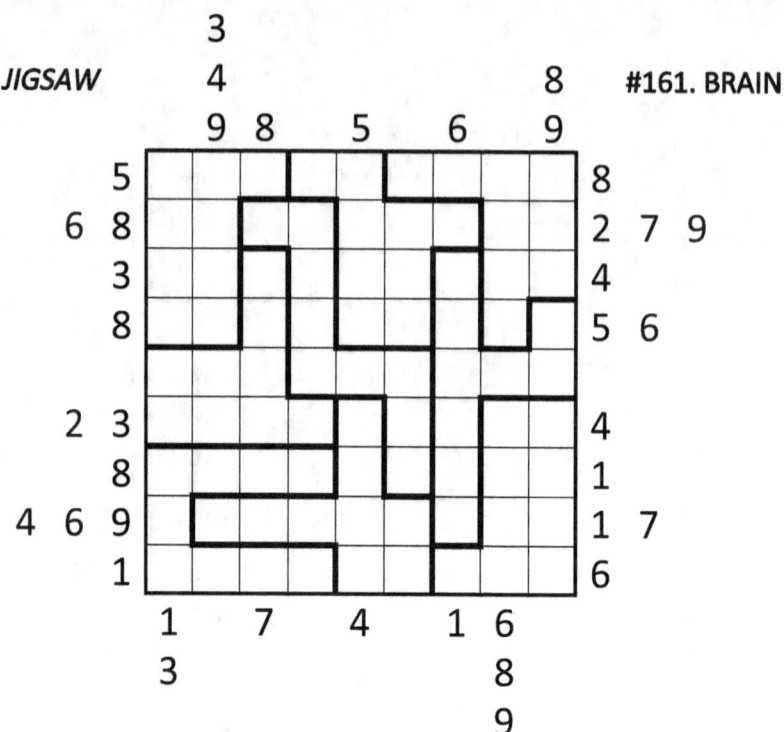

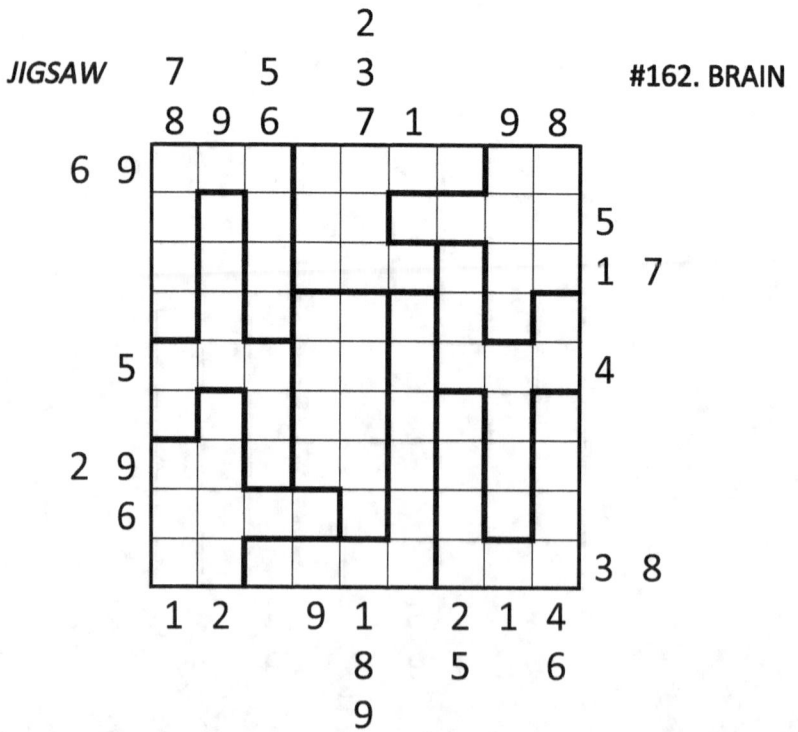

JIGSAW

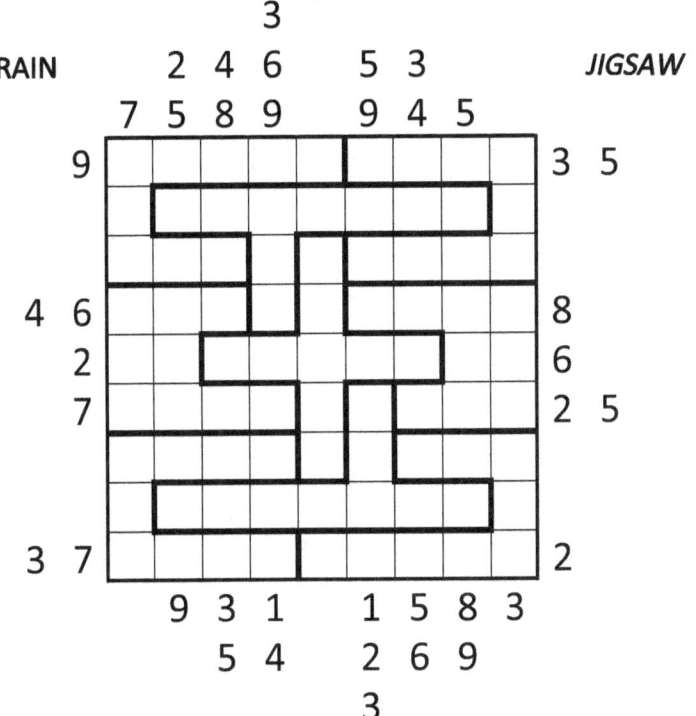

JIGSAW

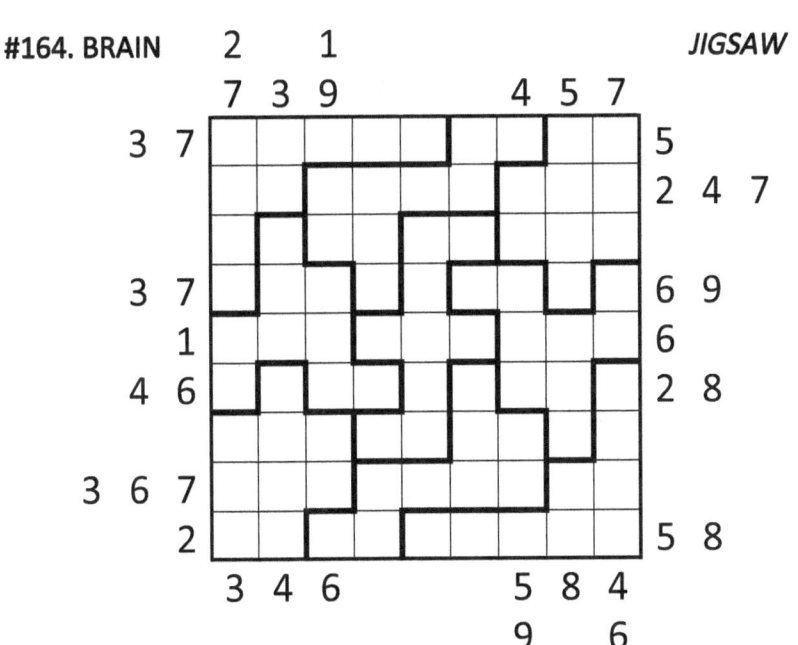

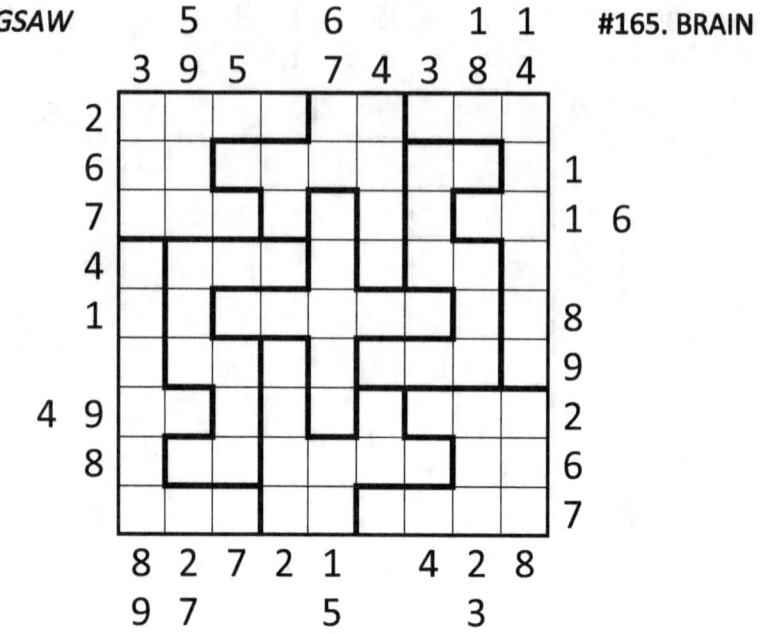

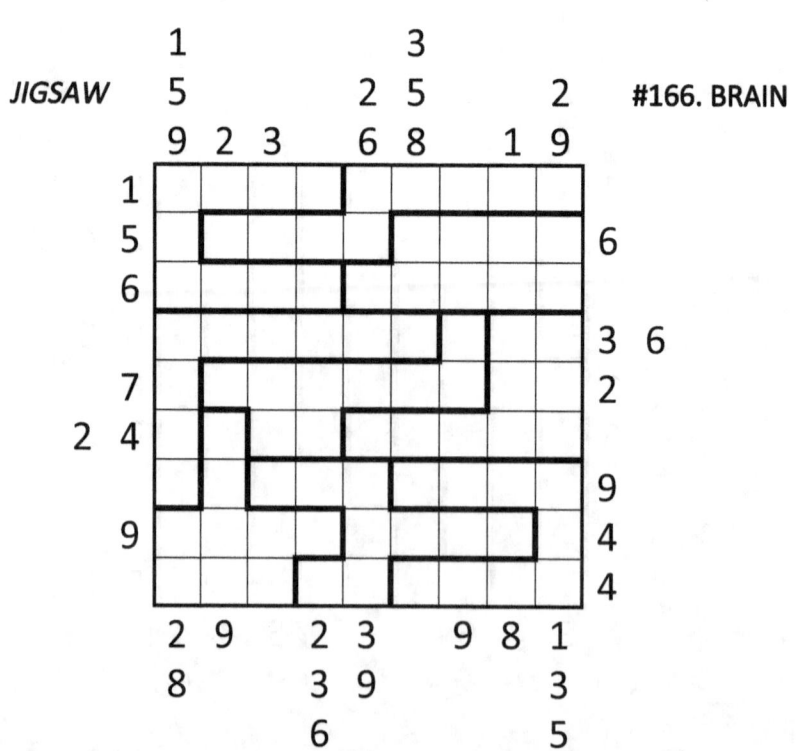

JIGSAW

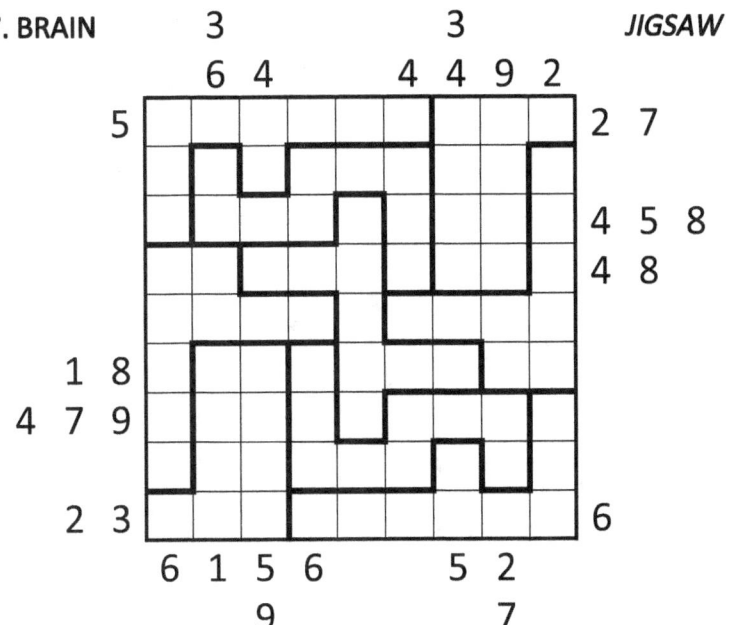

JIGSAW

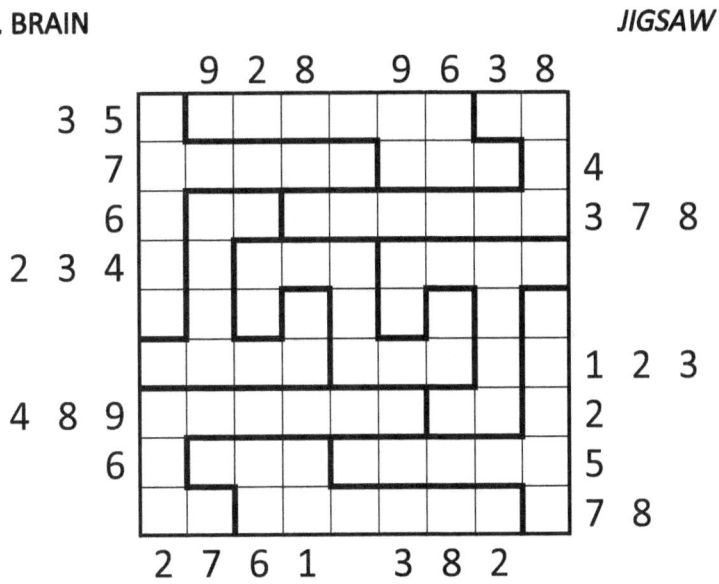

JIGSAW   #169. BRAIN

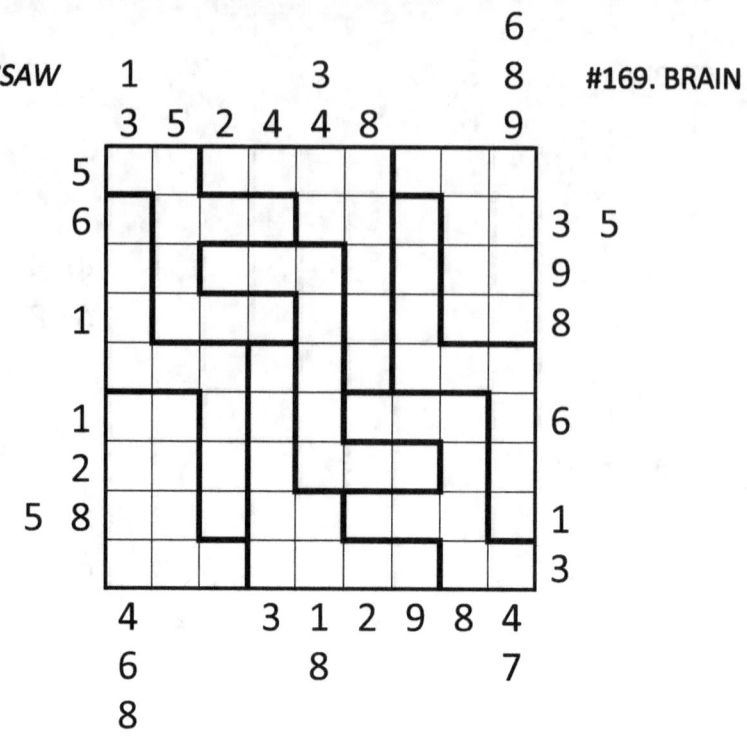

JIGSAW   #170. BRAIN

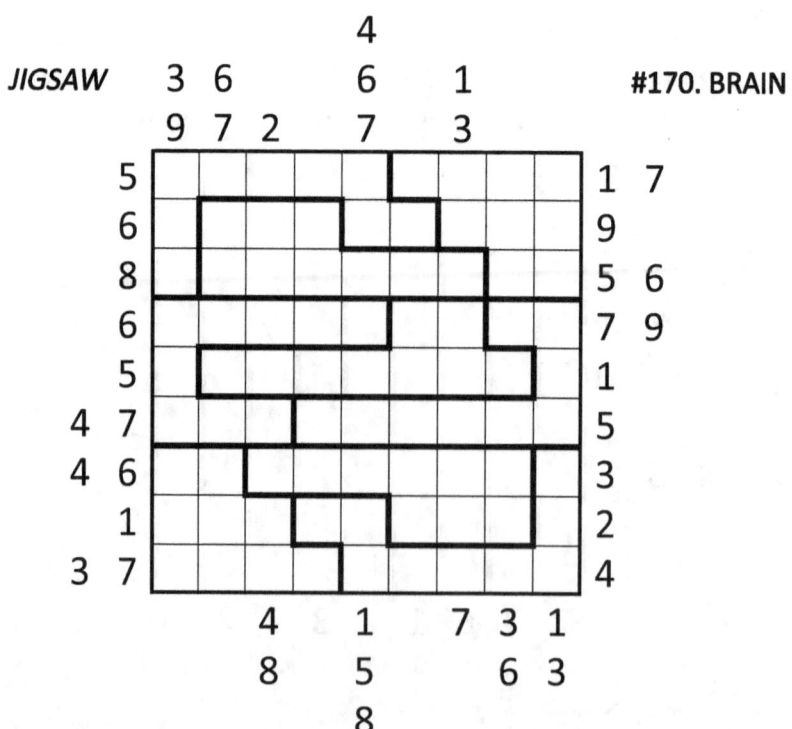

#171. BRAIN *JIGSAW*

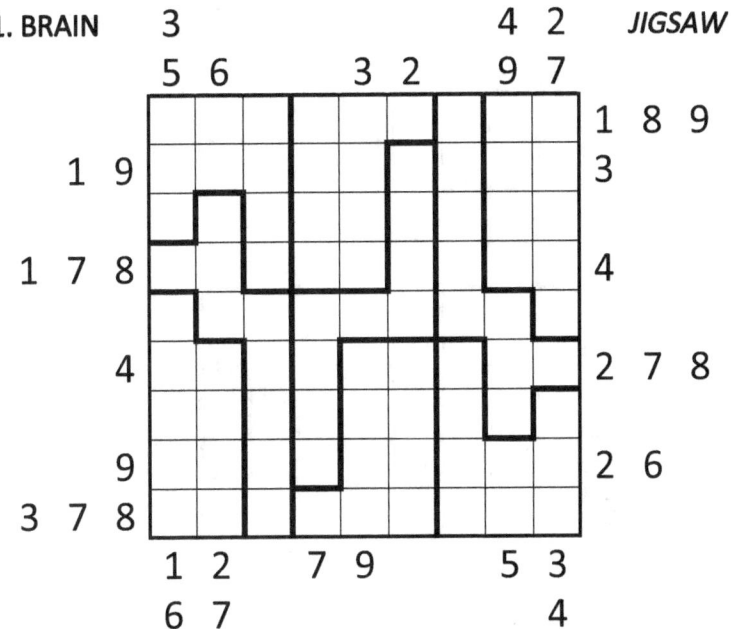

#172. BRAIN *JIGSAW*

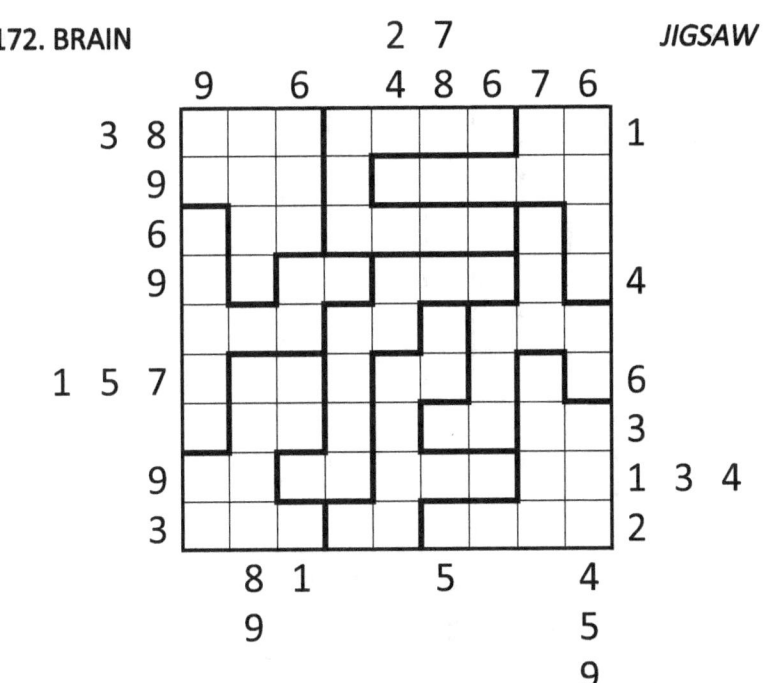

JIGSAW

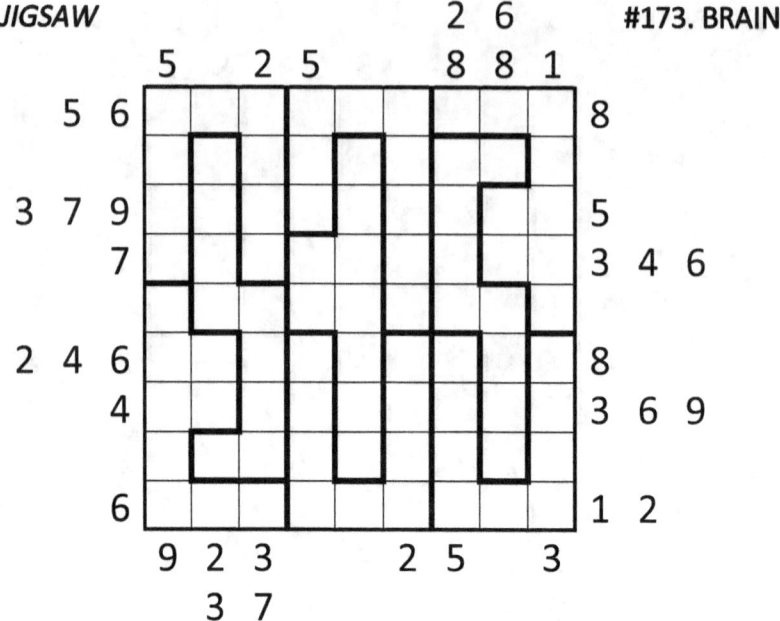

#173. BRAIN

JIGSAW

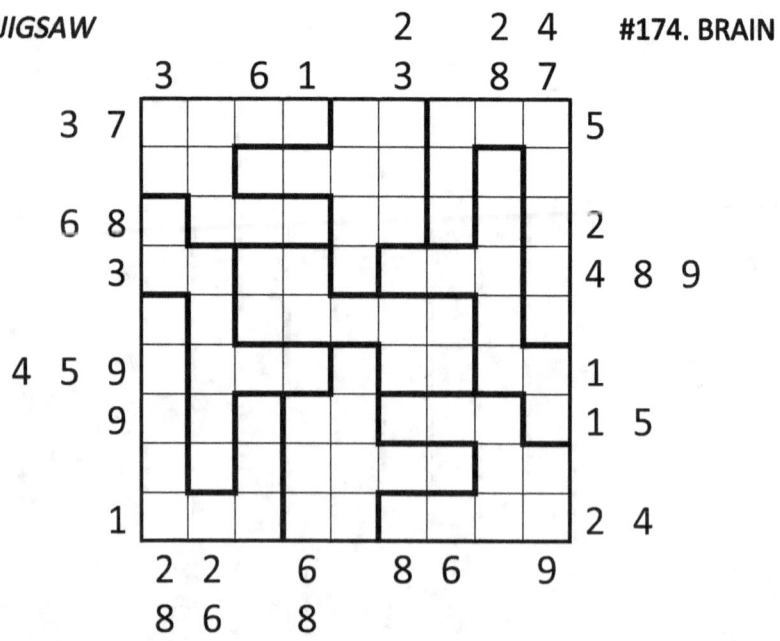

#174. BRAIN

#175. BRAIN

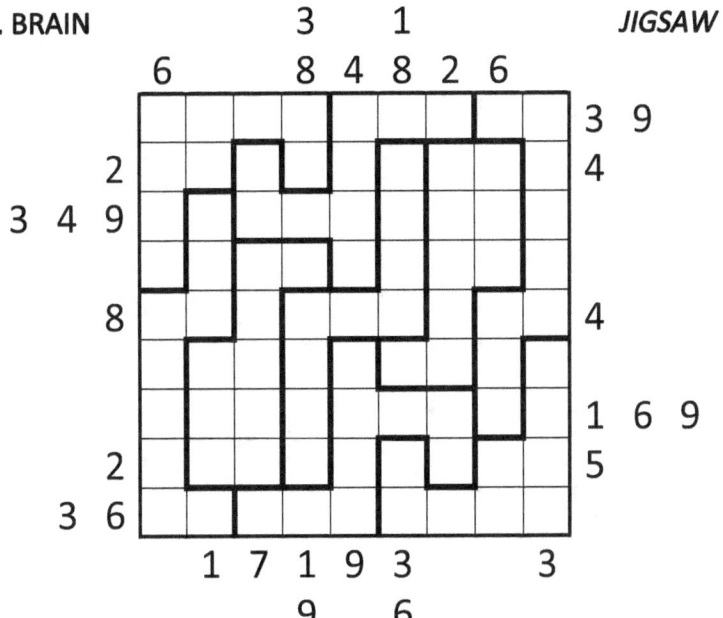

#176. BRAIN

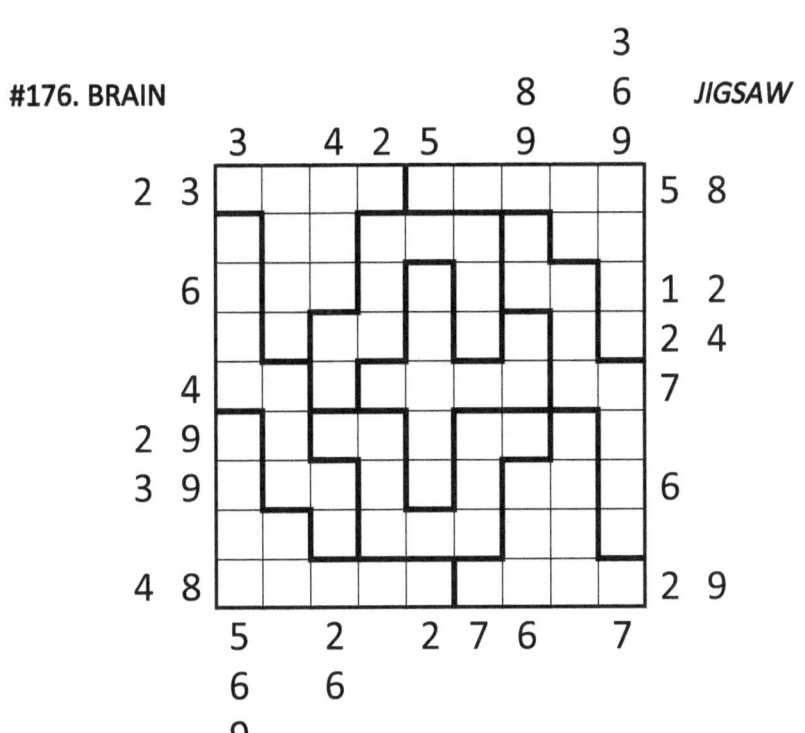

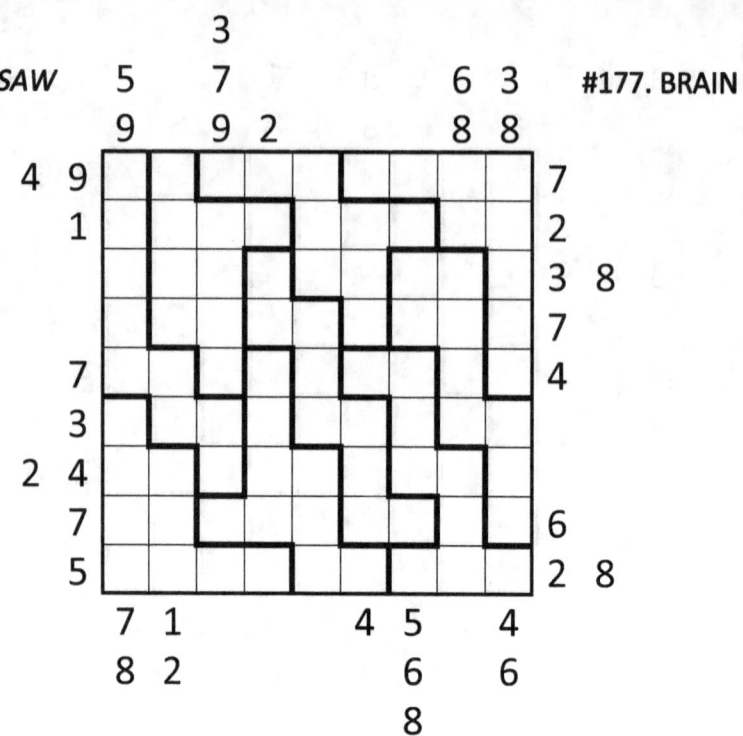

JIGSAW   #177. BRAIN

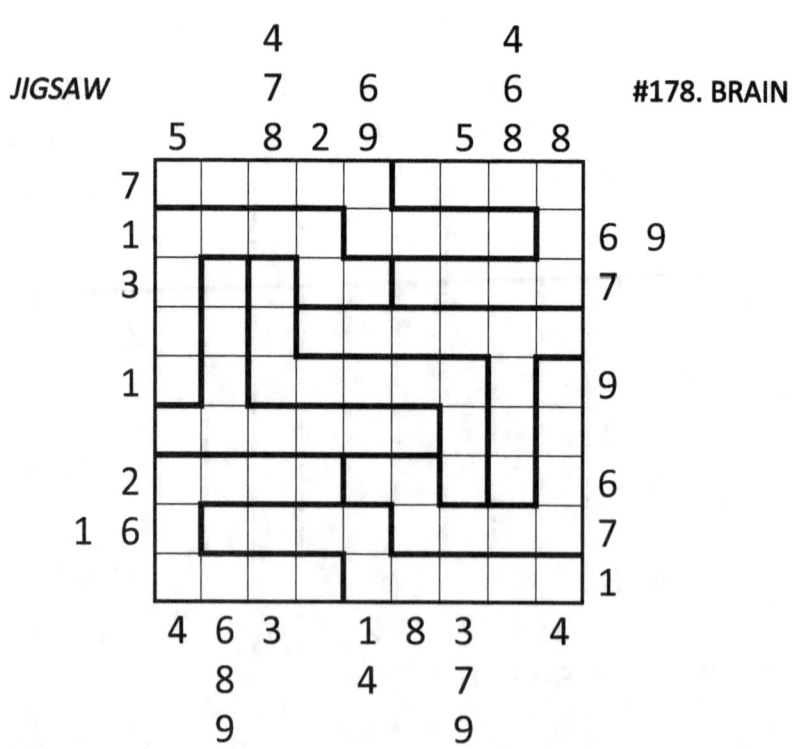

JIGSAW   #178. BRAIN

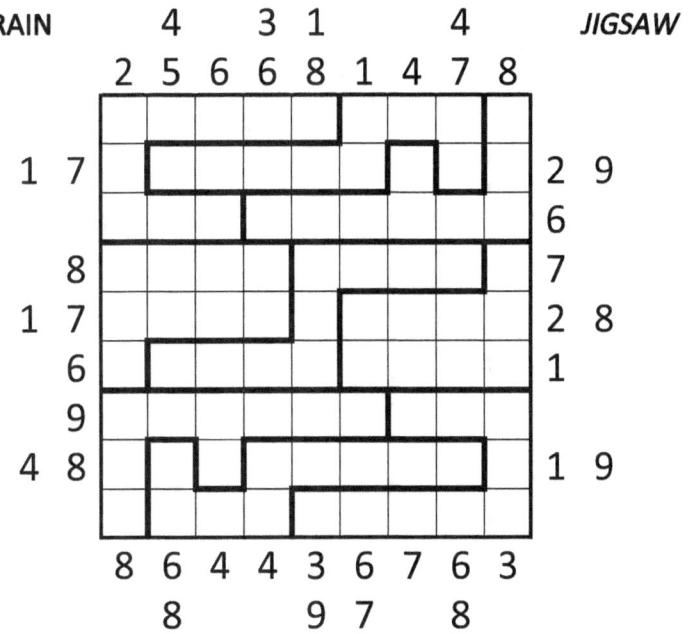

#179. BRAIN JIGSAW

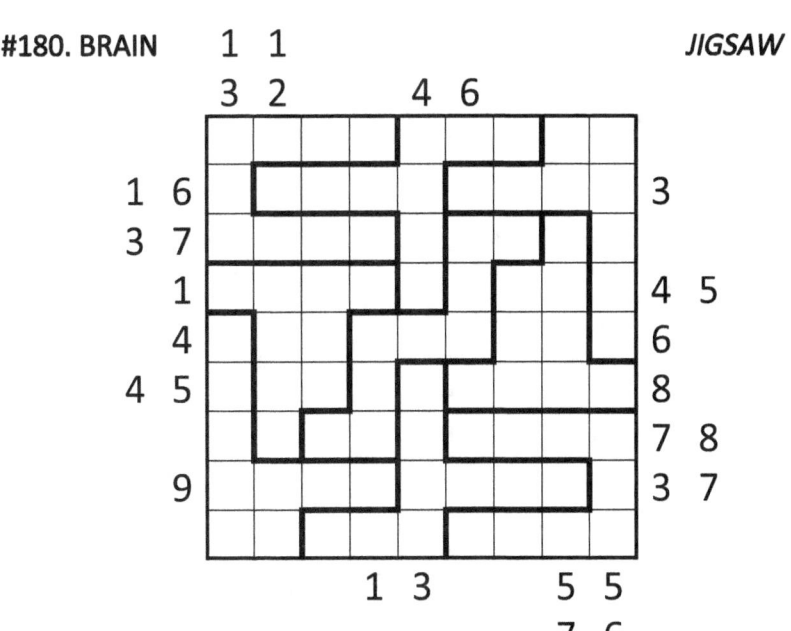

#180. BRAIN JIGSAW

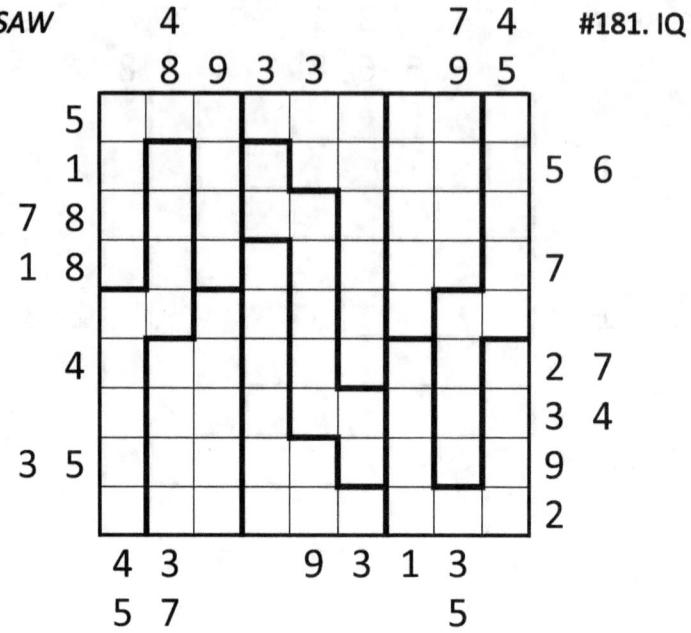

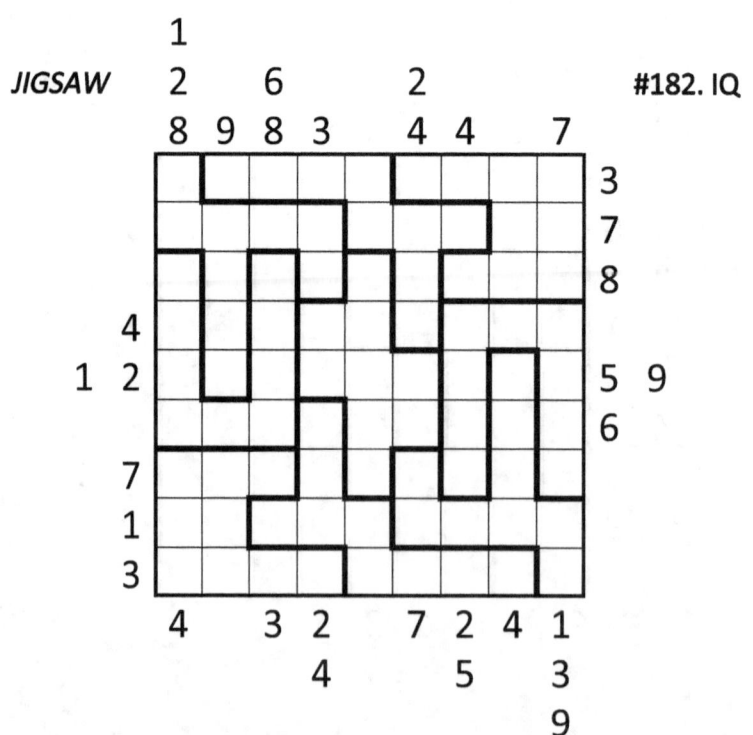

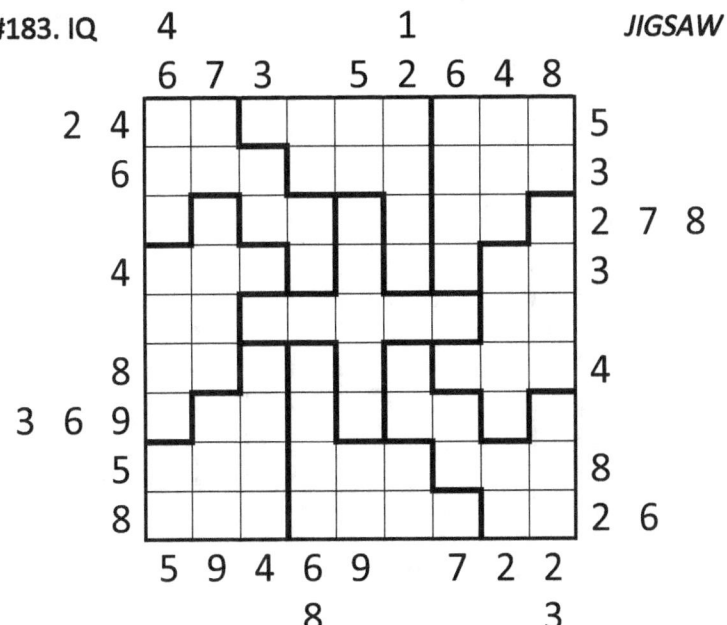

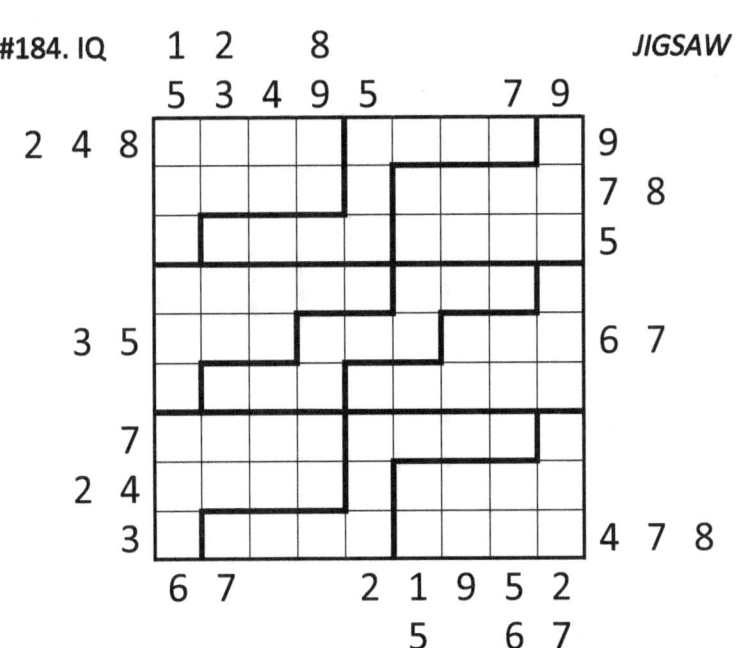

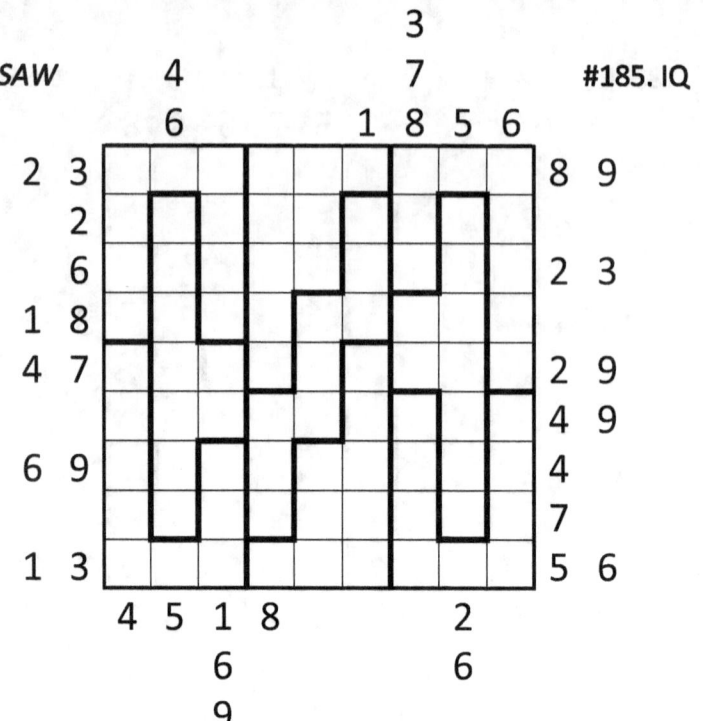

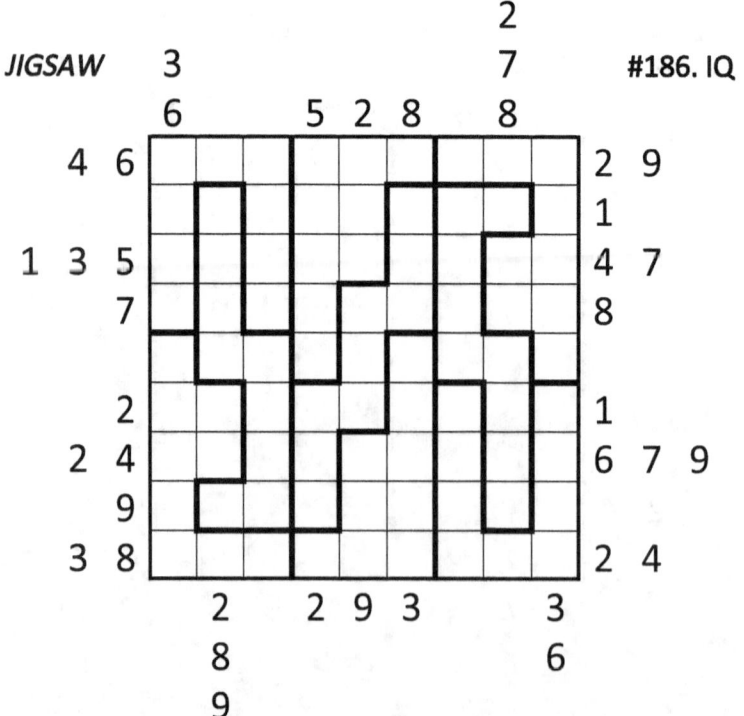

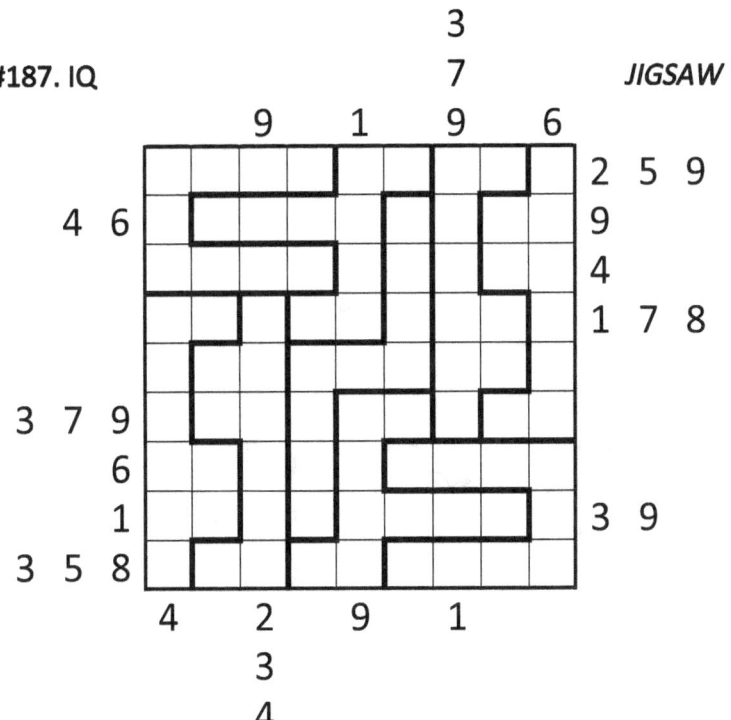

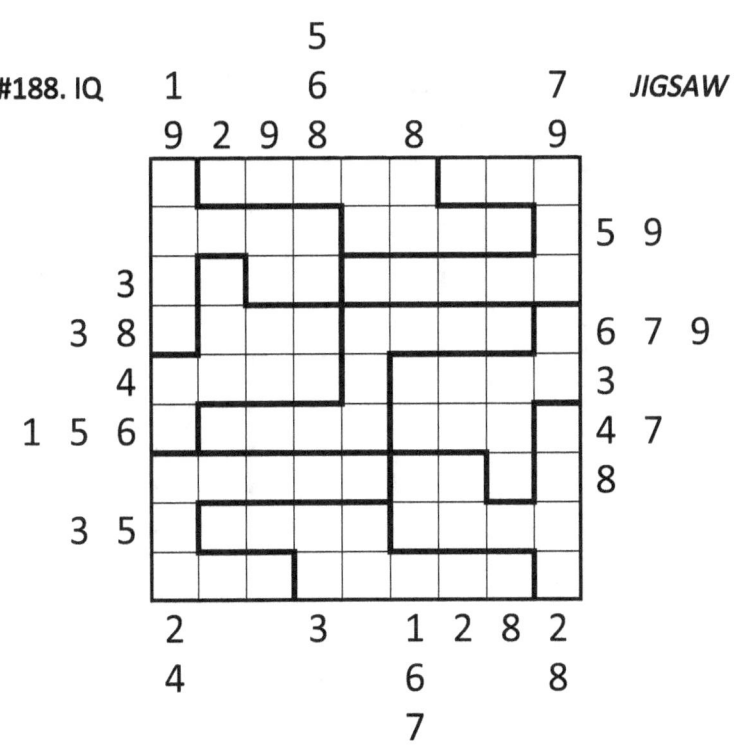

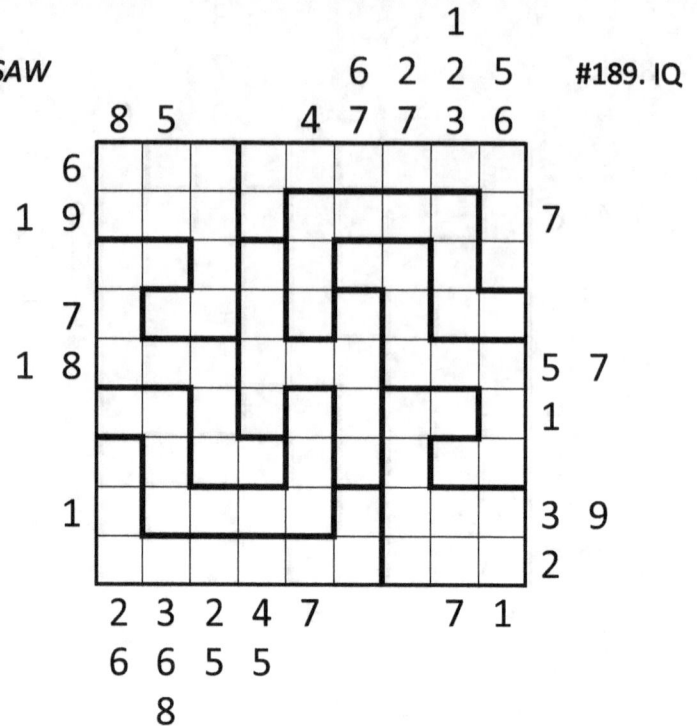

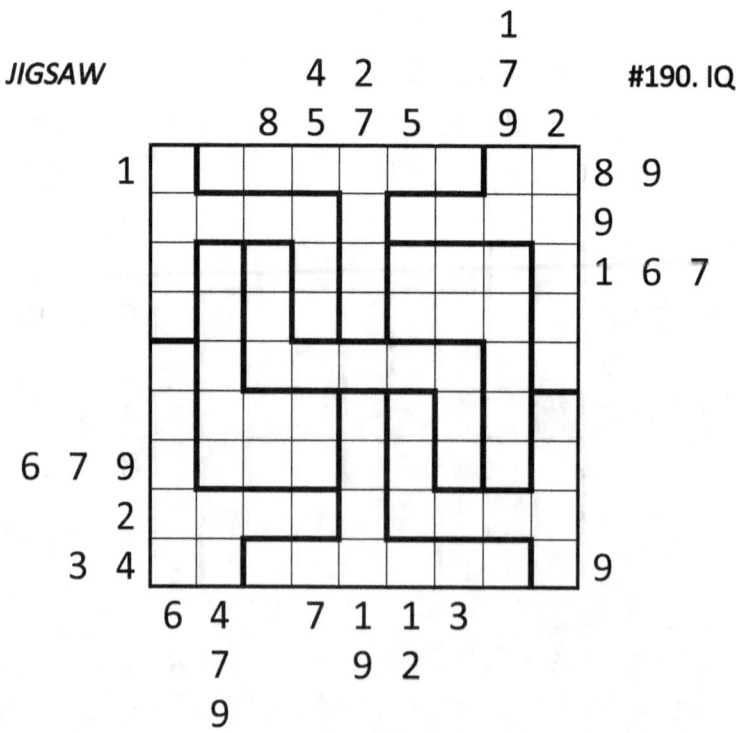

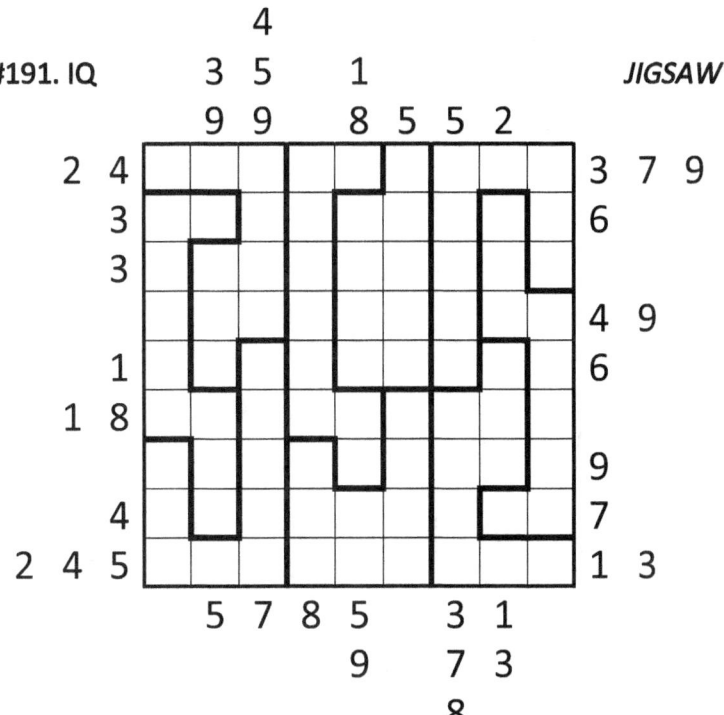

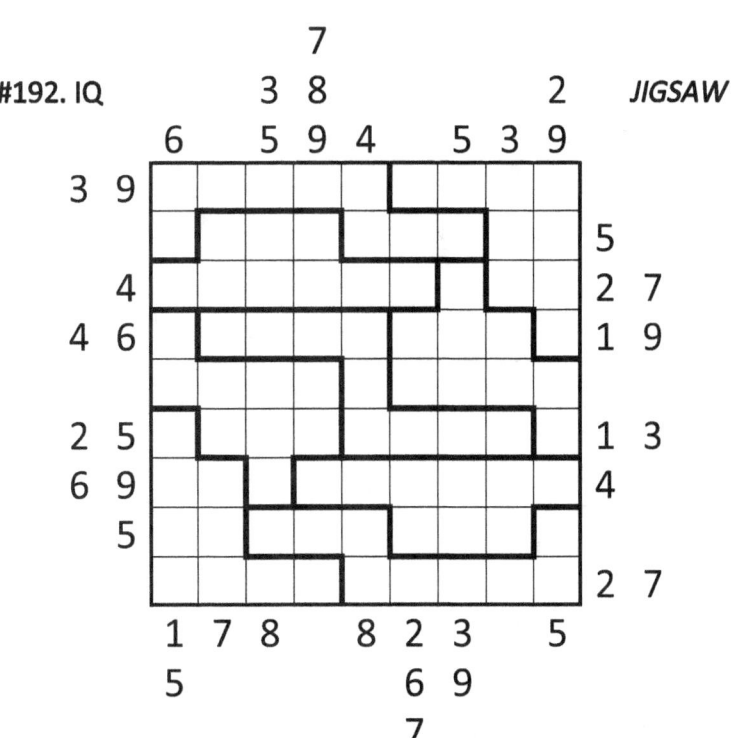

JIGSAW #193. IQ

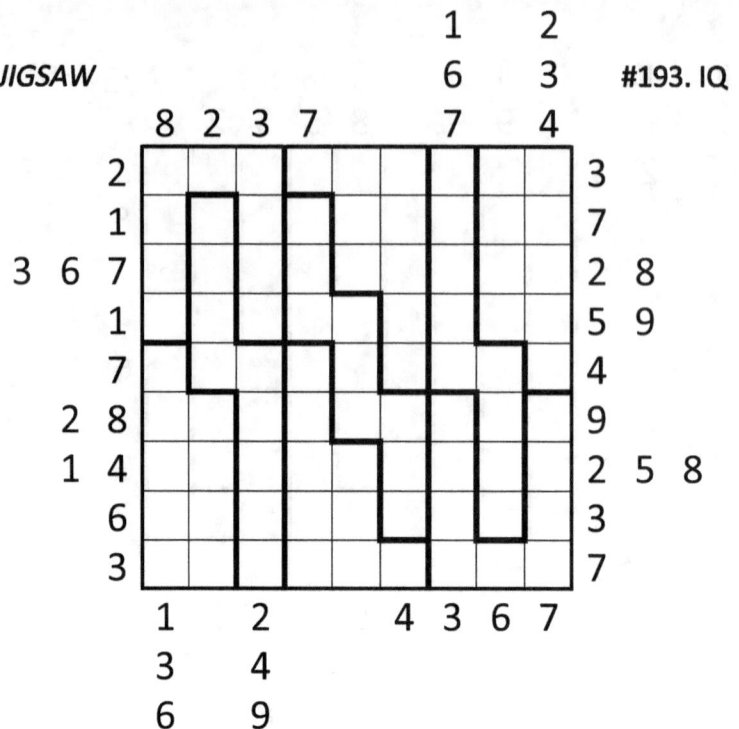

JIGSAW #194. IQ

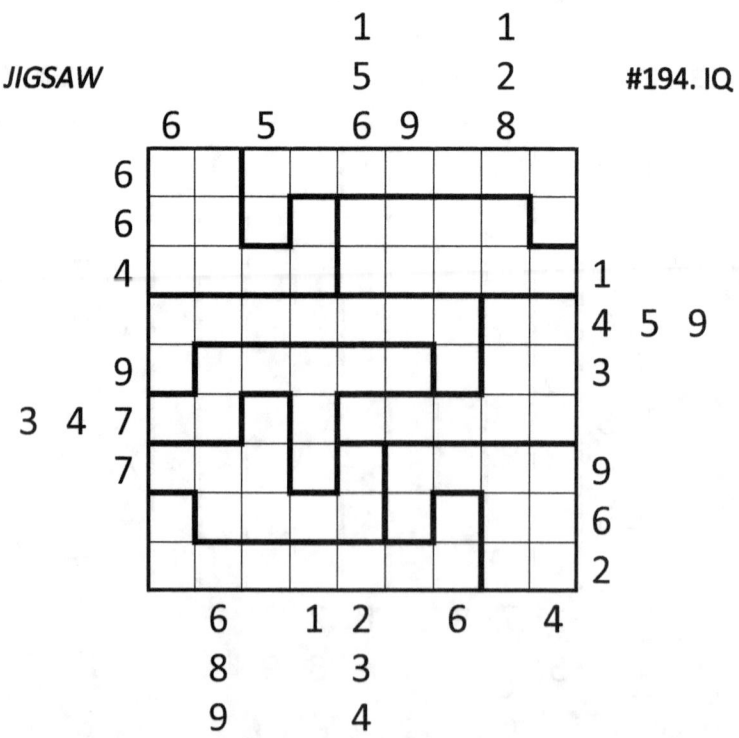

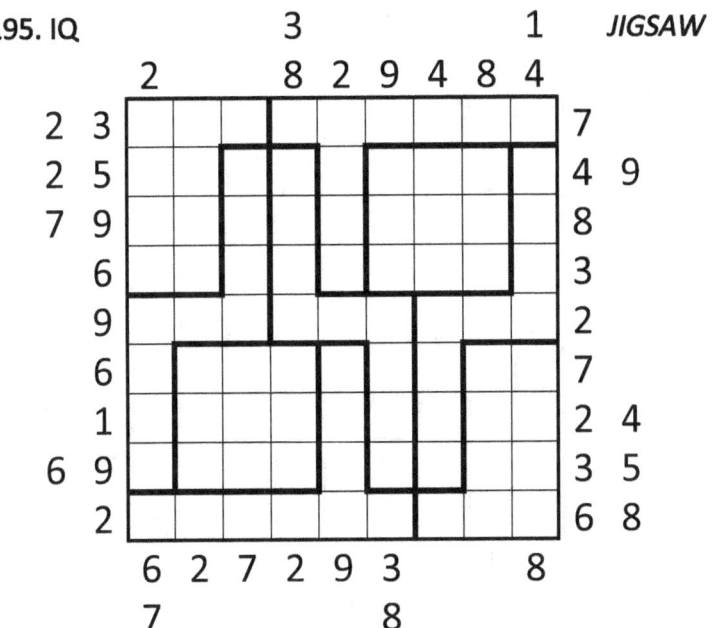

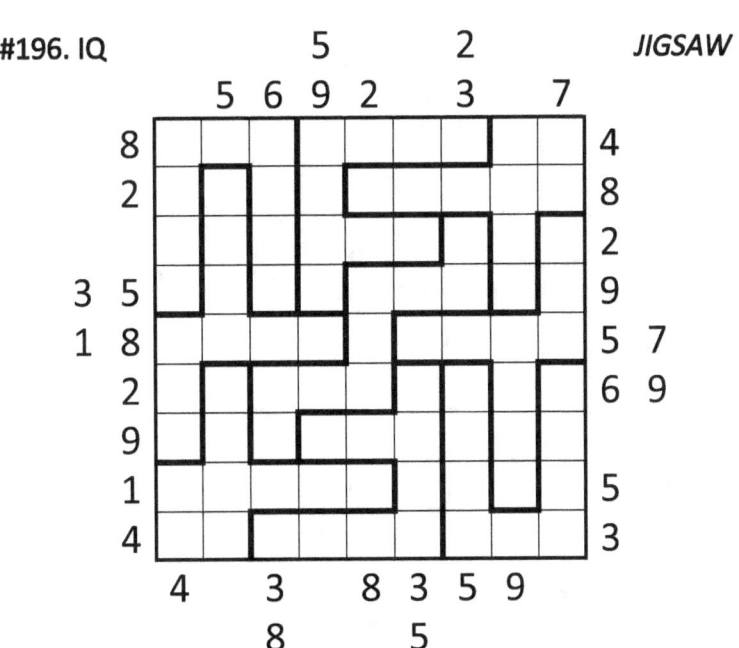

JIGSAW   #197. IQ

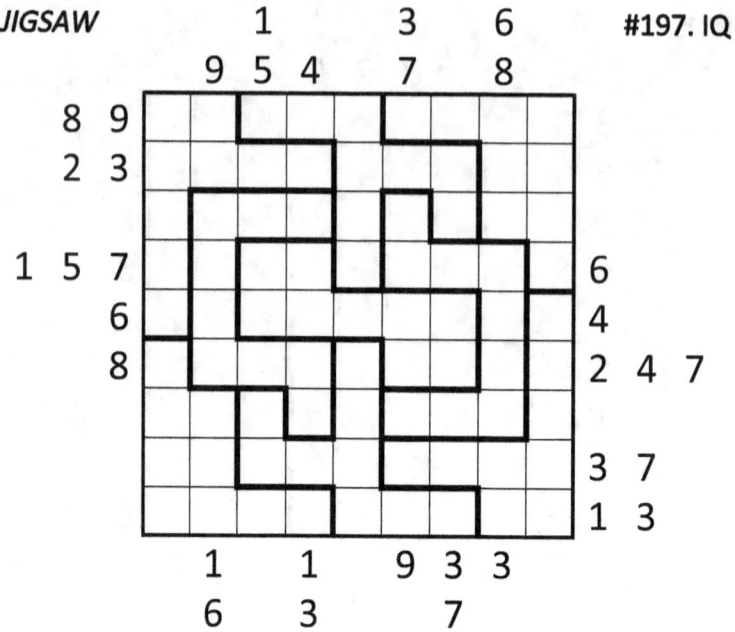

JIGSAW   #198. IQ

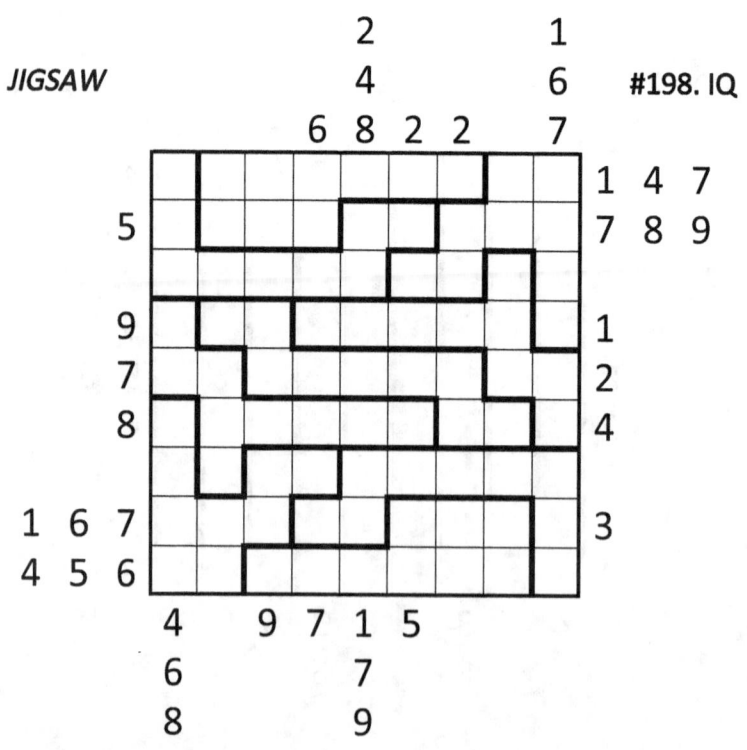

JIGSAW

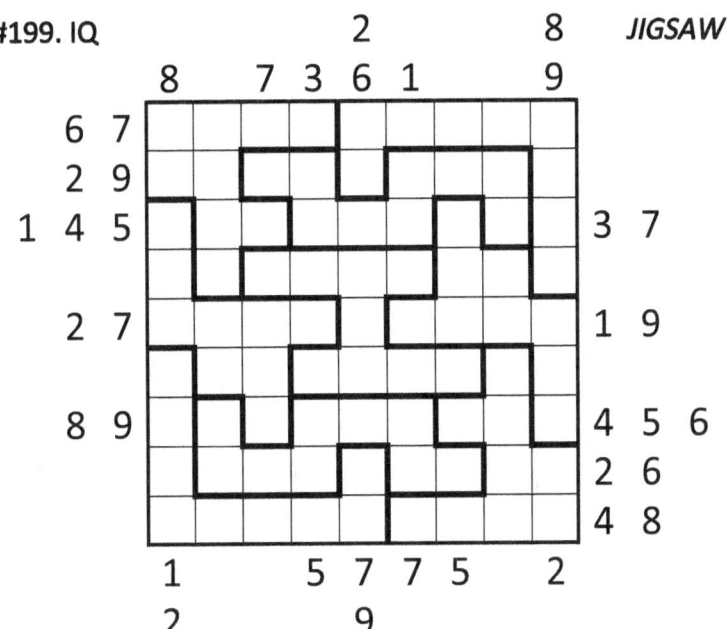

#200. IQ

JIGSAW

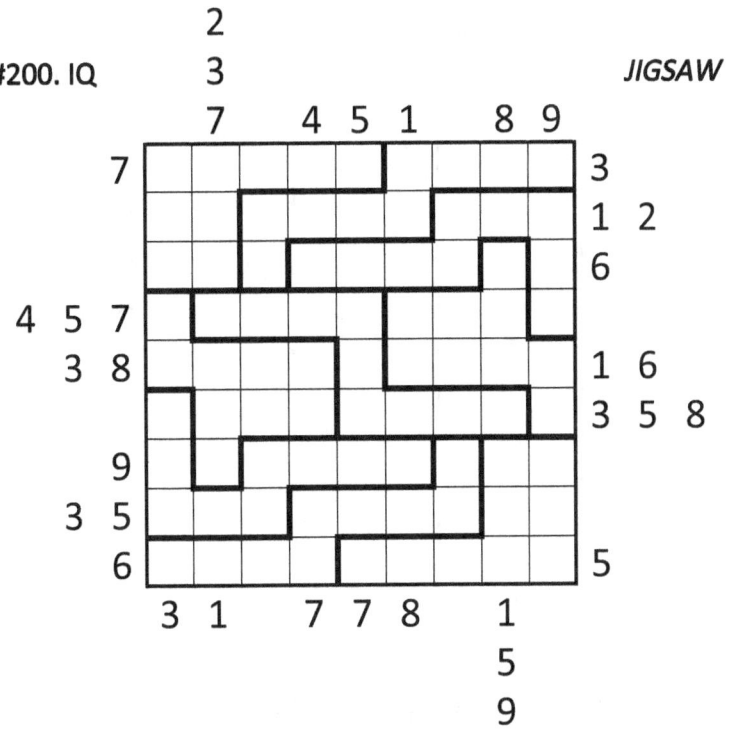

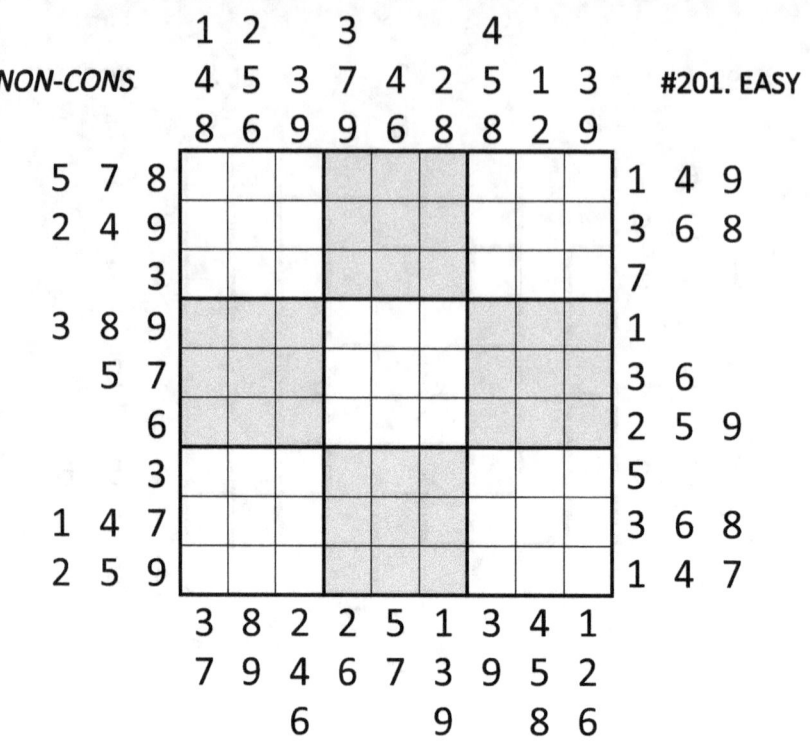

NON-CONS #201. EASY

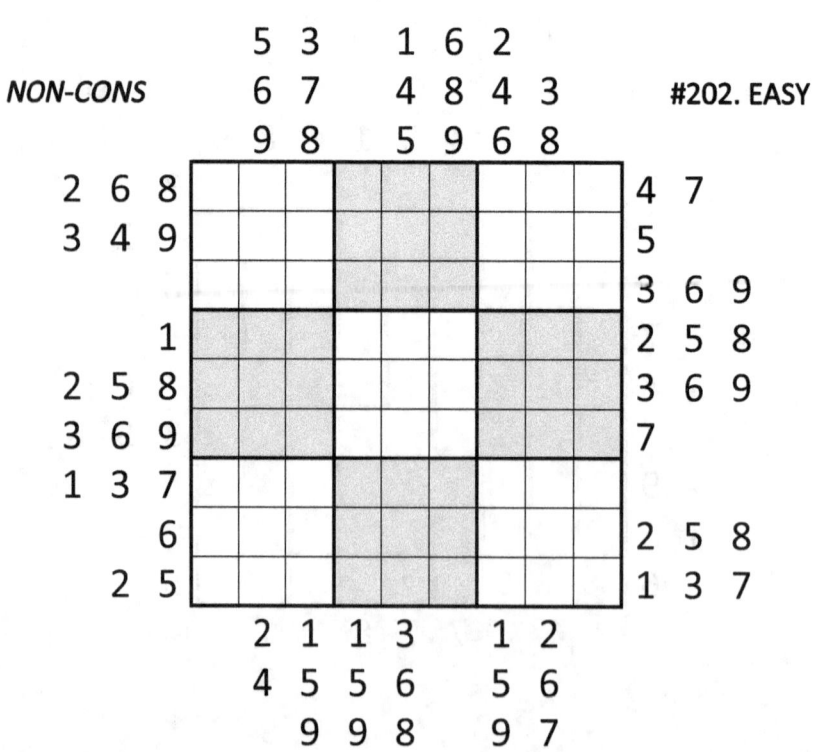

NON-CONS #202. EASY

#203. EASY

NON-CONS

	3 6 9	1 4 7	2 5 8		4	3 6 9	9	2 5 8	1 4 7	
2 7 9										4 6 8
4 6 8										1 3 5
4 6 8										1 3 5
1 3 5										2 7 9
3 5 7										2 4 9
1 6 8										3 5 7
	2 5 8	3 6 9	1	1 4 7	6		1 4 7	3 6 9	2 5 8	

#204. EASY

NON-CONS

	1 4	2 5 7	3 6 8	3 6 8	1 4 7	2 5 9	2	3 8	1 4 7	
3 7										2 4 6
2 4 6										1 5 8
2 4 6										1 5 8
3 7 9										2 4 6
1 5 8										3 7 9
1 5 8										3 7 9
3 7 9										4 6
	3 6 8	1 9	7	2 5 9	3 6 8	1 4 7	1 4 7	2 5 9	6 8	

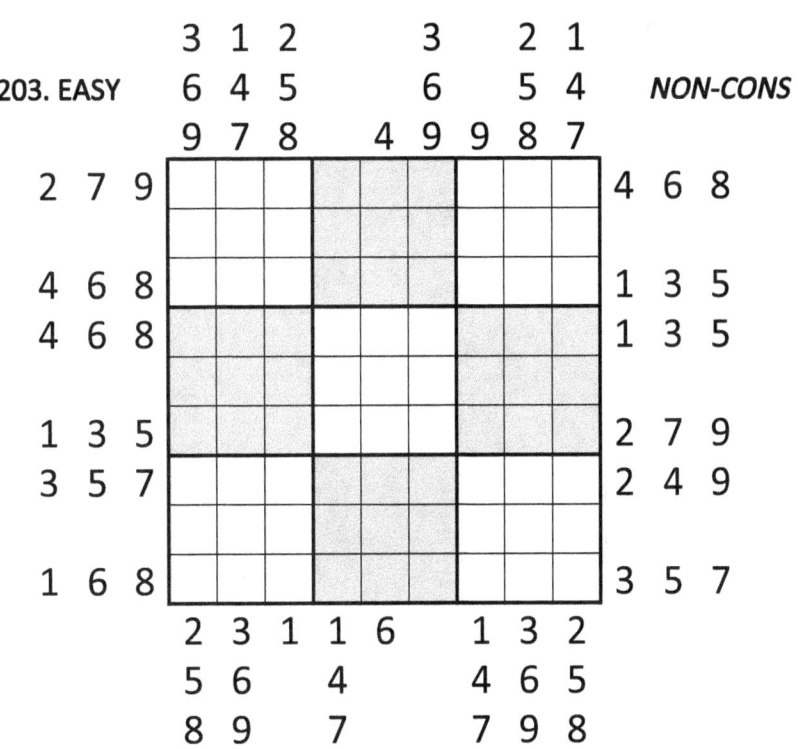

NON-CONS

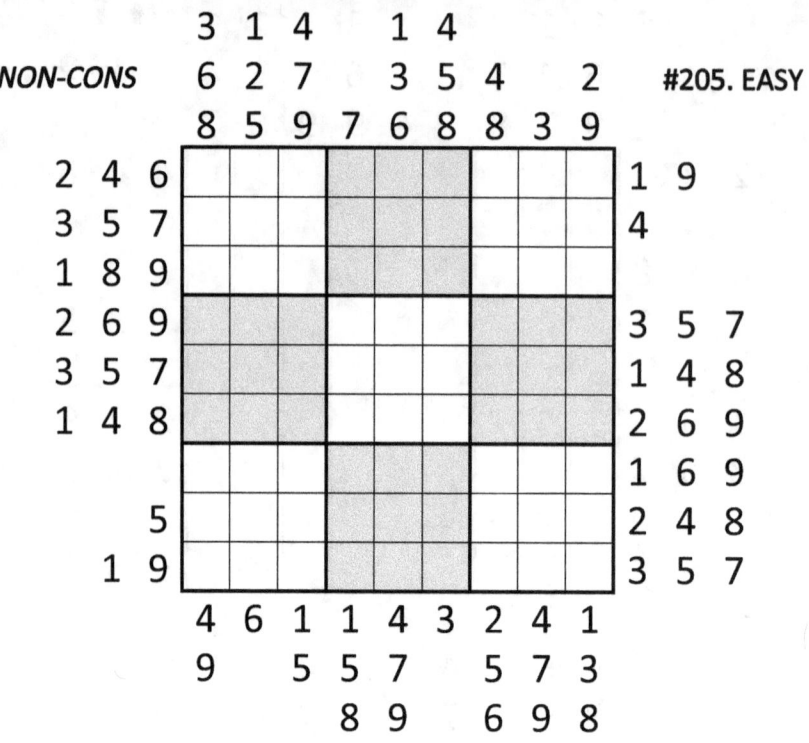

NON-CONS

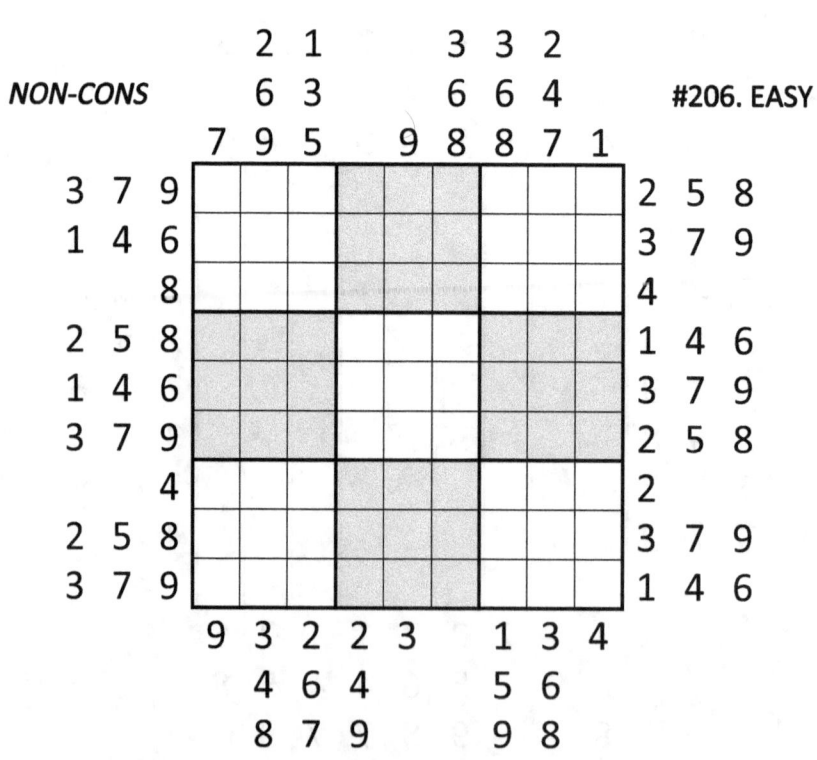

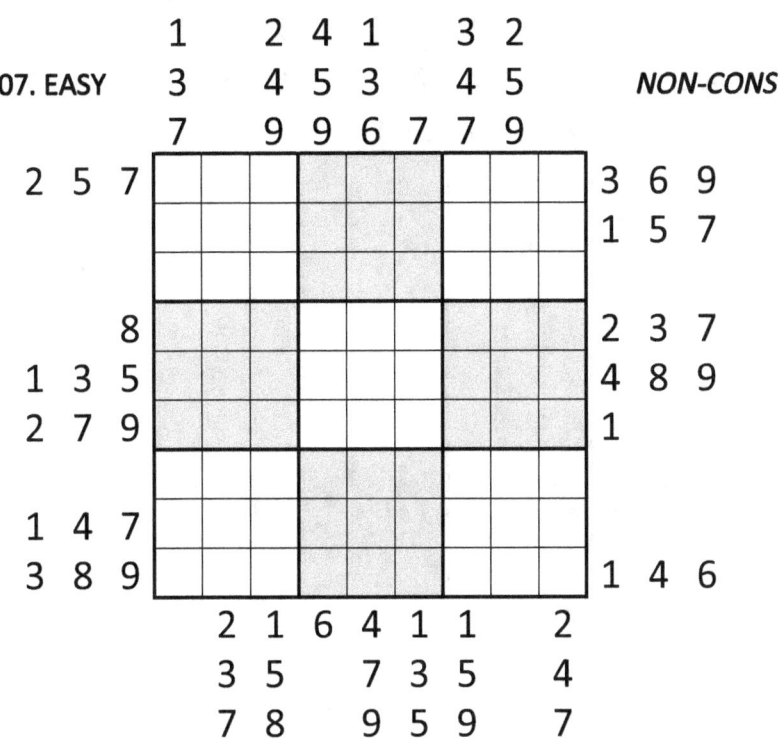

#207. EASY — NON-CONS

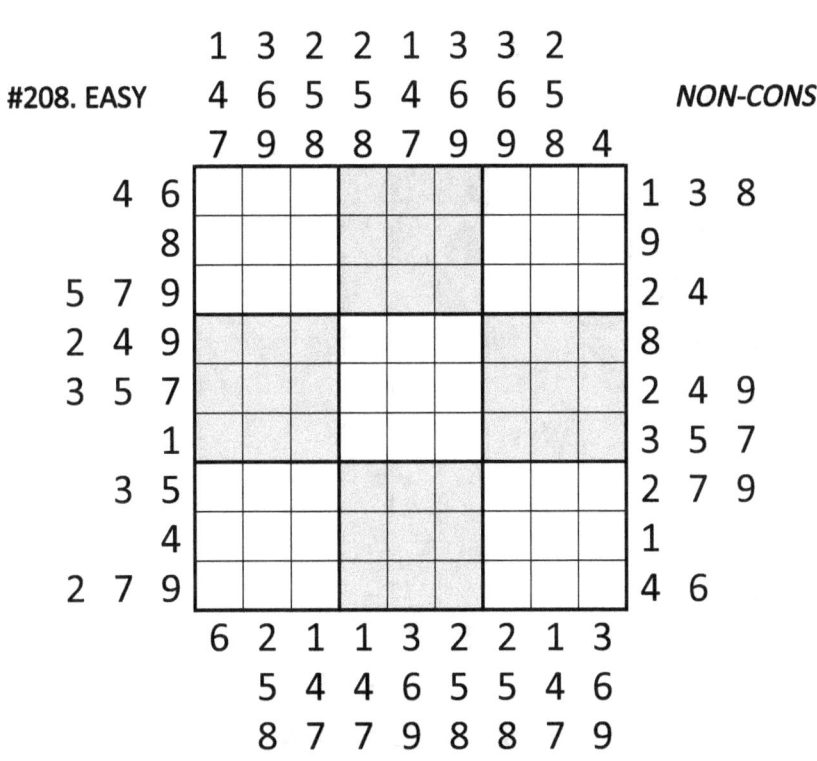

Outside Sudoku by amazon.com/djape, page 109

#208. EASY — NON-CONS

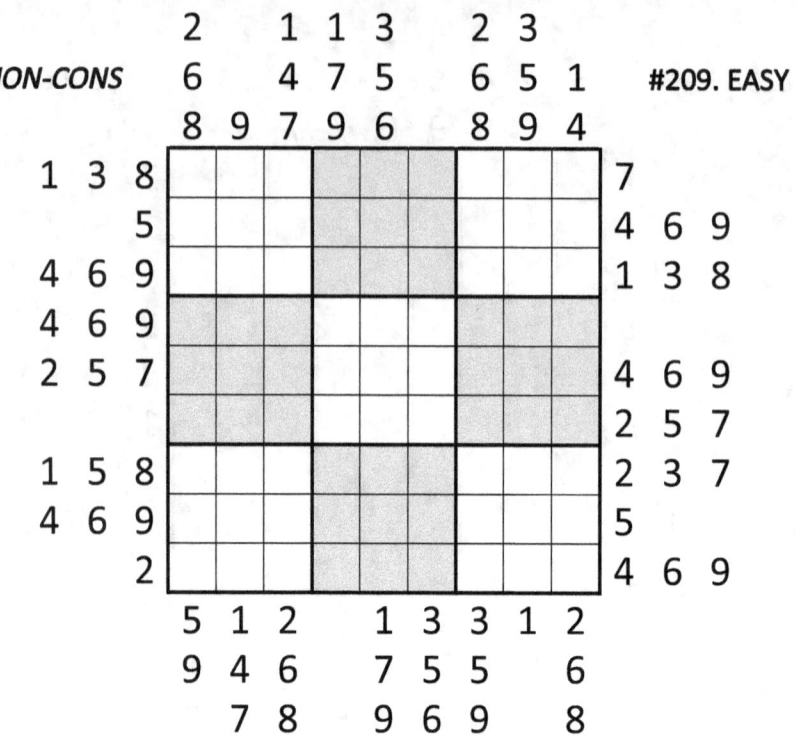

NON-CONS   #209. EASY

Outside Sudoku by amazon.com/djape, page 110

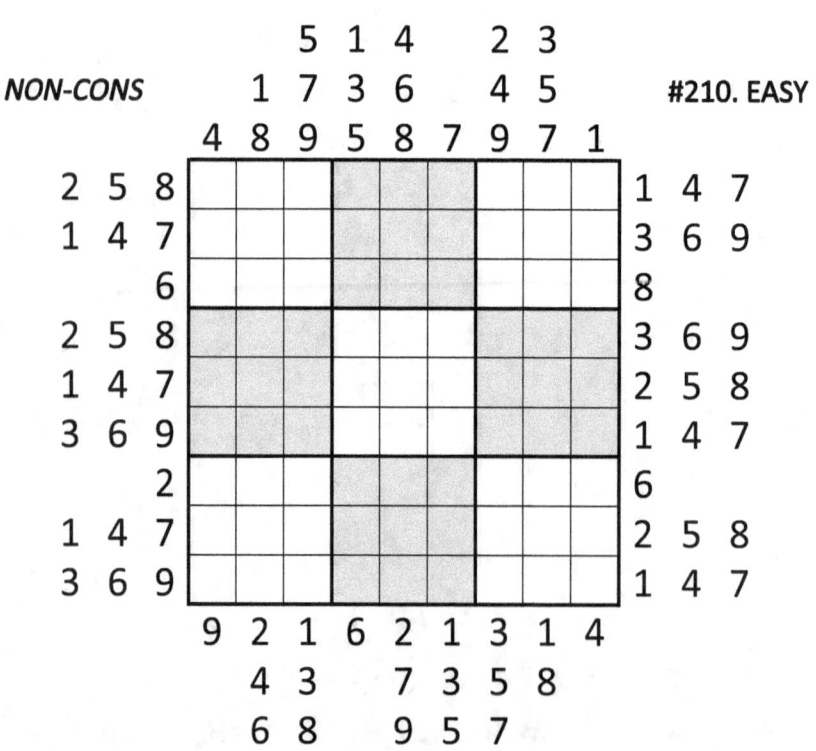

NON-CONS   #210. EASY

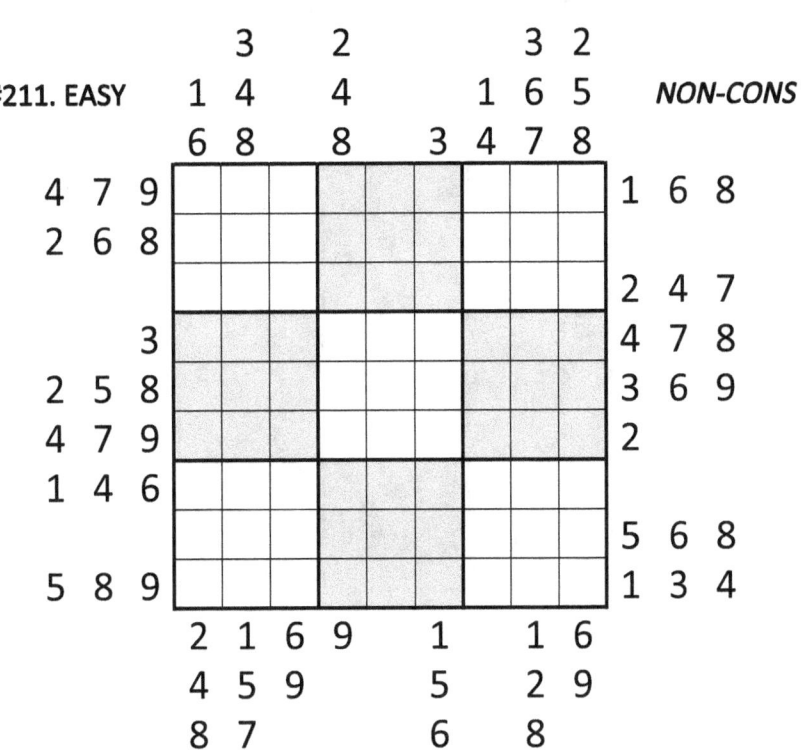

#211. EASY — NON-CONS

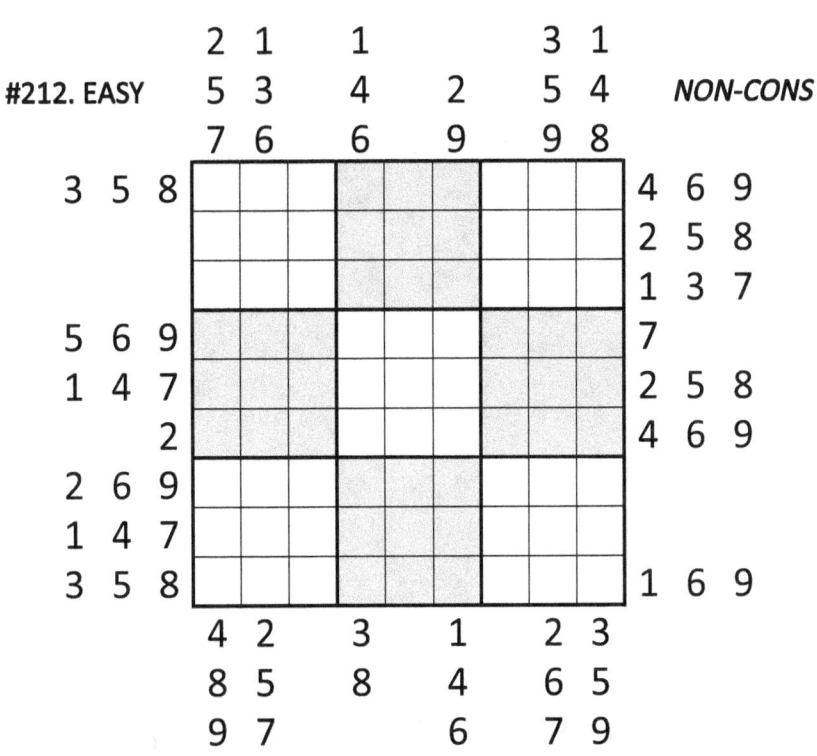

Outside Sudoku by amazon.com/djape, page 111 _____

#212. EASY — NON-CONS

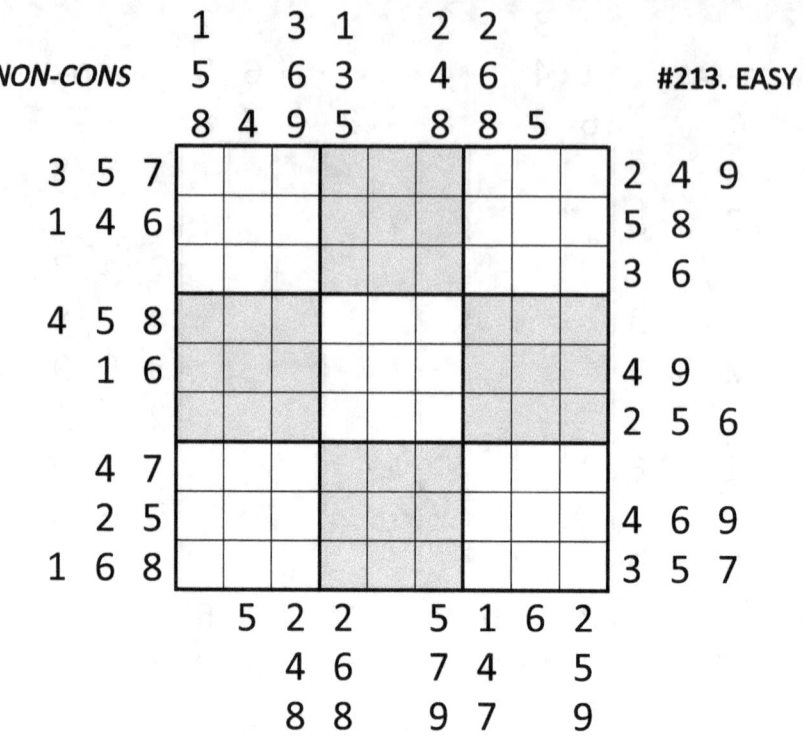

NON-CONS

#213. EASY

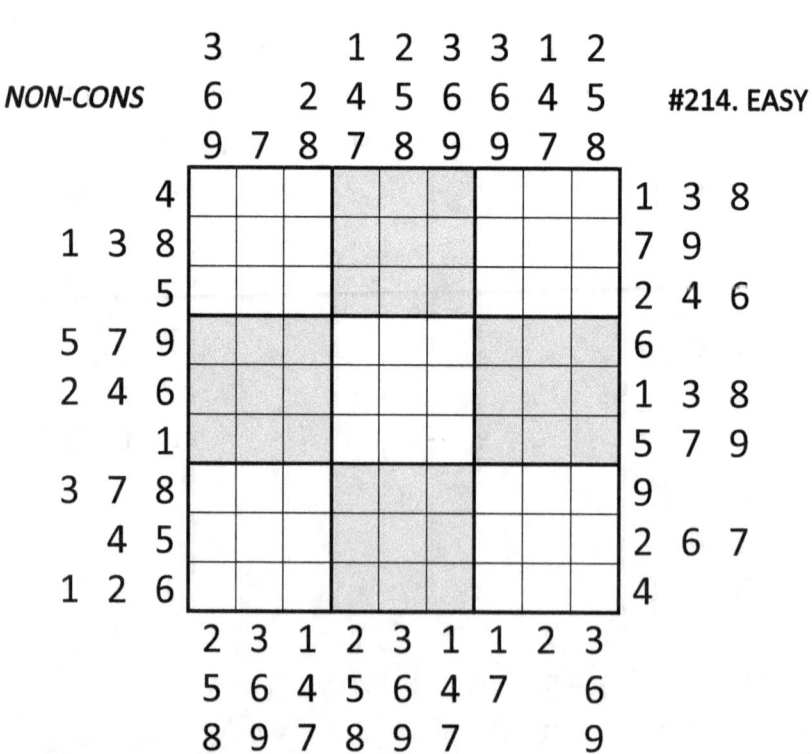

NON-CONS

#214. EASY

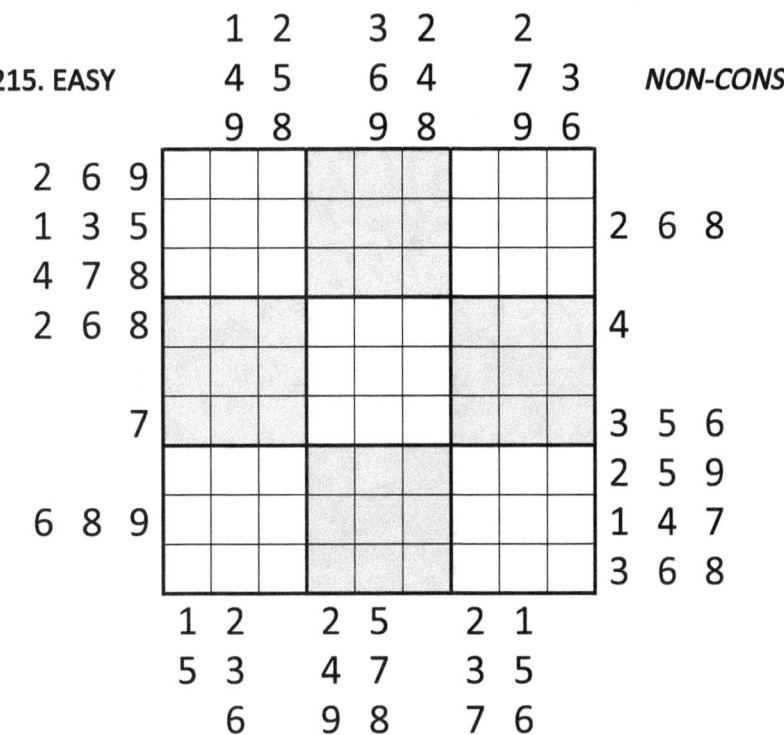

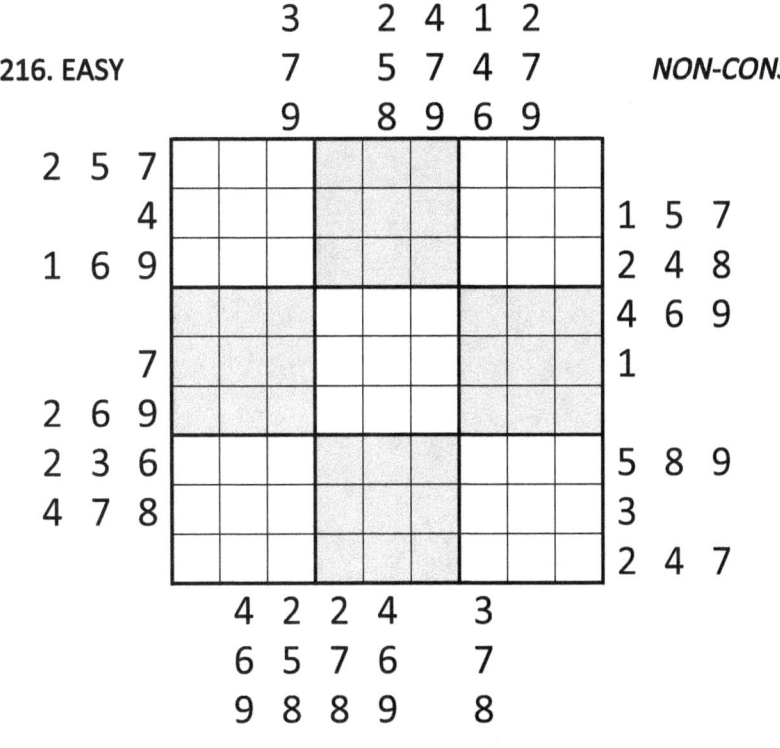

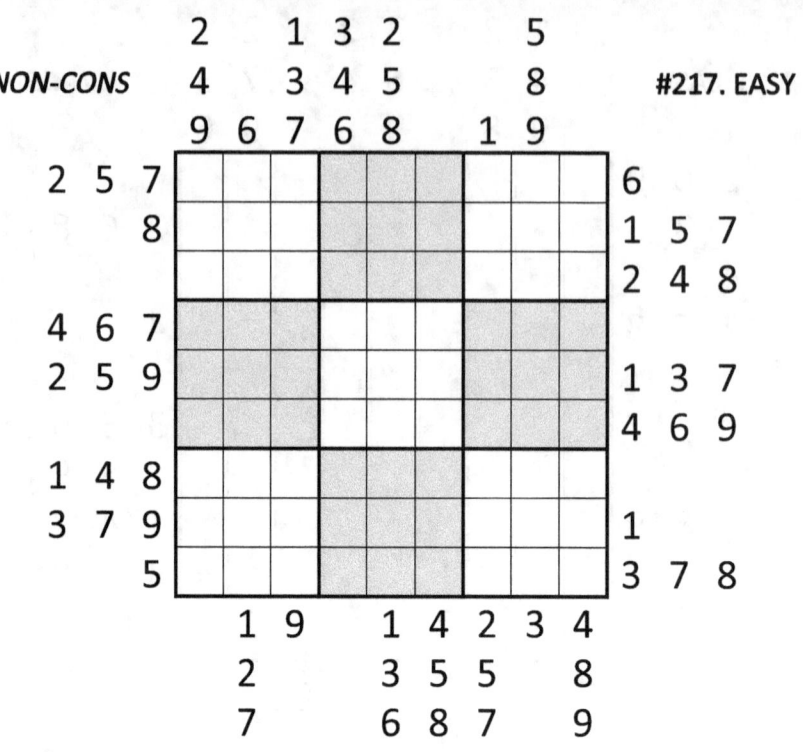

NON-CONS #217. EASY

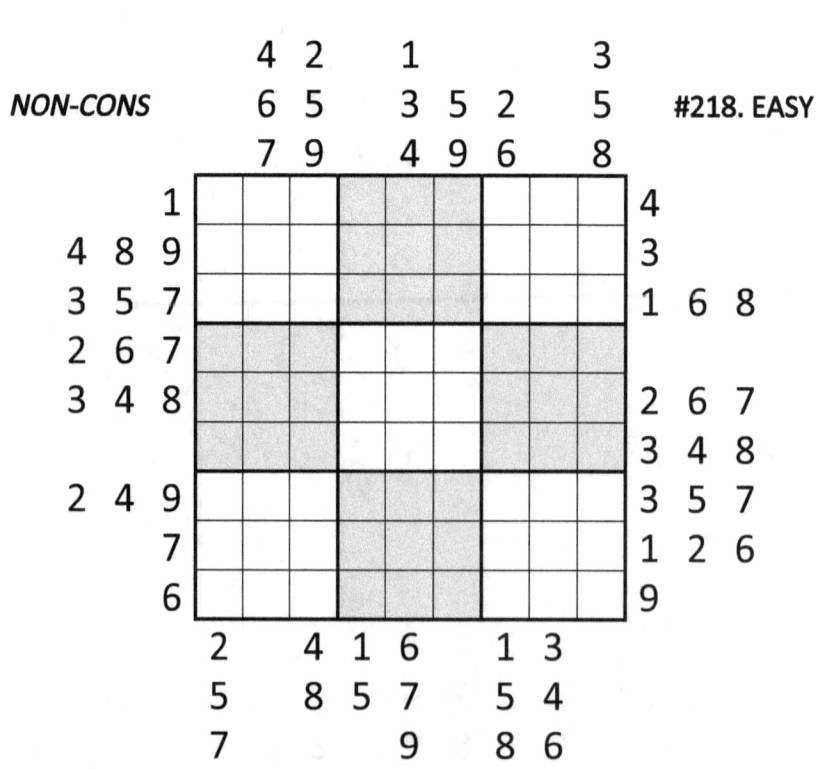

NON-CONS #218. EASY

#219. EASY NON-CONS

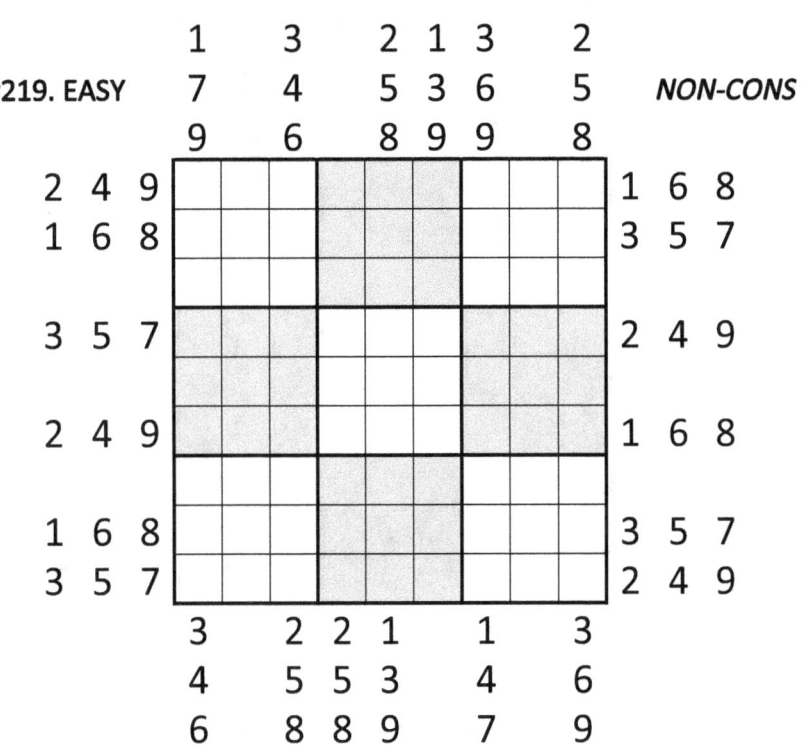

#220. EASY NON-CONS

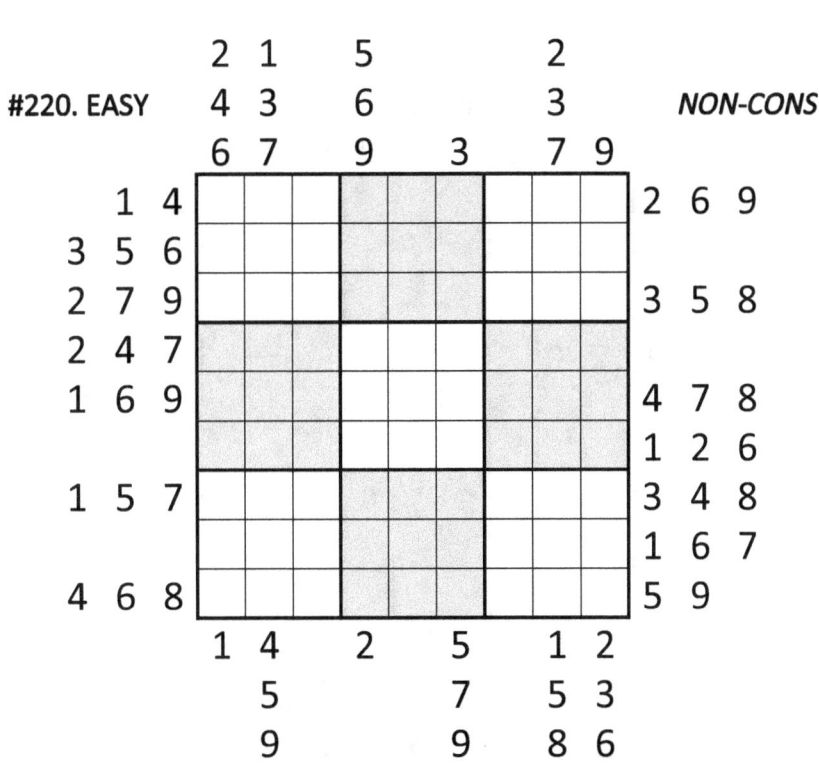

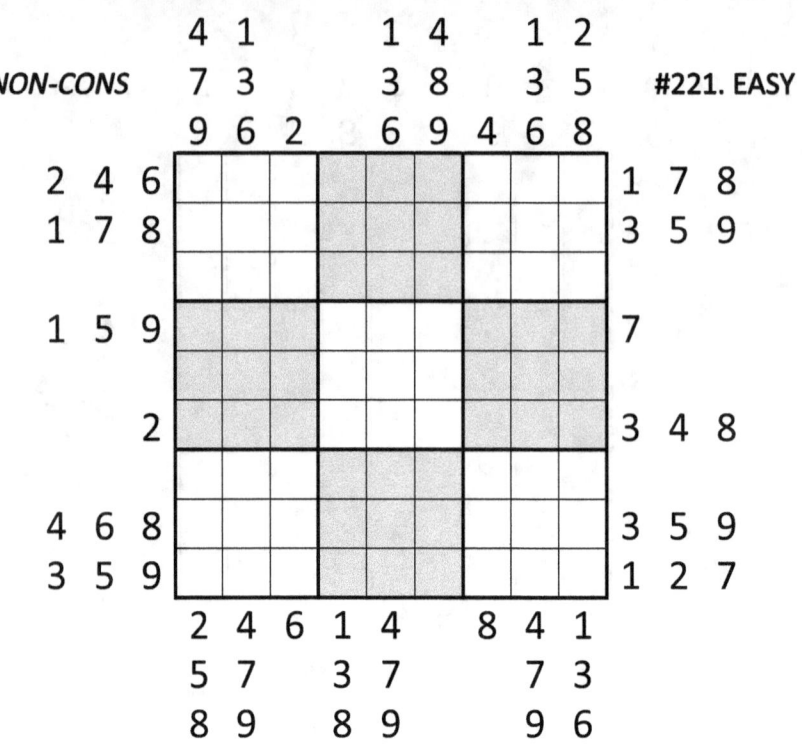

NON-CONS   #221. EASY

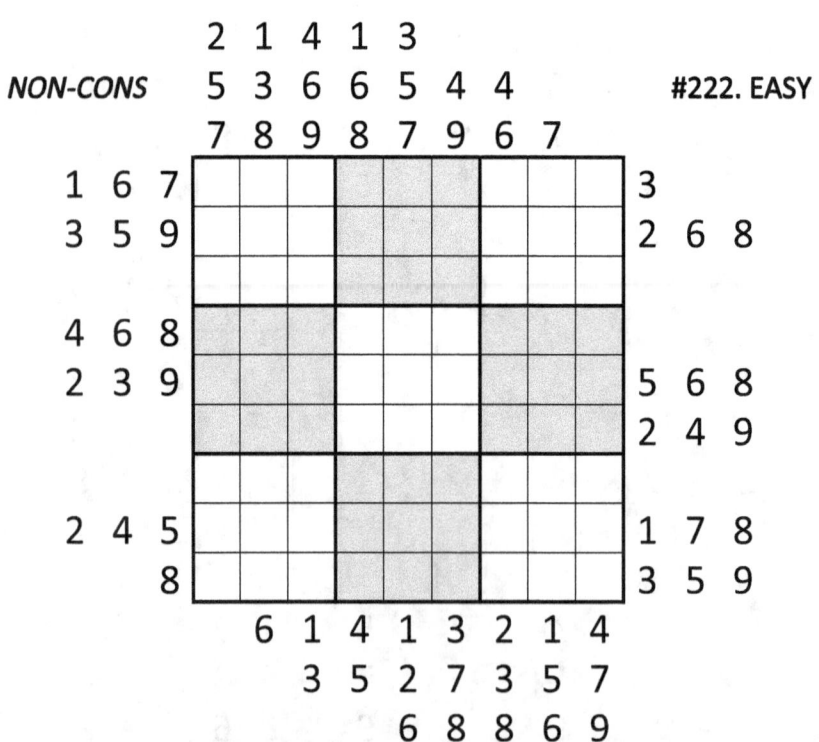

NON-CONS   #222. EASY

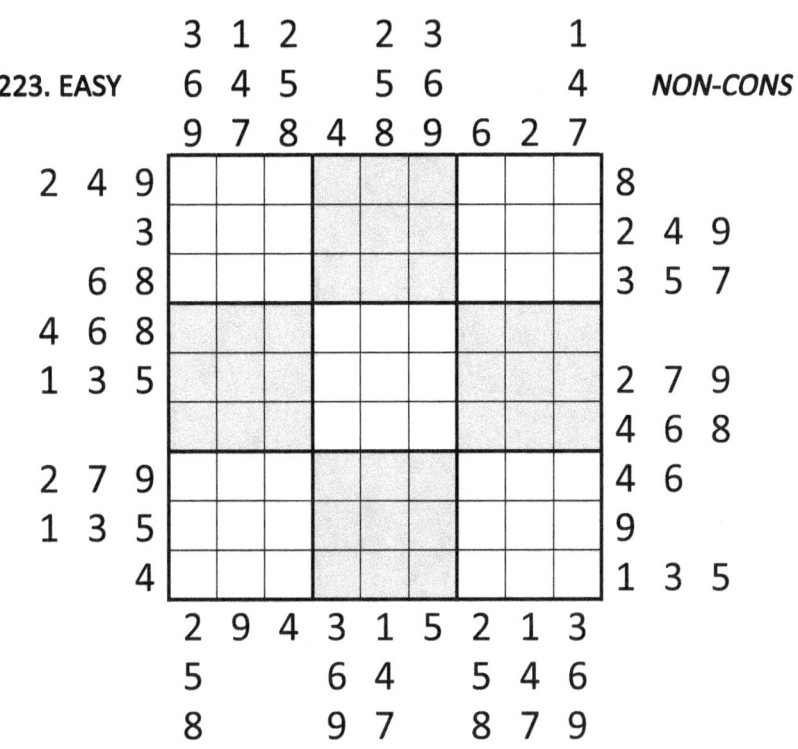

#223. EASY — NON-CONS

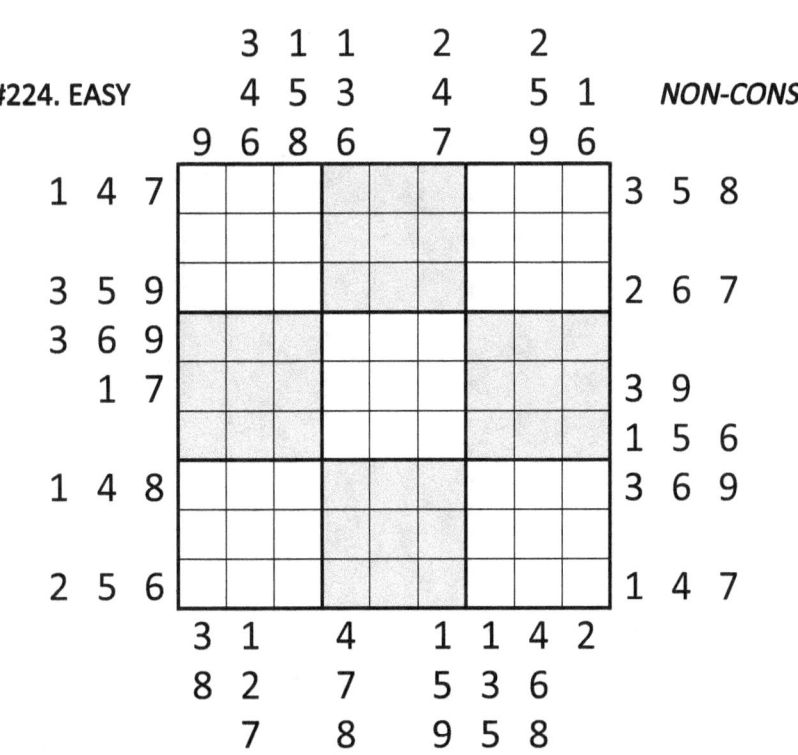

#224. EASY — NON-CONS

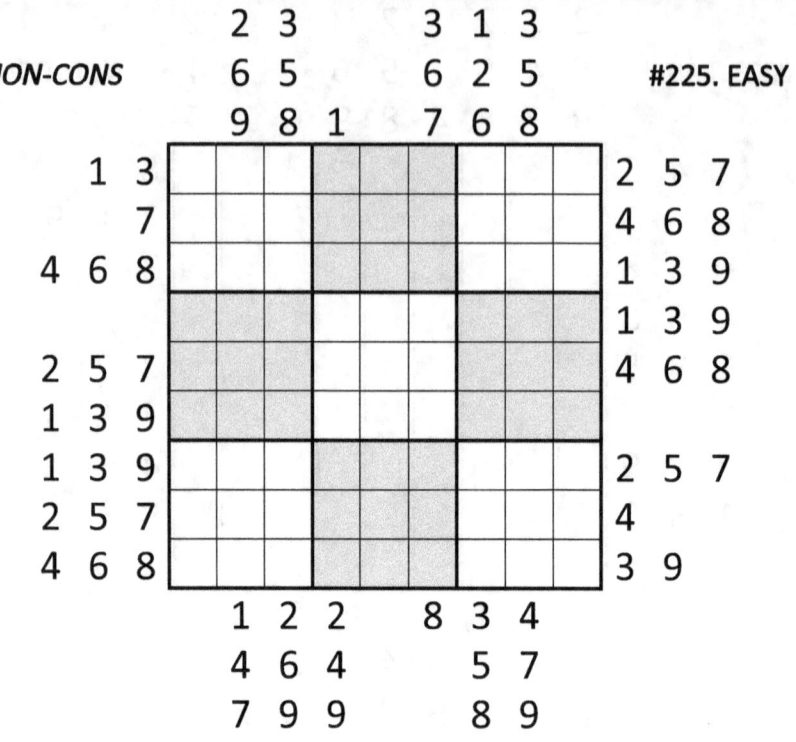

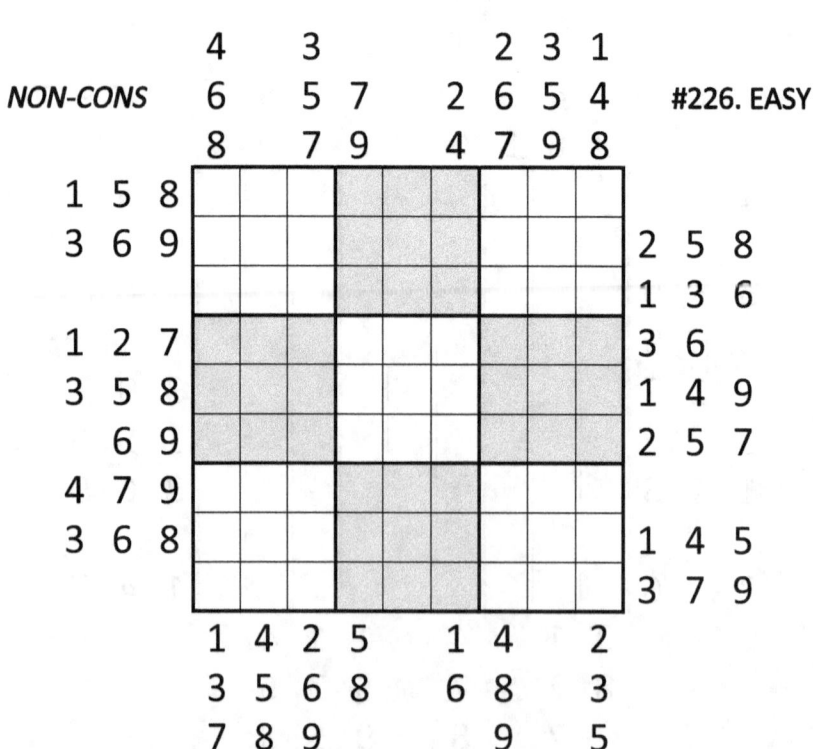

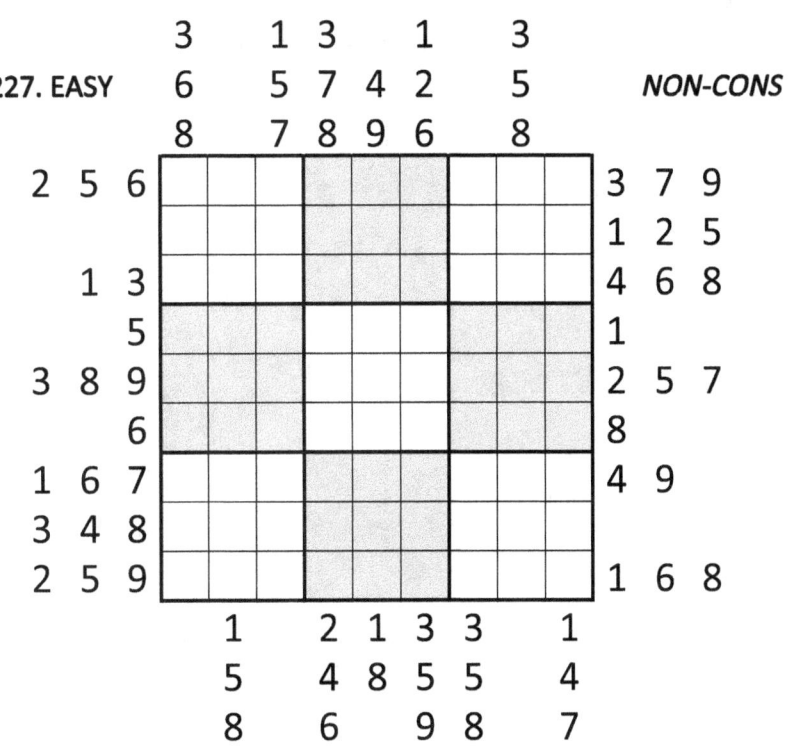

```
              3   1 3   1   3
 #227. EASY   6   5 7 4 2   5        NON-CONS
              8   7 8 9 6   8

   2  5  6  ┌──────┬──────┬──────┐   3  7  9
            │      │░░░░░░│      │   1  2  5
      1  3  │      │░░░░░░│      │   4  6  8
         5  ├──────┼──────┼──────┤   1
   3  8  9  │░░░░░░│      │░░░░░░│   2  5  7
         6  │░░░░░░│      │░░░░░░│   8
   1  6  7  ├──────┼──────┼──────┤   4  9
   3  4  8  │      │░░░░░░│      │
   2  5  9  │      │░░░░░░│      │   1  6  8
           └──────┴──────┴──────┘
              1   2 1 3 3   1
              5   4 8 5 5   4
              8   6   9 8   7
```

Outside Sudoku by amazon.com/djape, page 119 _____

```
              2     1   1   2
 #228. EASY   5 8 2 6 5 5   4        NON-CONS
              6 9 8 7 9 9   6

   1  5  8  ┌──────┬──────┬──────┐   4  7  9
            │      │░░░░░░│      │
   3  6  9  │      │░░░░░░│      │   1  2  8
   2  6  8  ├──────┼──────┼──────┤   3  5  9
            │░░░░░░│      │░░░░░░│
   3  7  9  │░░░░░░│      │░░░░░░│   1  4  6
   2  3  7  ├──────┼──────┼──────┤   5  6  9
            │      │░░░░░░│      │
   4  6  8  │      │░░░░░░│      │   1  3  7
           └──────┴──────┴──────┘
              2   1 1 3 2 2 1
              5   3 6 5 7 6 4
              8   6   8   9
```

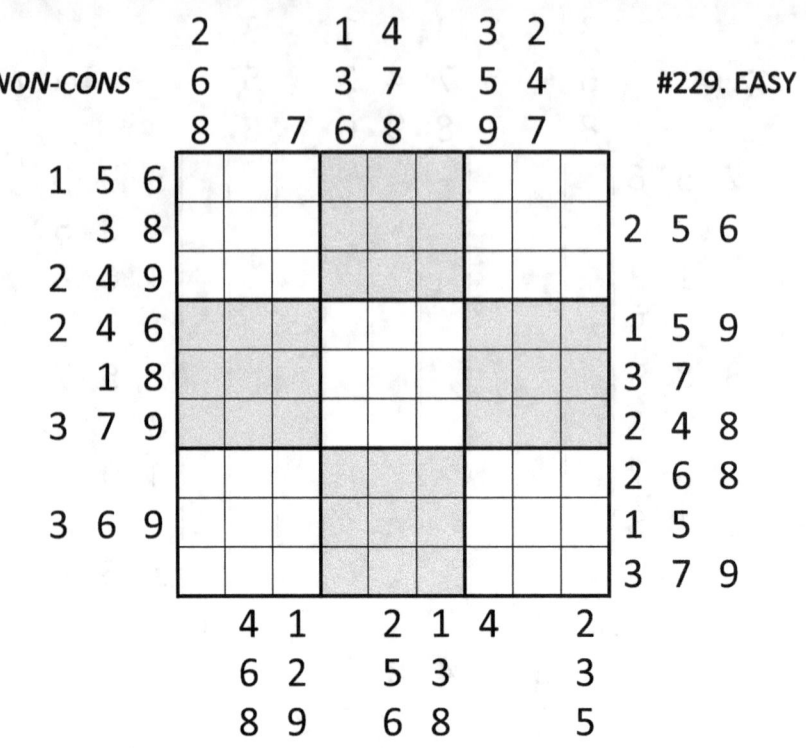

NON-CONS

#229. EASY

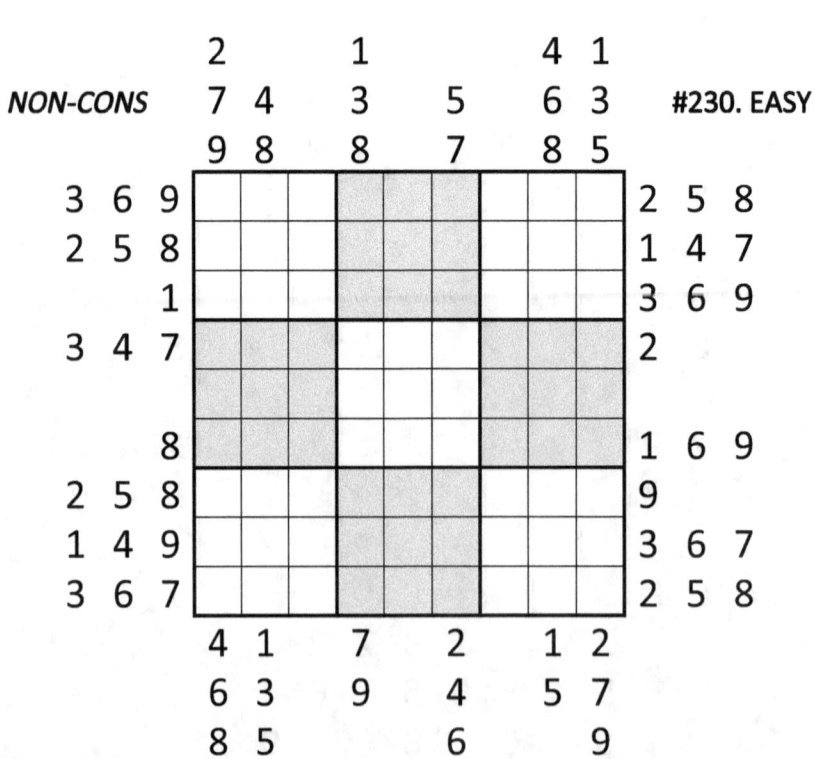

NON-CONS

#230. EASY

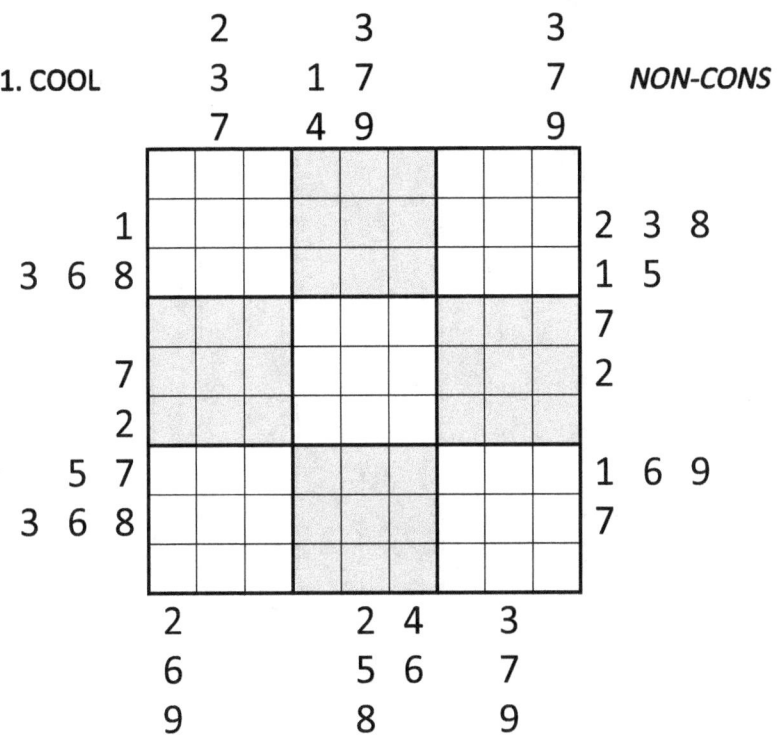

#231. COOL   NON-CONS

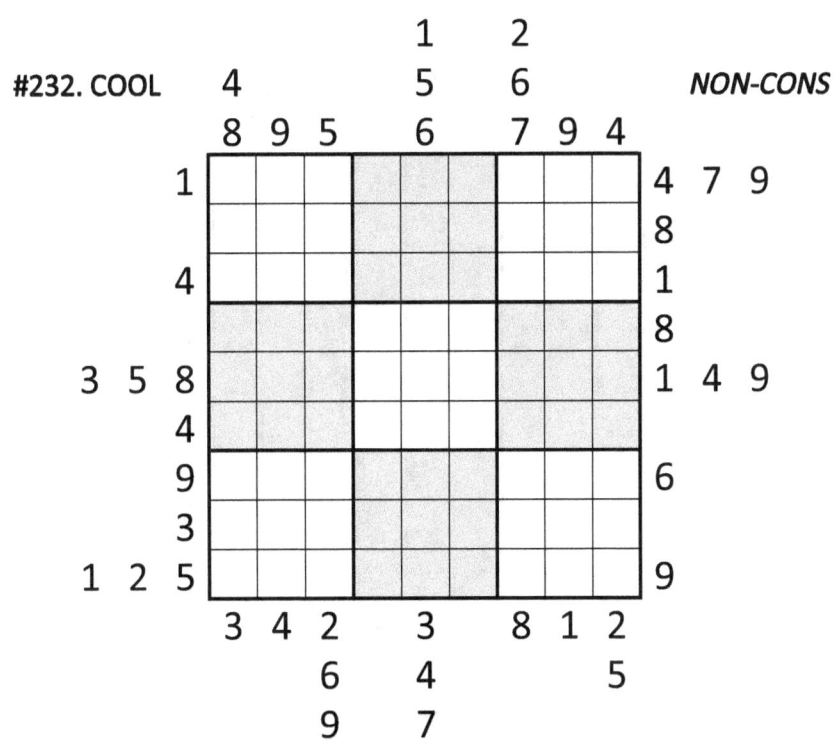

#232. COOL   NON-CONS

NON-CONS

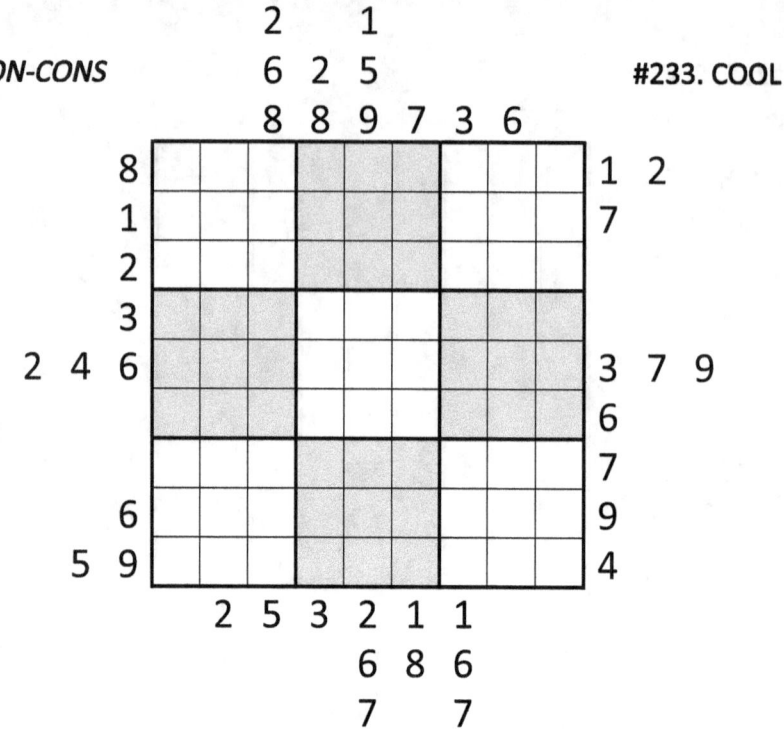

NON-CONS

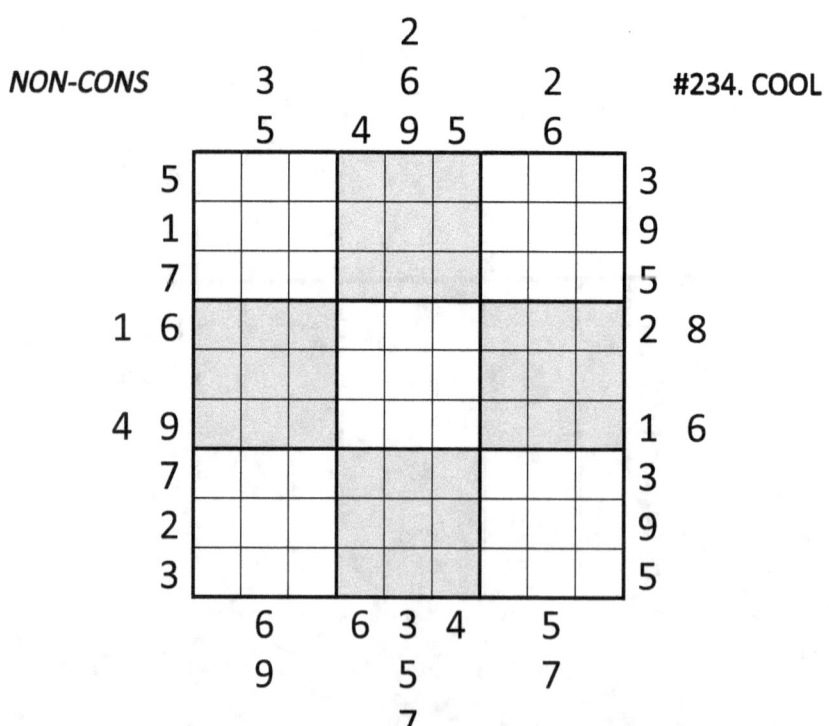

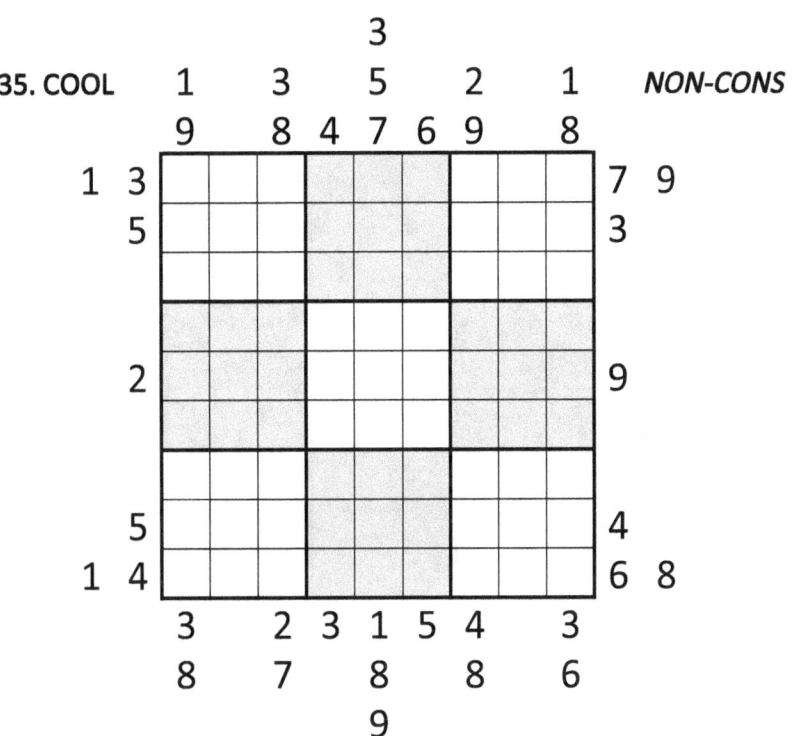

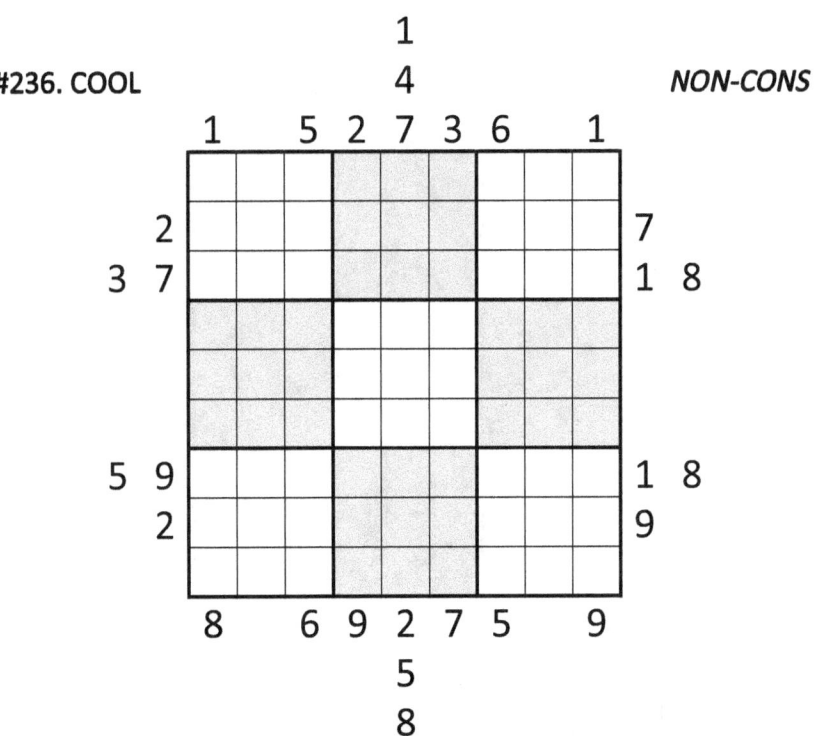

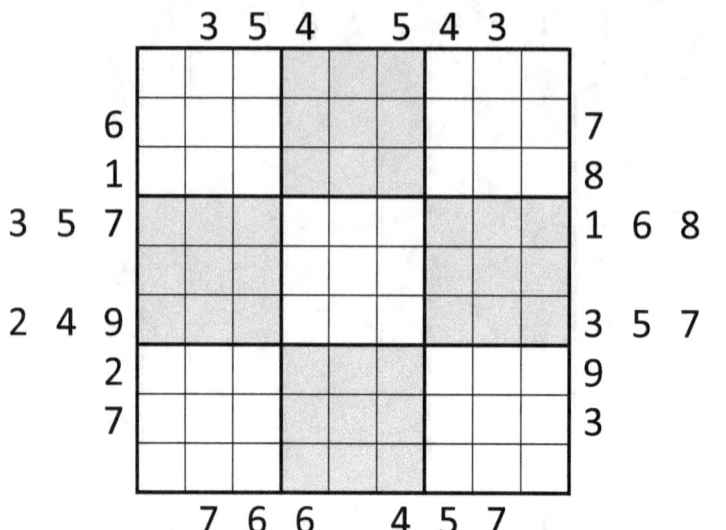

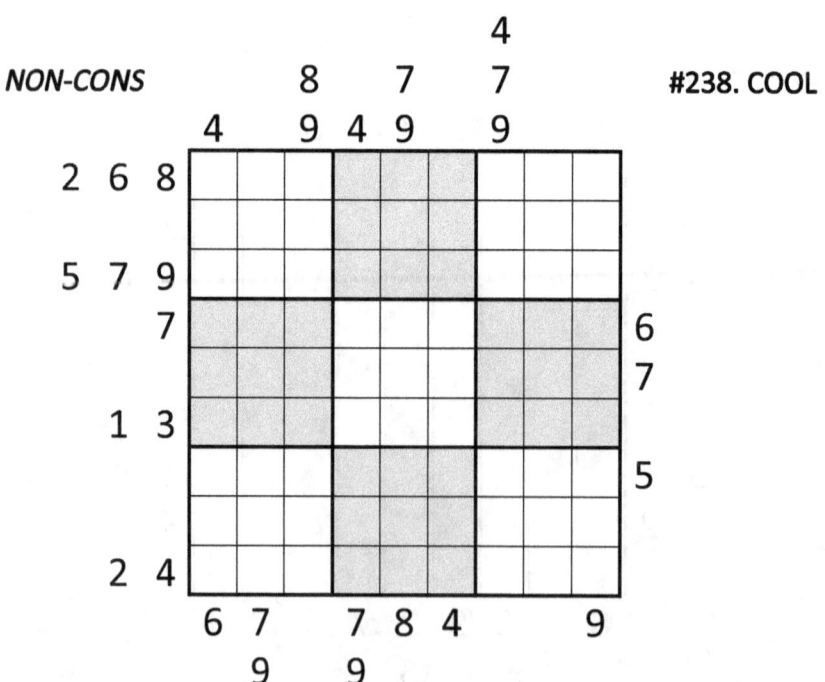

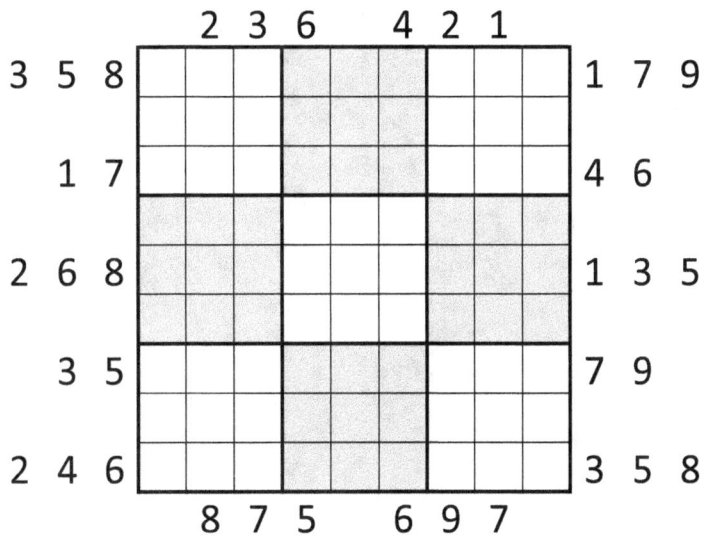

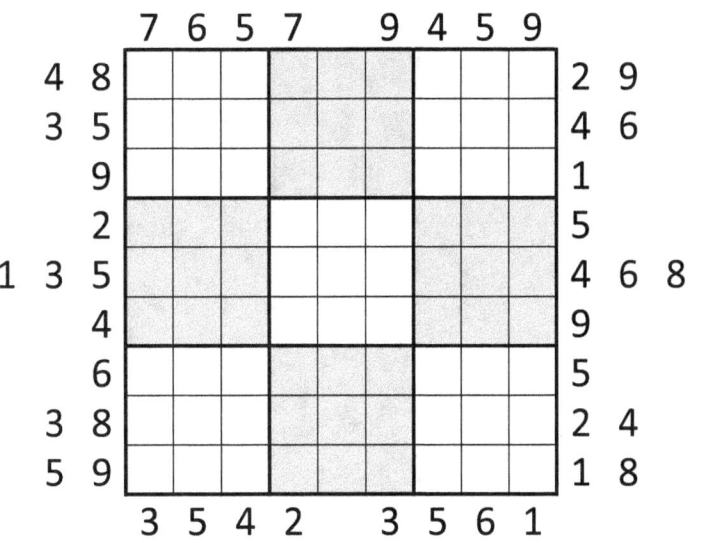

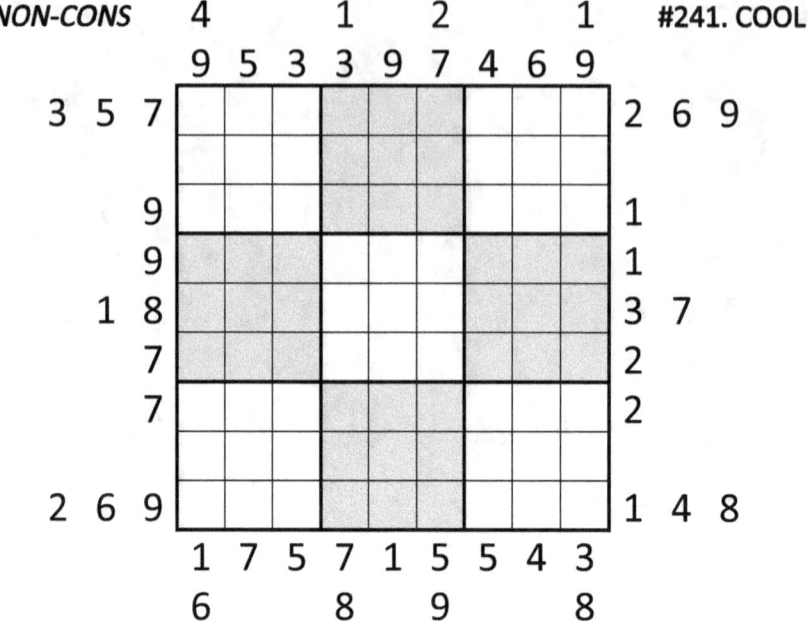

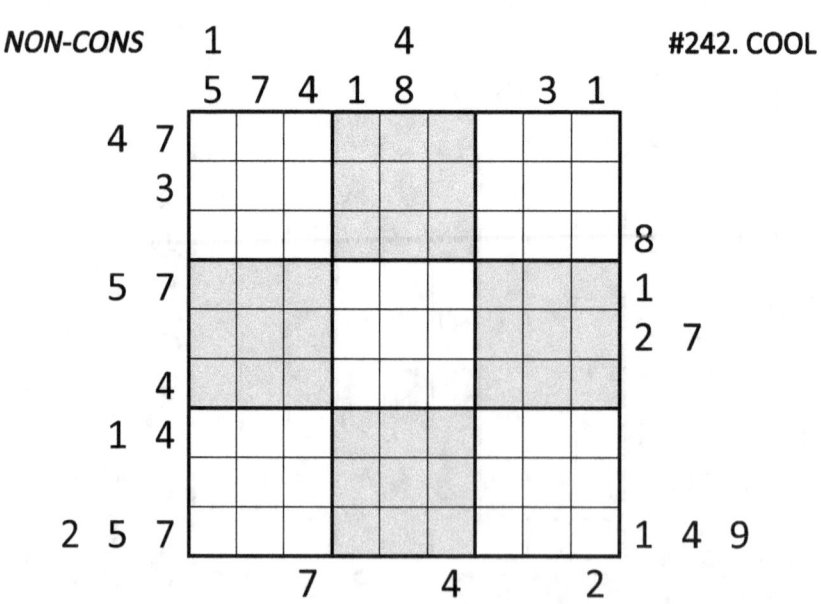

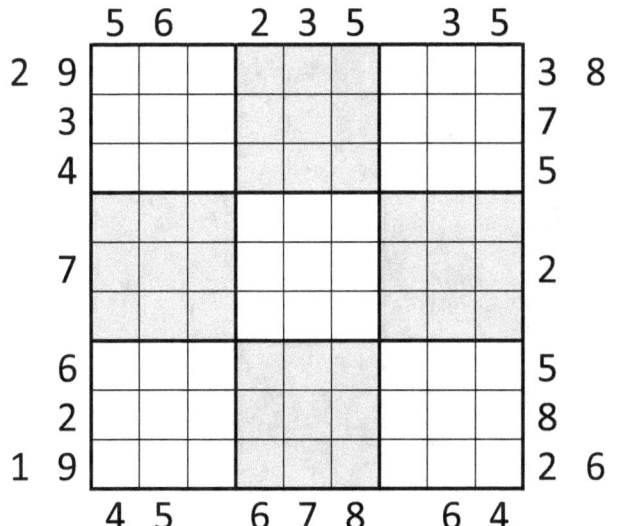

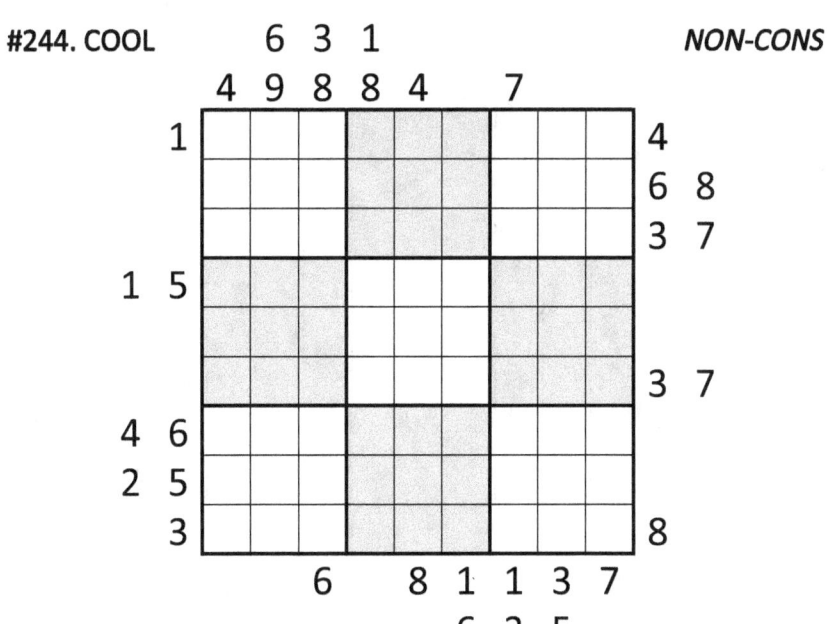

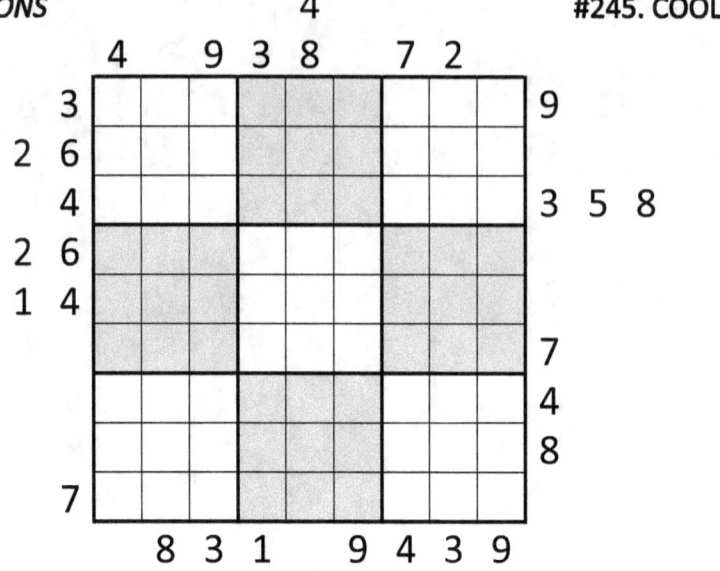

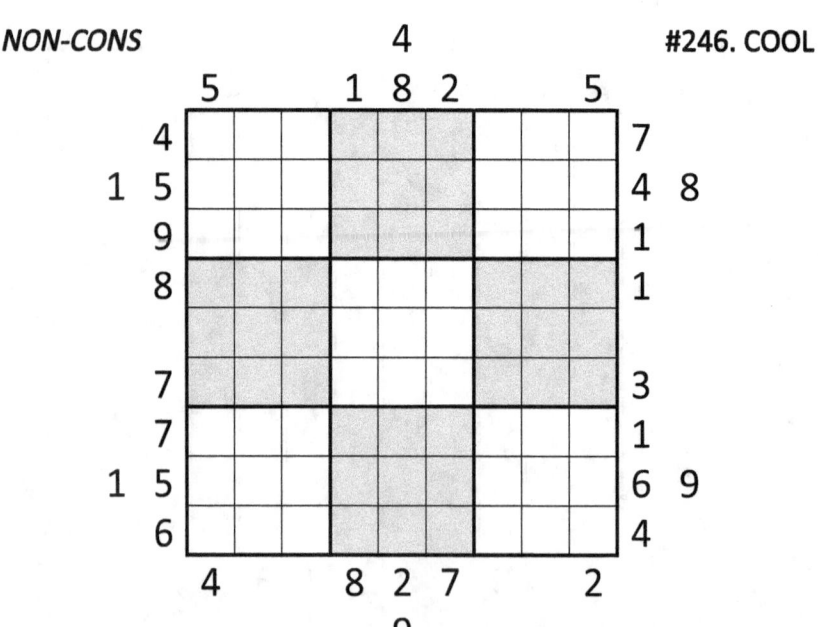

NON-CONS

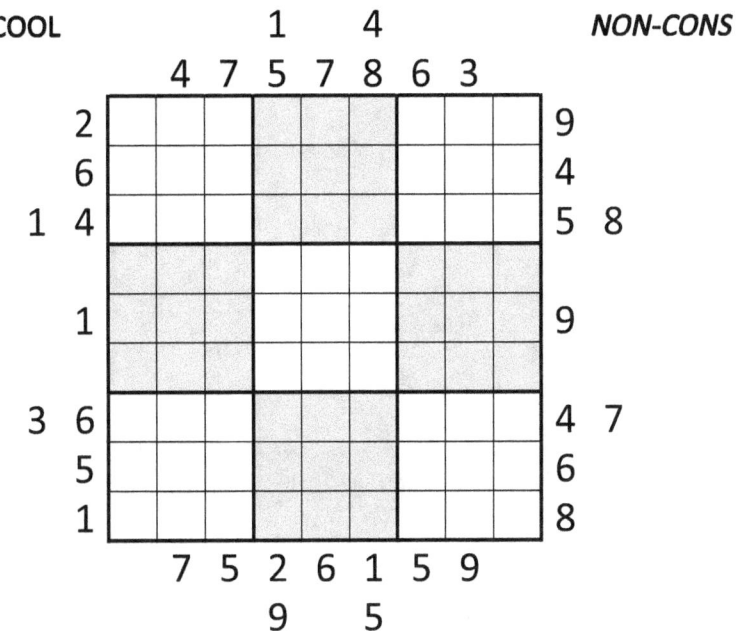

Outside Sudoku by amazon.com/djape, page 129 _____

NON-CONS

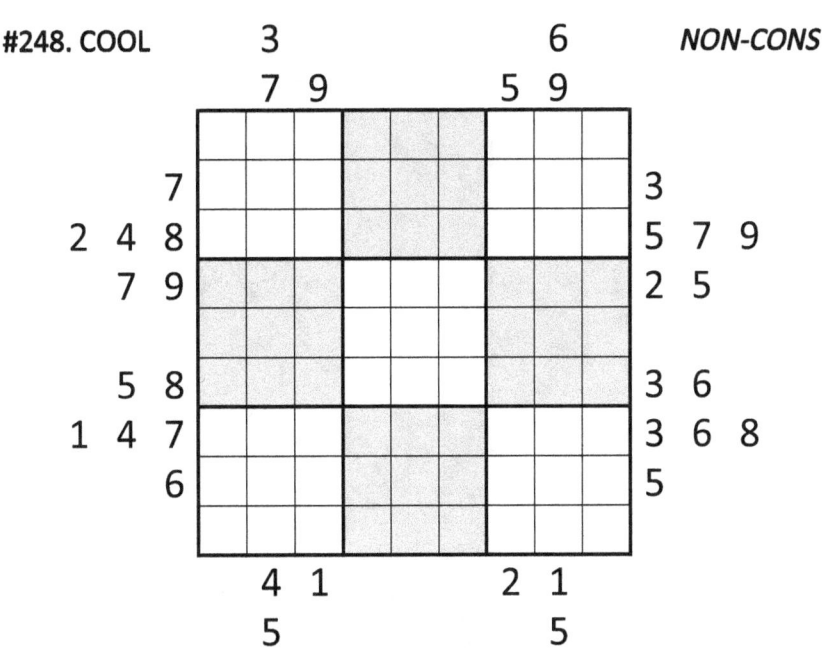

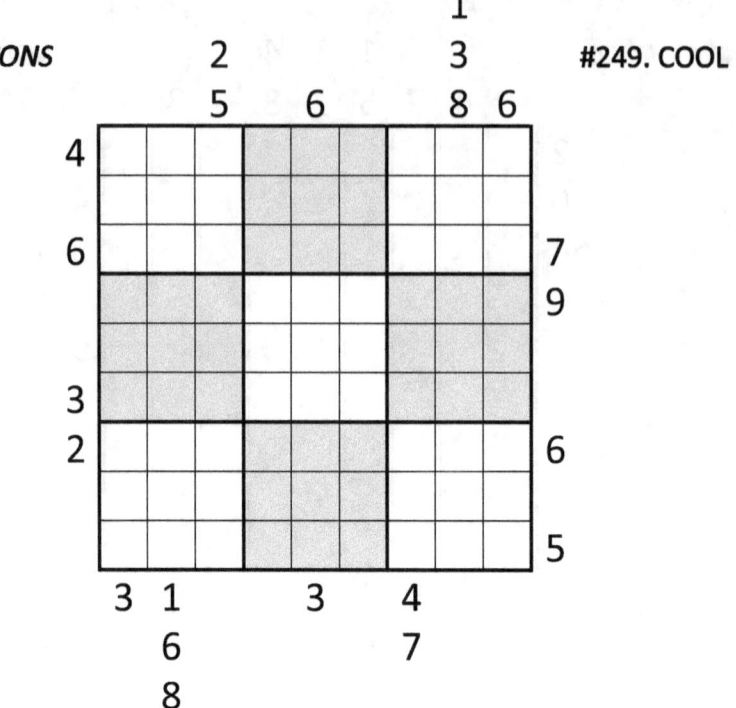

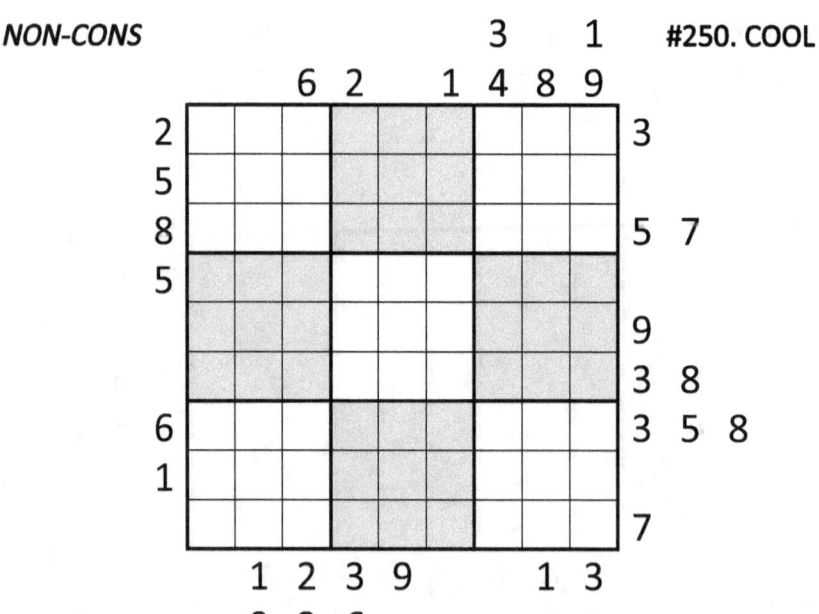

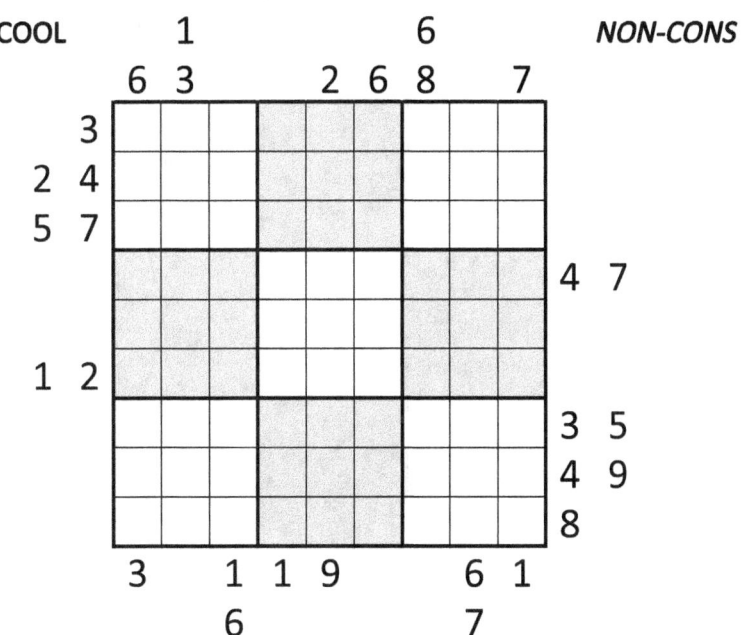

#251. COOL — NON-CONS

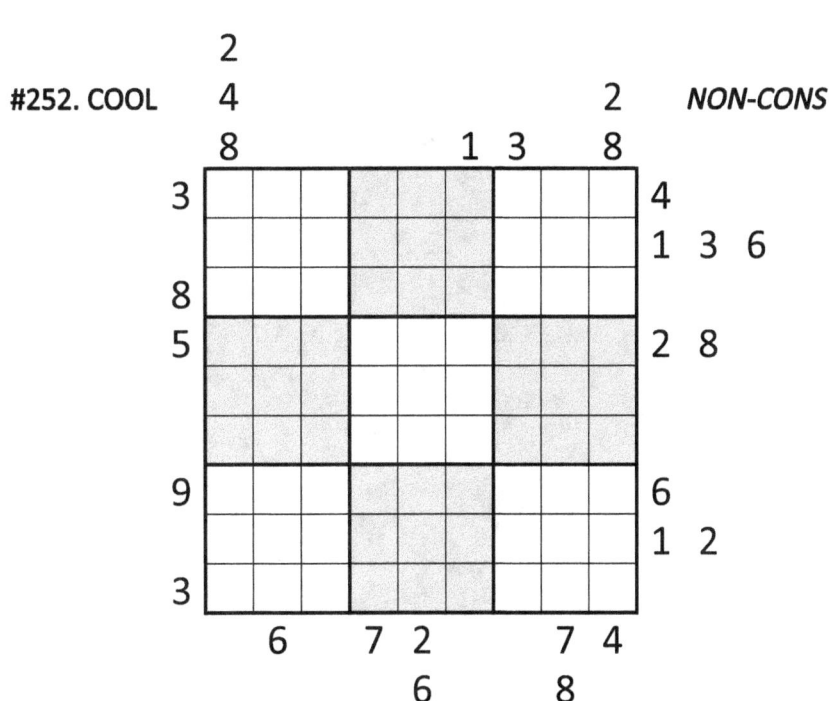

Outside Sudoku by amazon.com/djape, page 131 _____

#252. COOL — NON-CONS

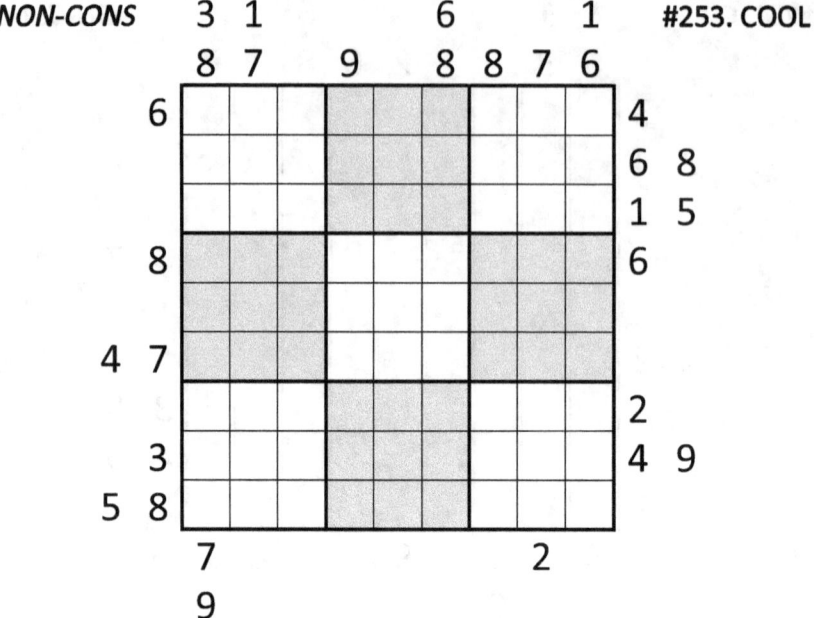

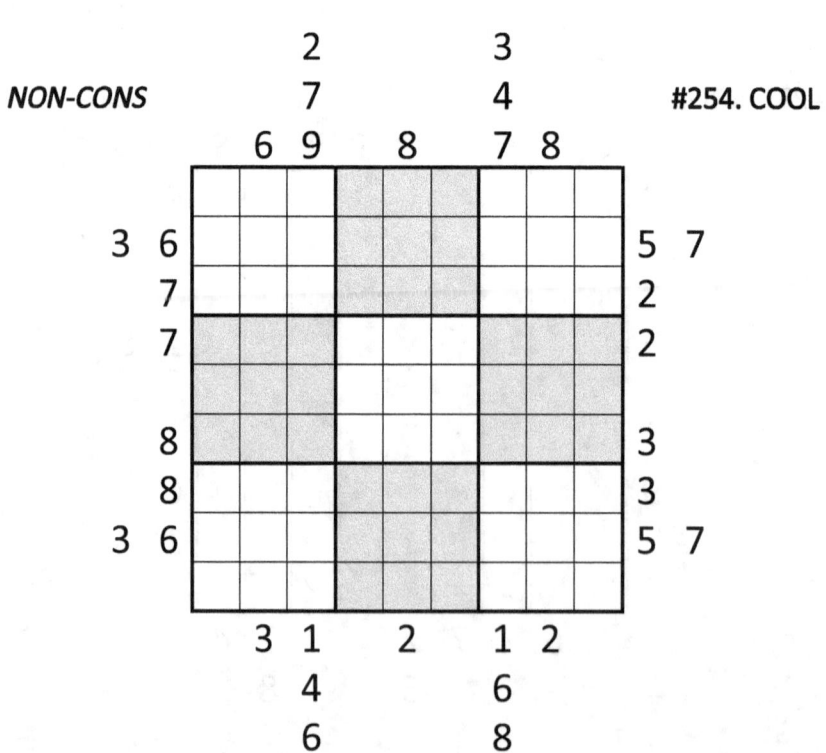

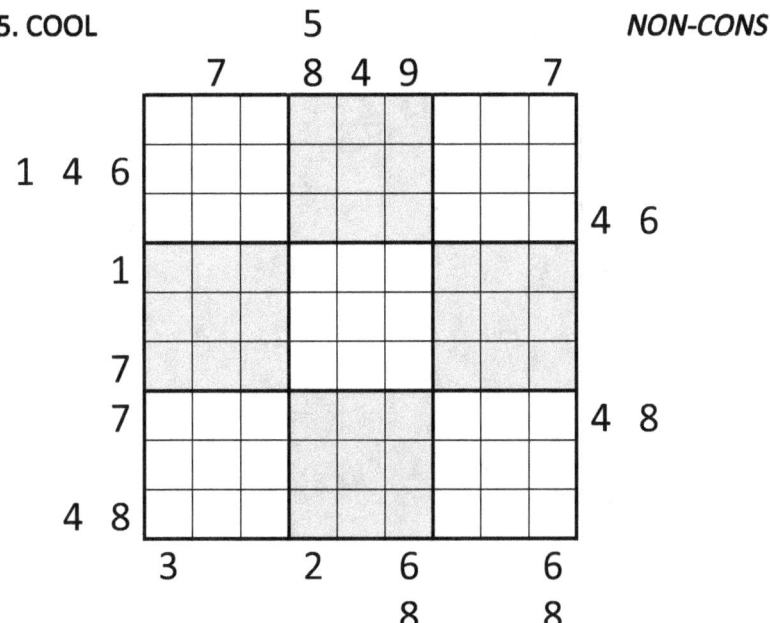

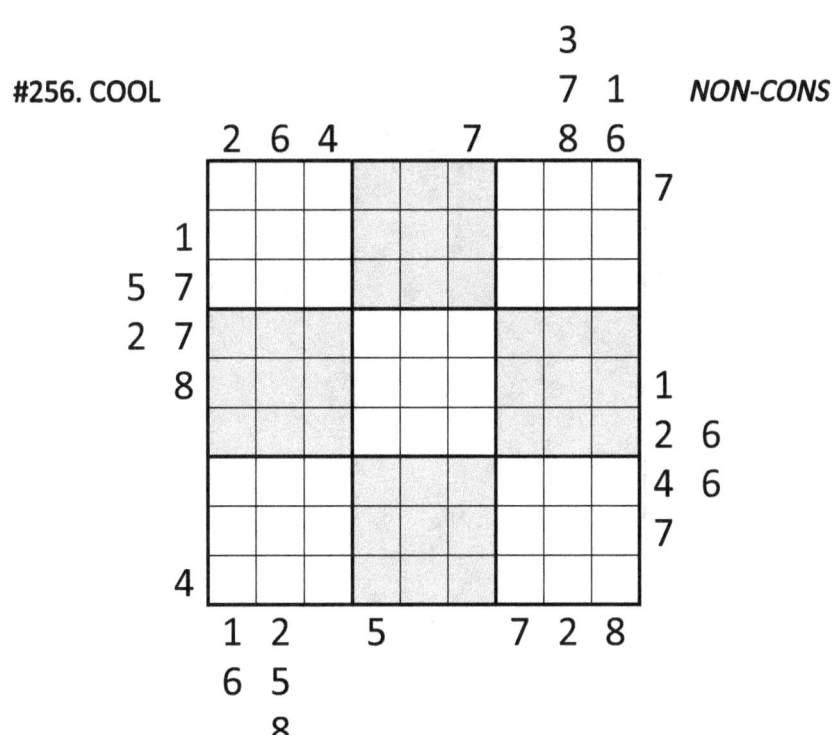

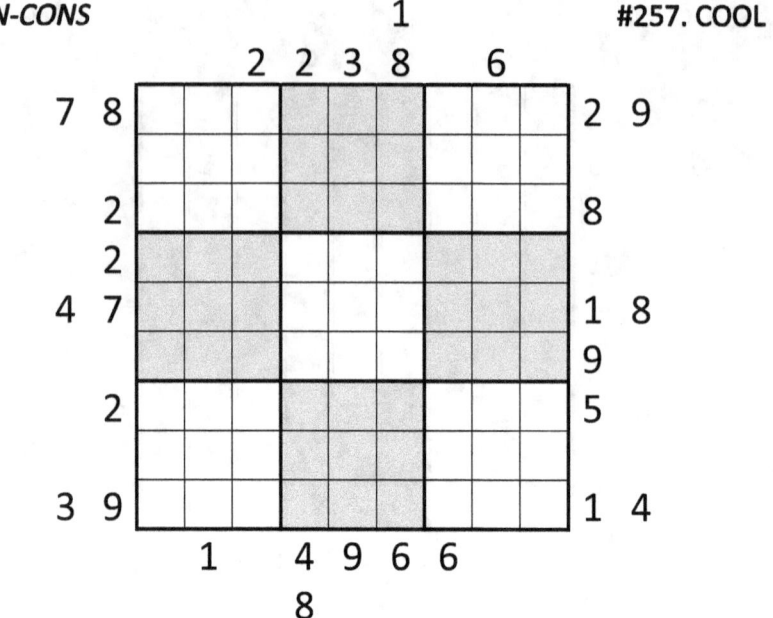

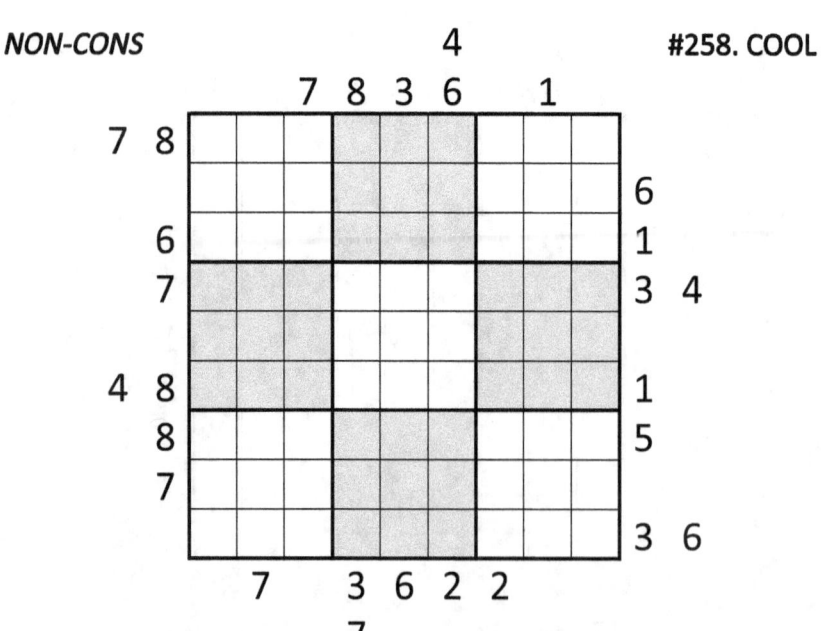

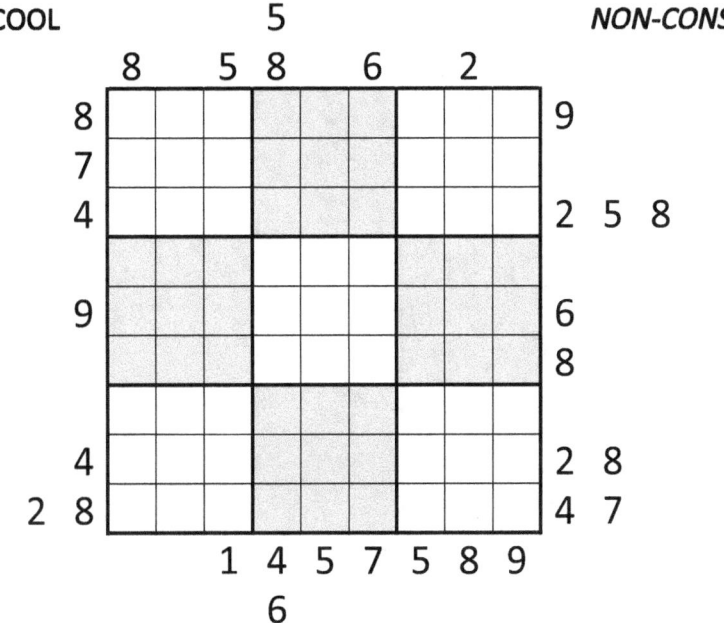

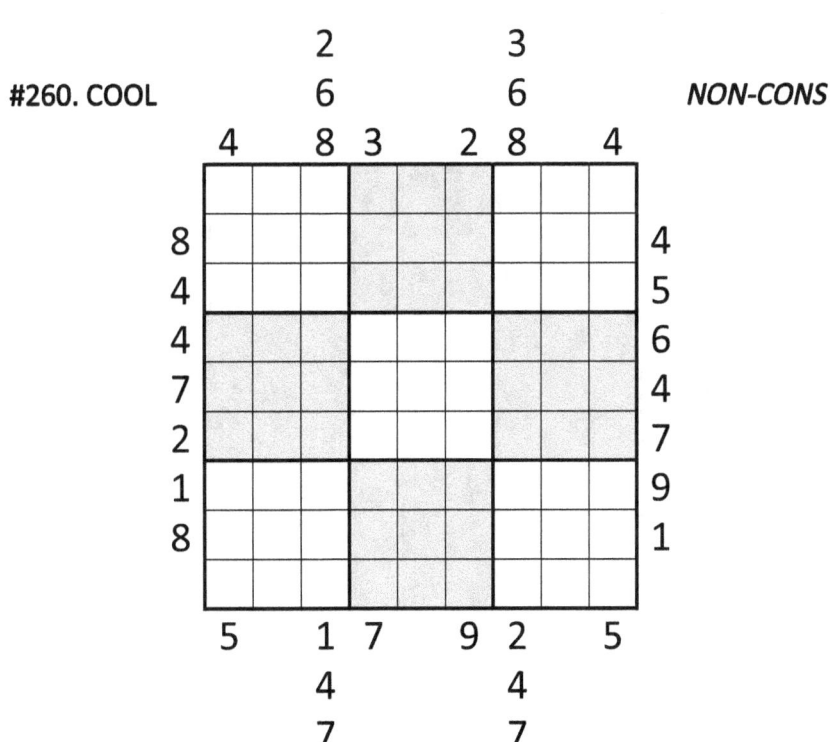

#261. THINKER

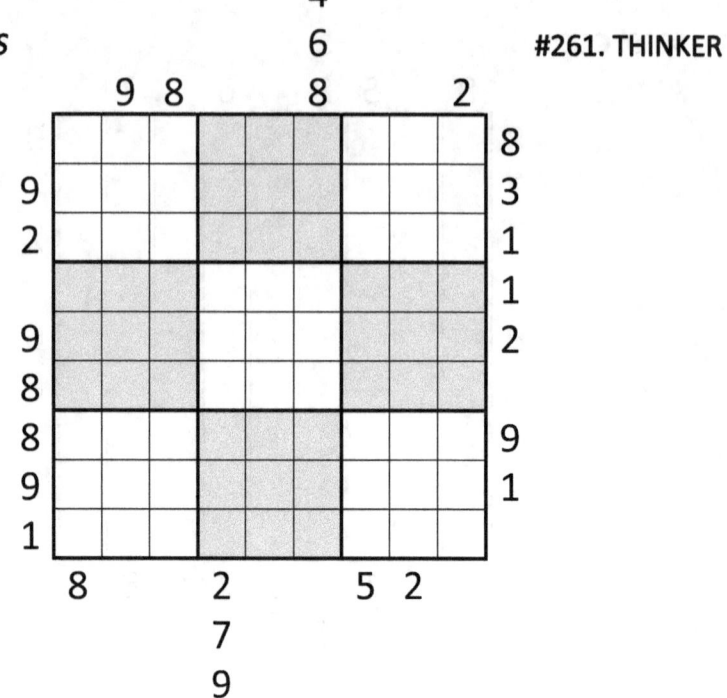

#262. THINKER

NON-CONS

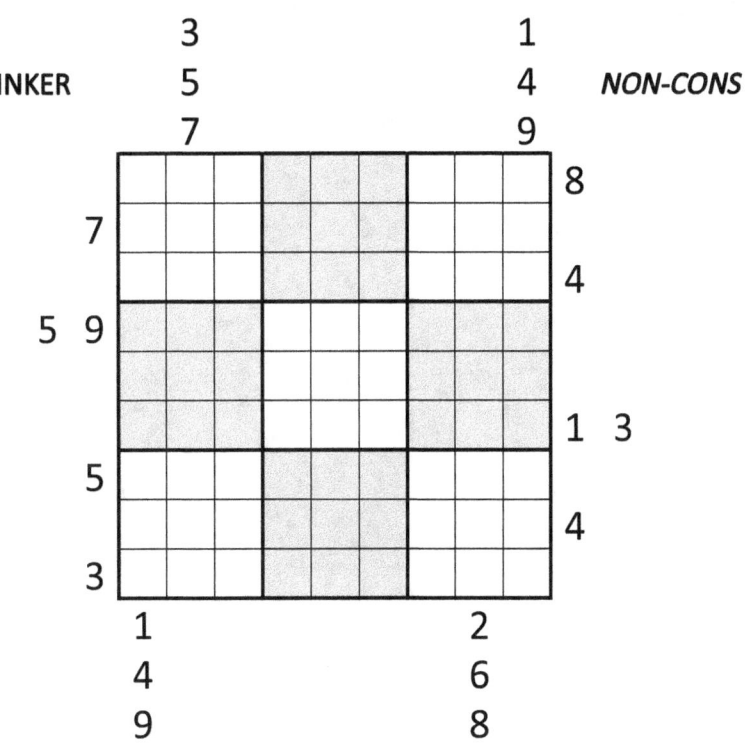

NON-CONS

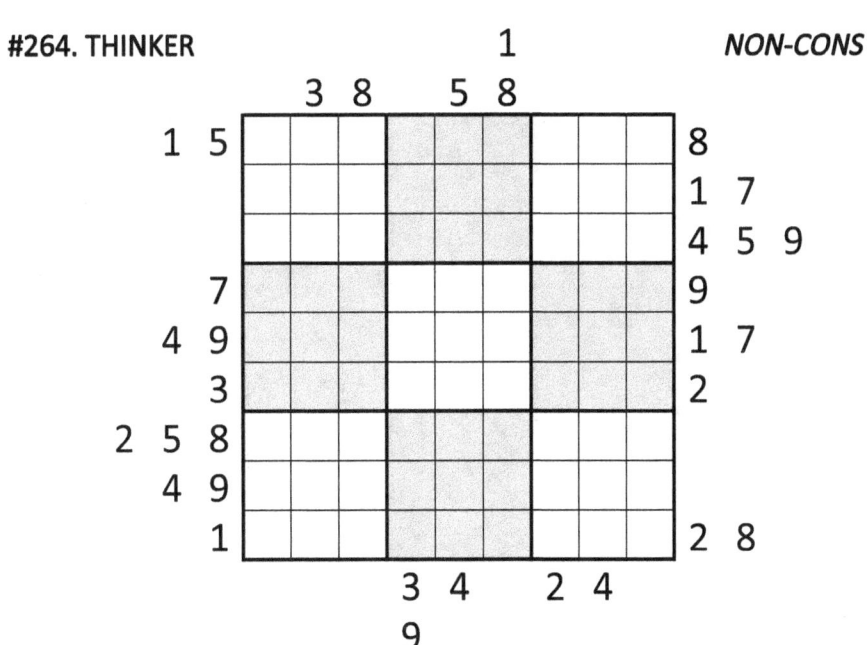

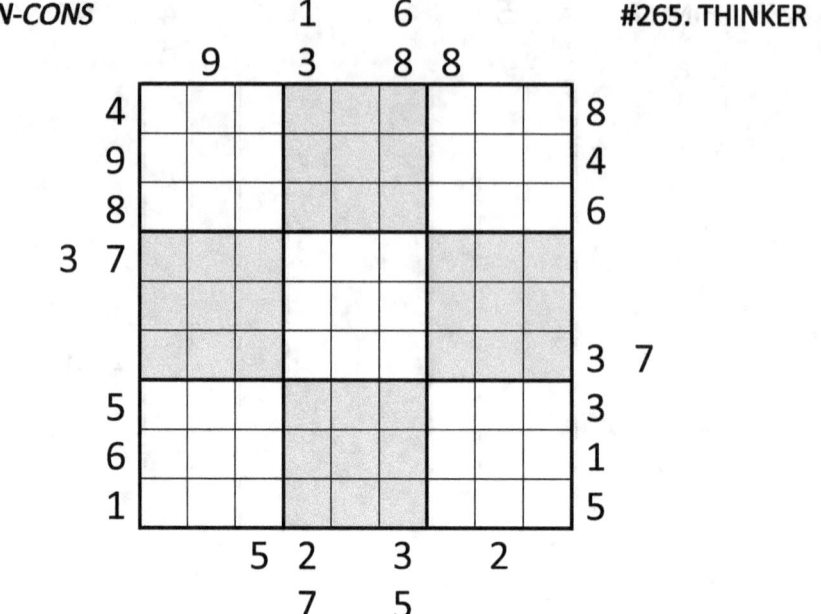

NON-CONS

#265. THINKER

_____ Outside Sudoku by amazon.com/djape, page 138

NON-CONS

#266. THINKER

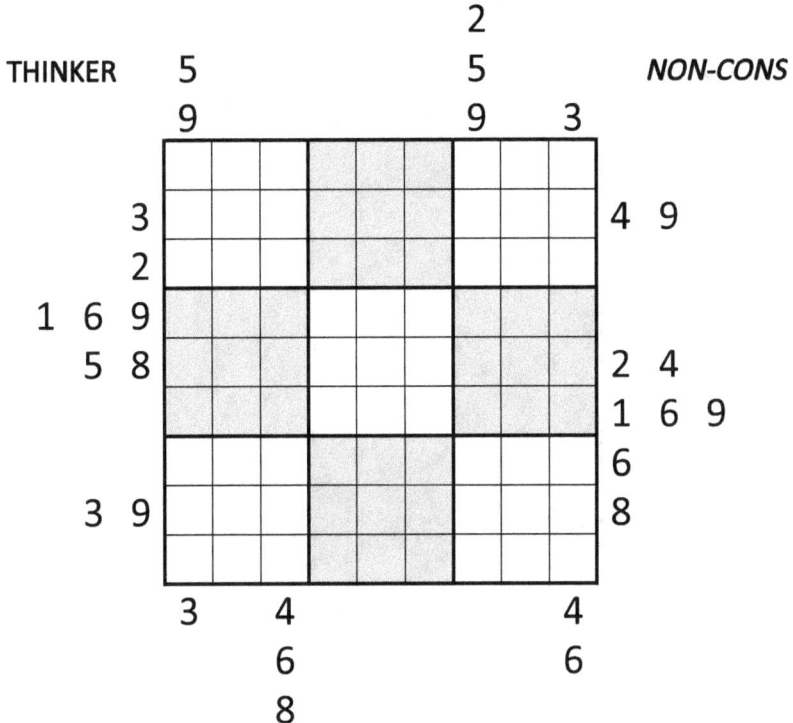

Outside Sudoku by amazon.com/djape, page 139 _____

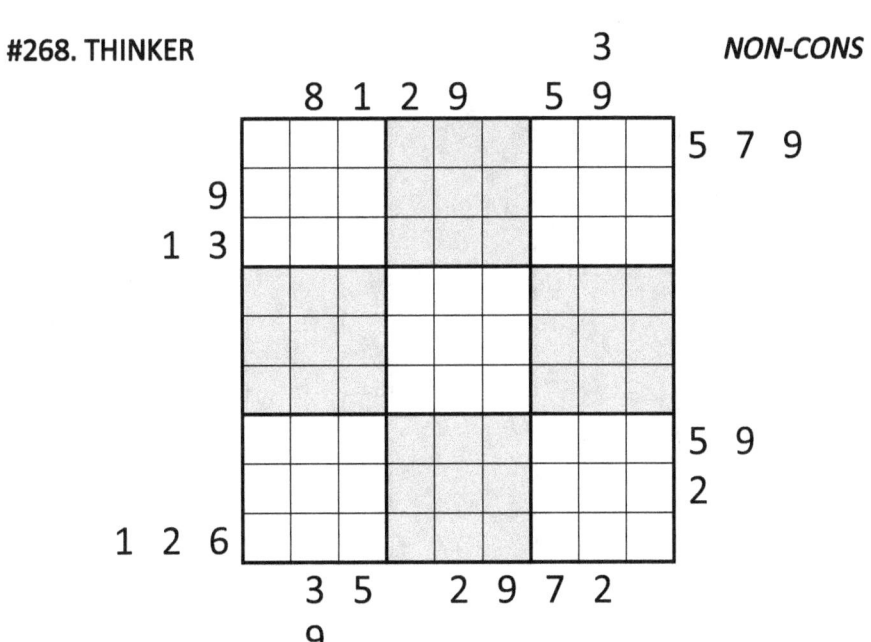

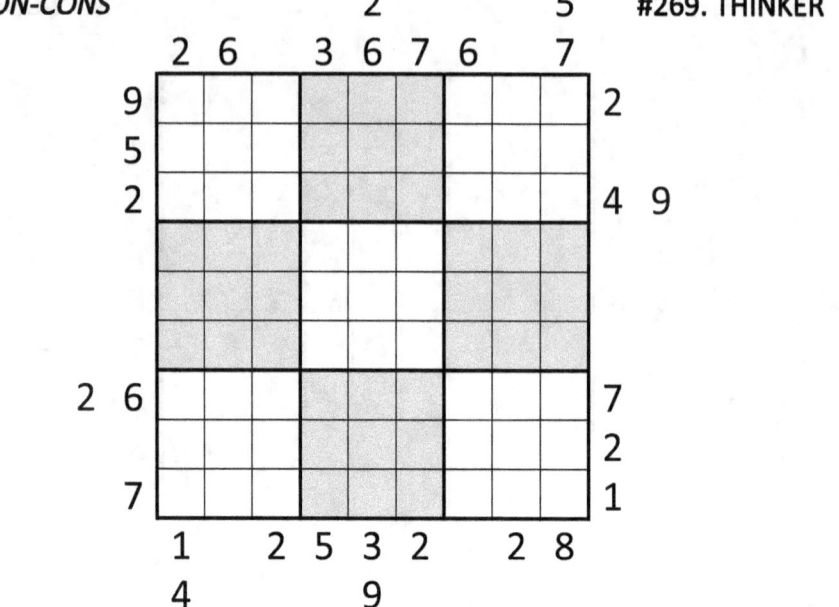

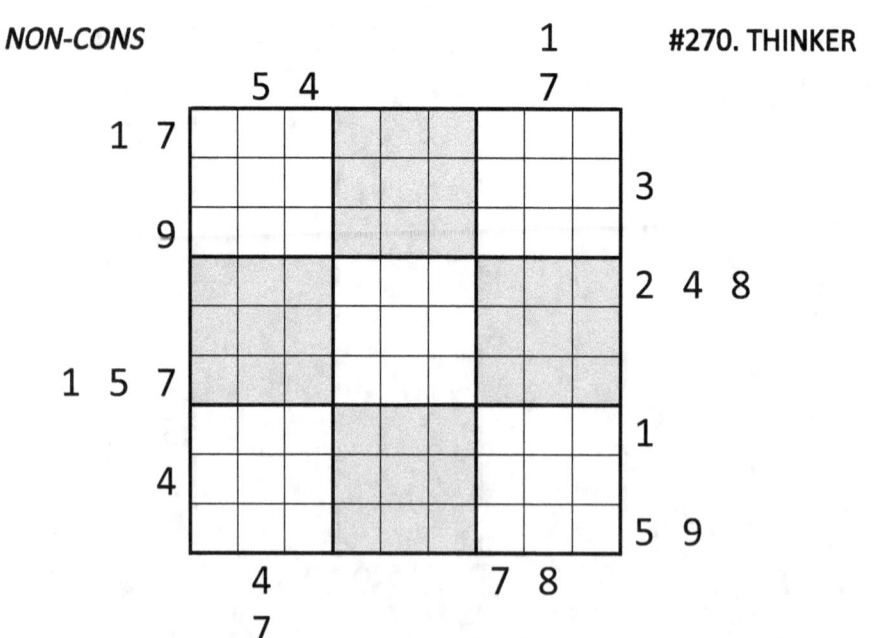

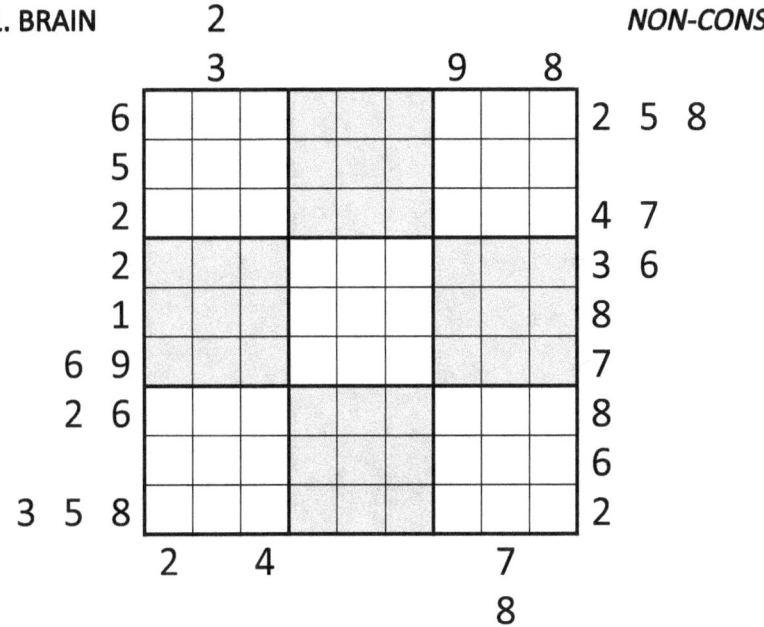

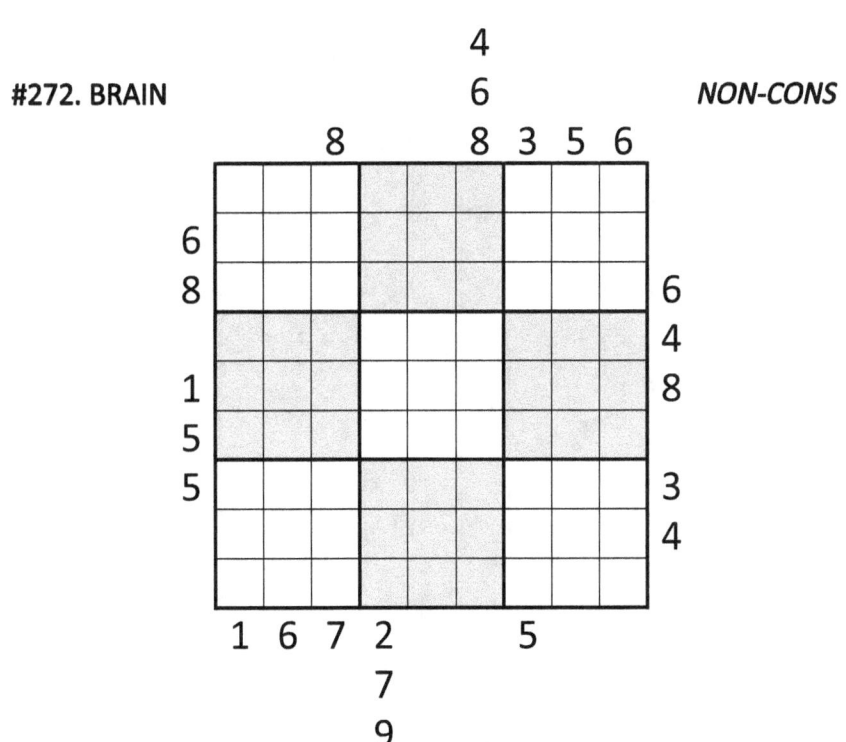

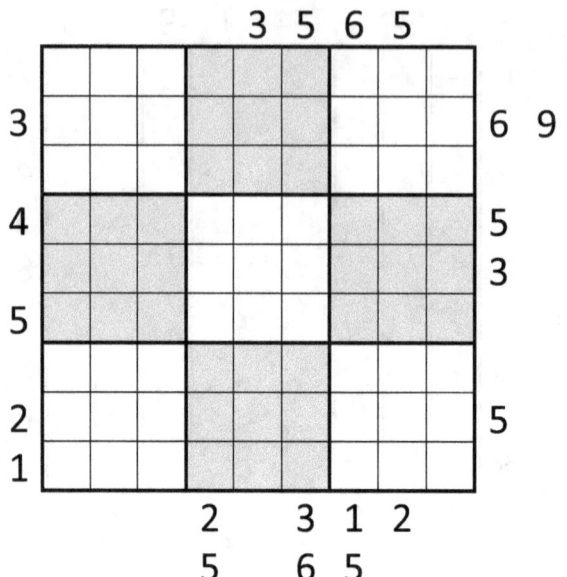

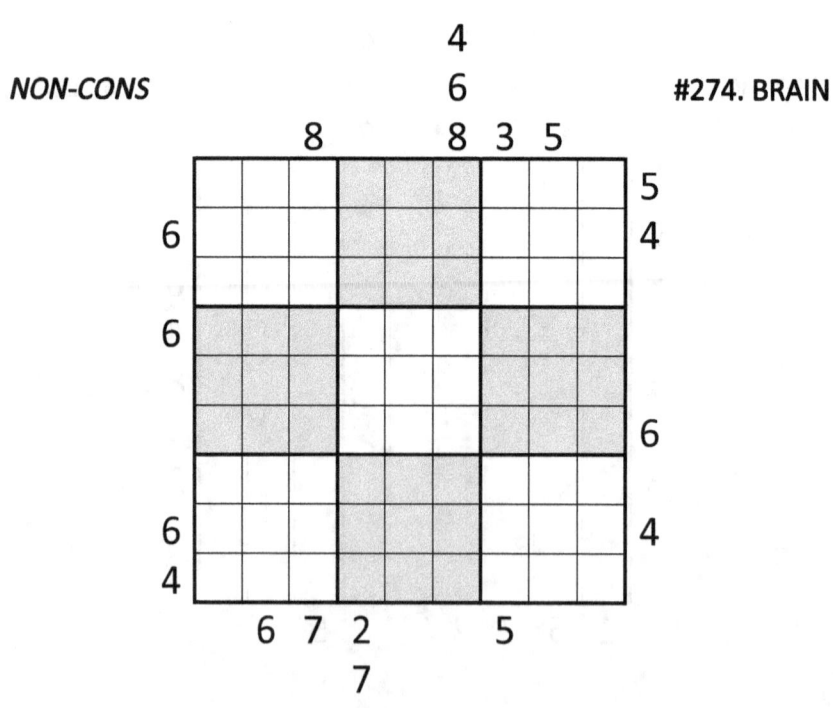

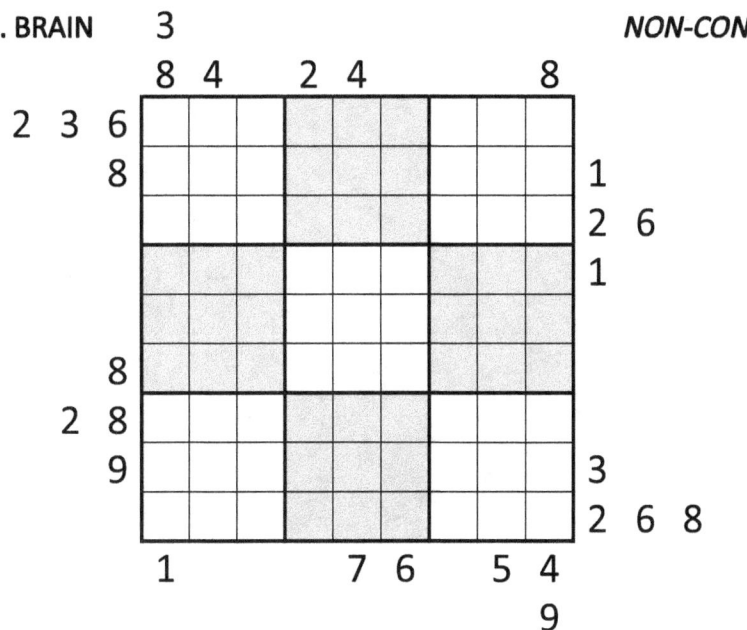

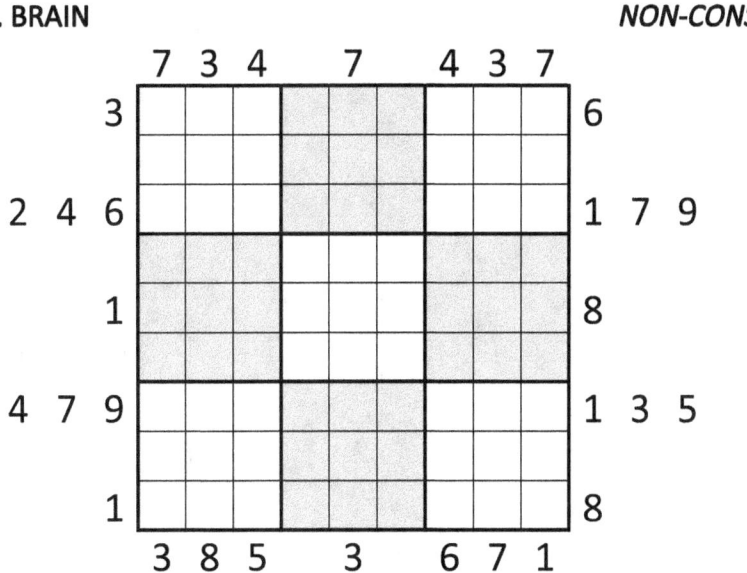

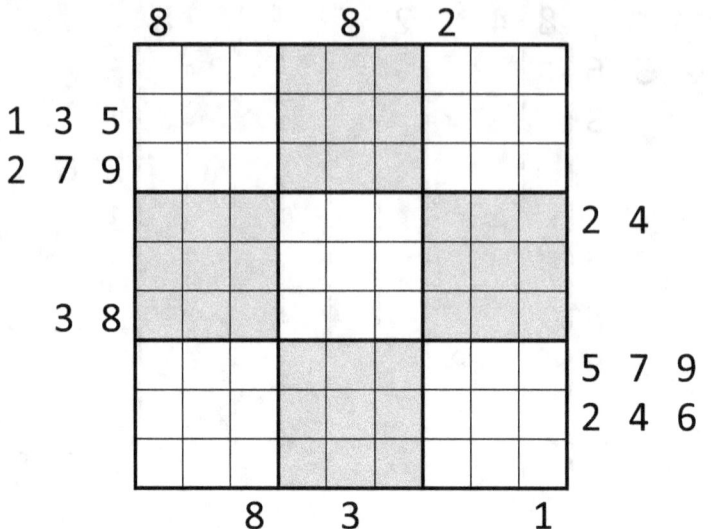

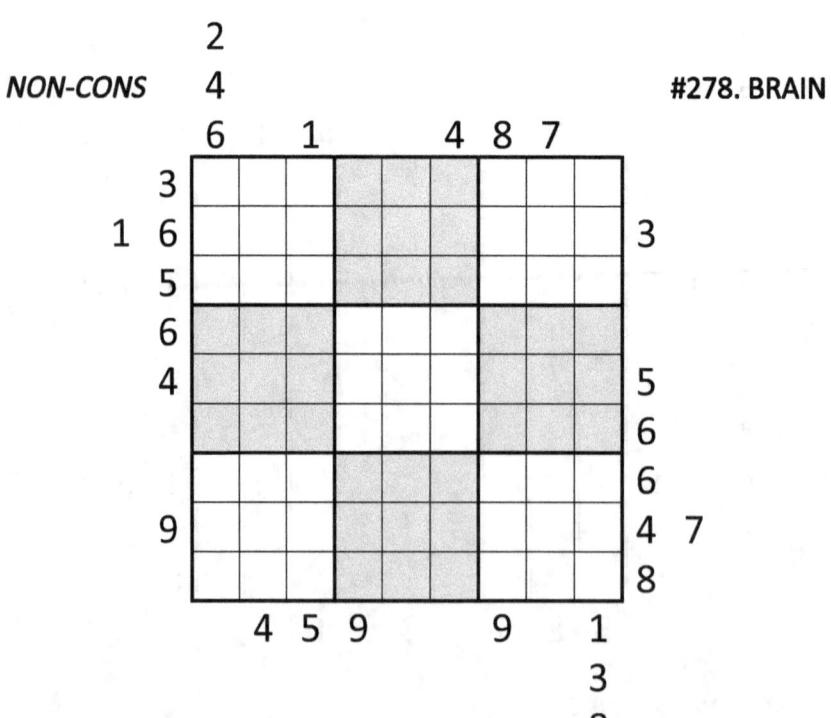

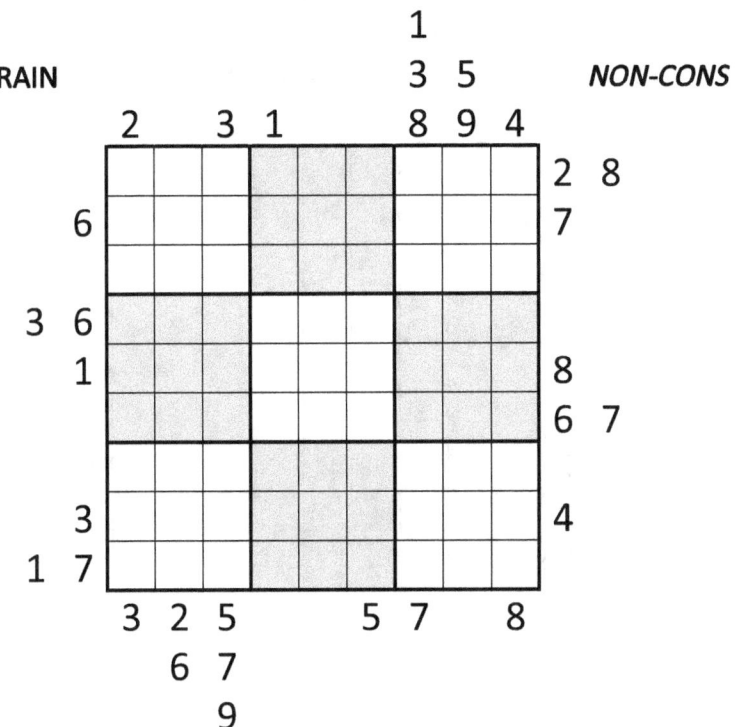

#279. BRAIN   NON-CONS

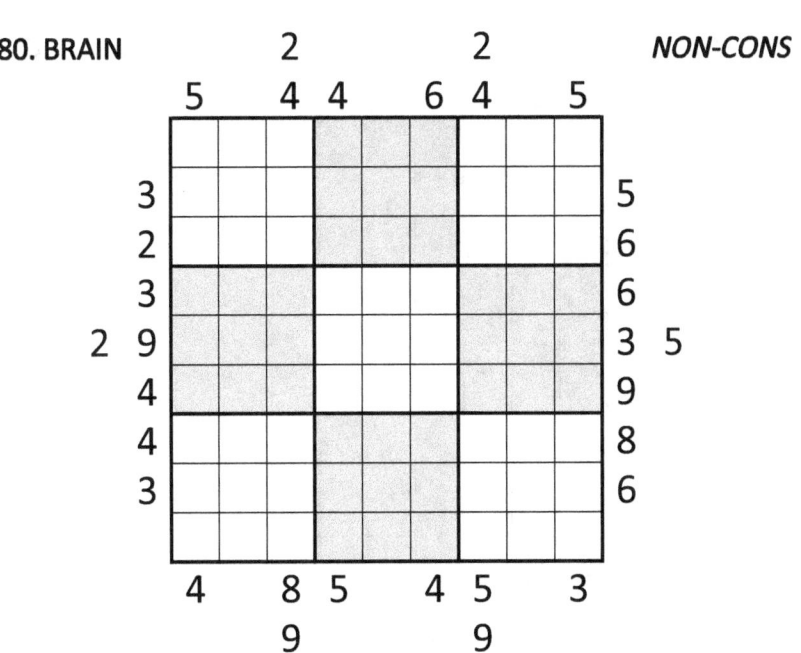

#280. BRAIN   NON-CONS

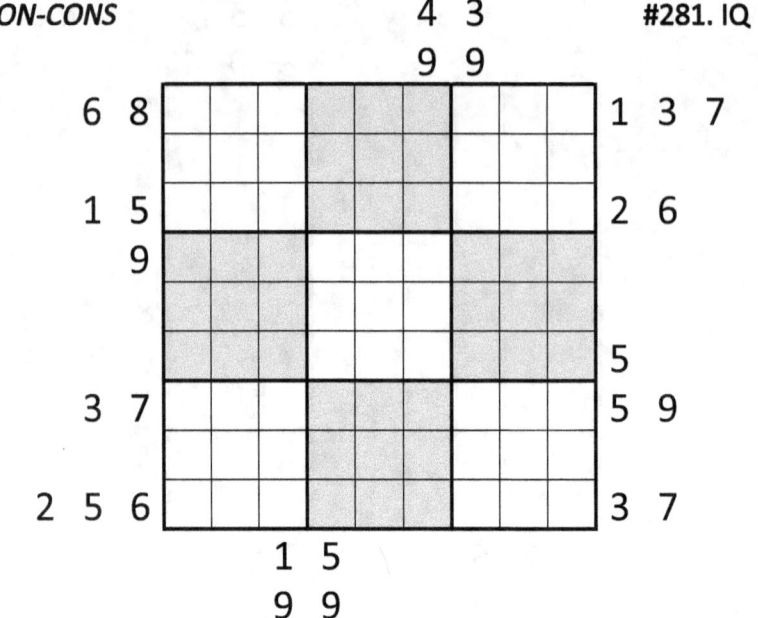

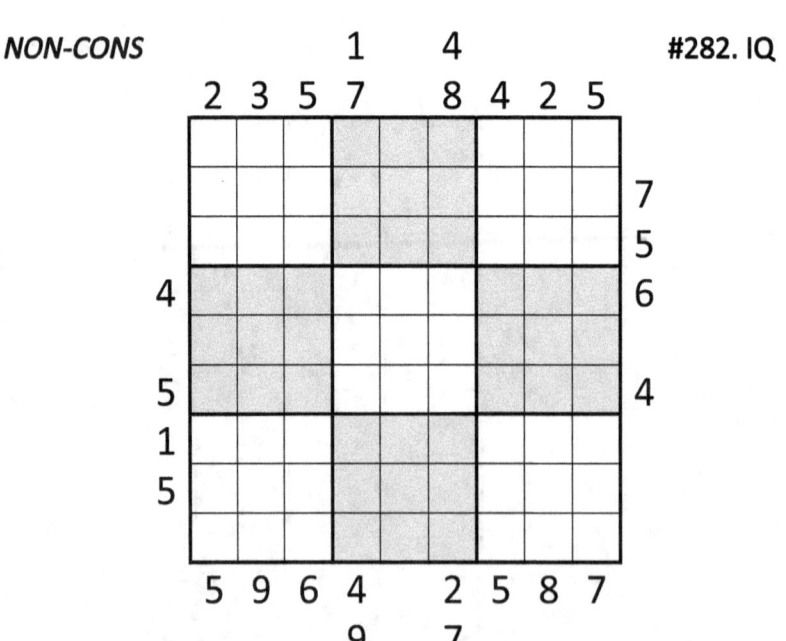

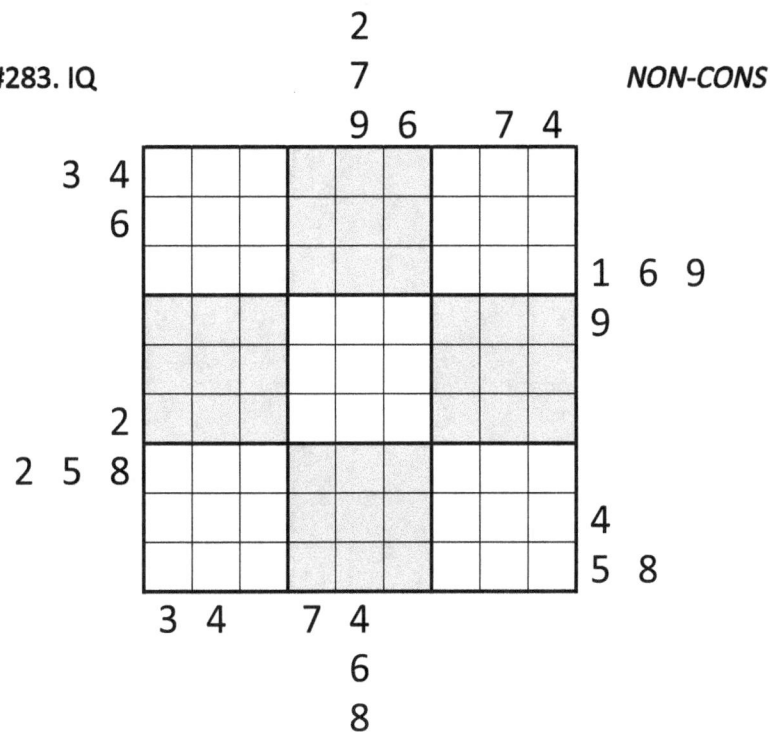

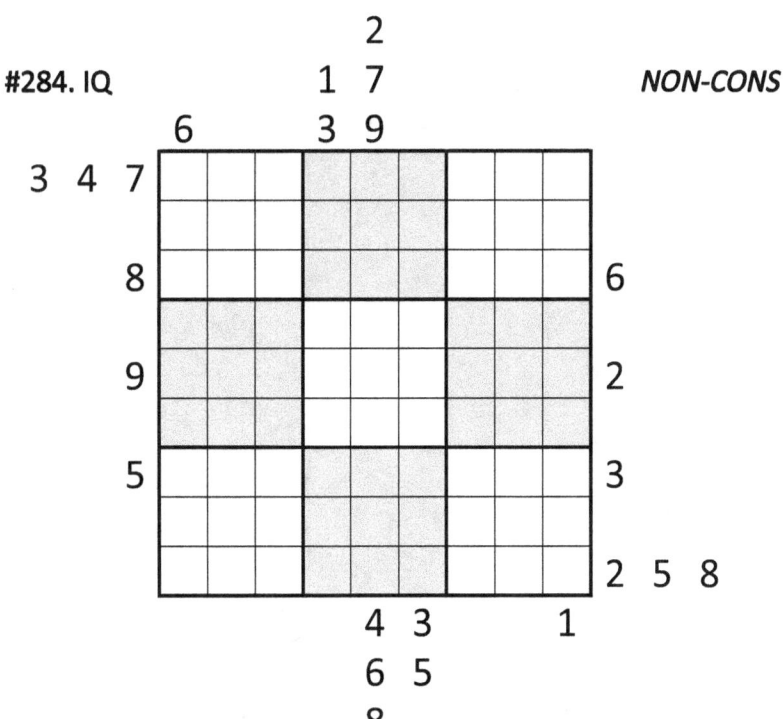

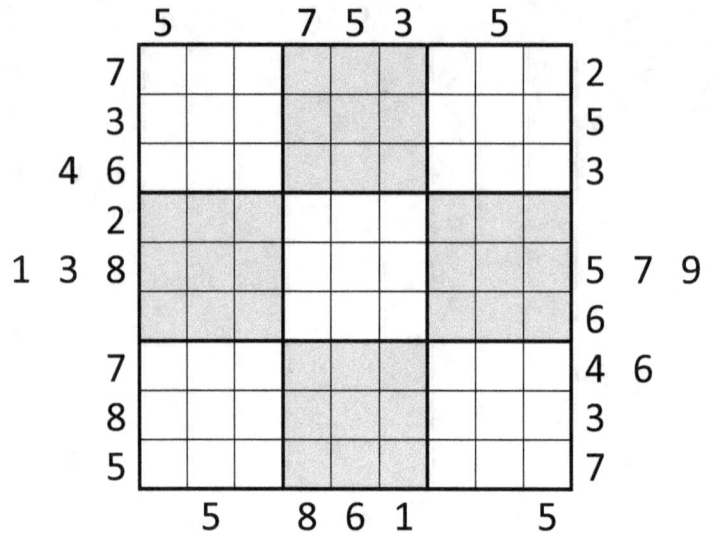

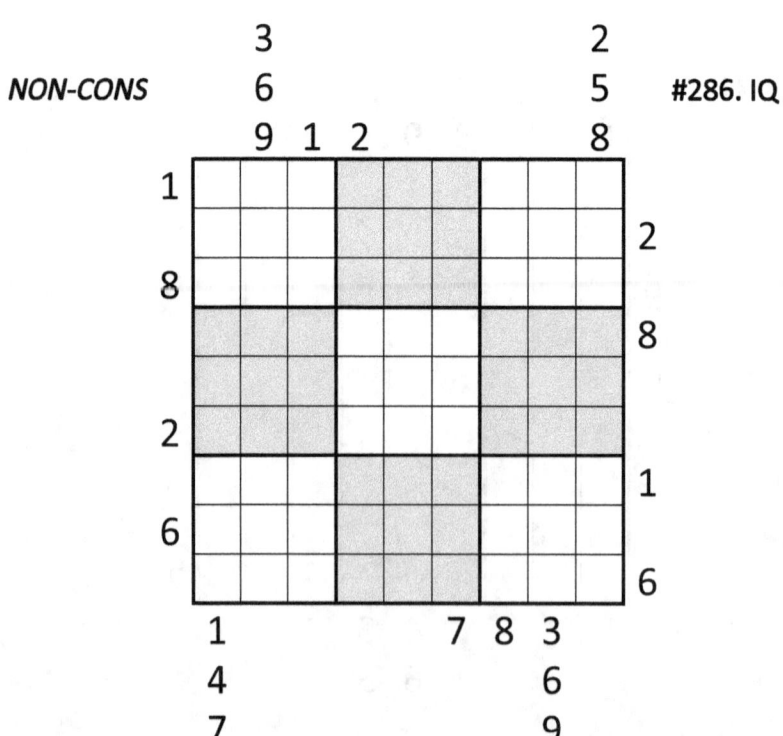

NON-CONS

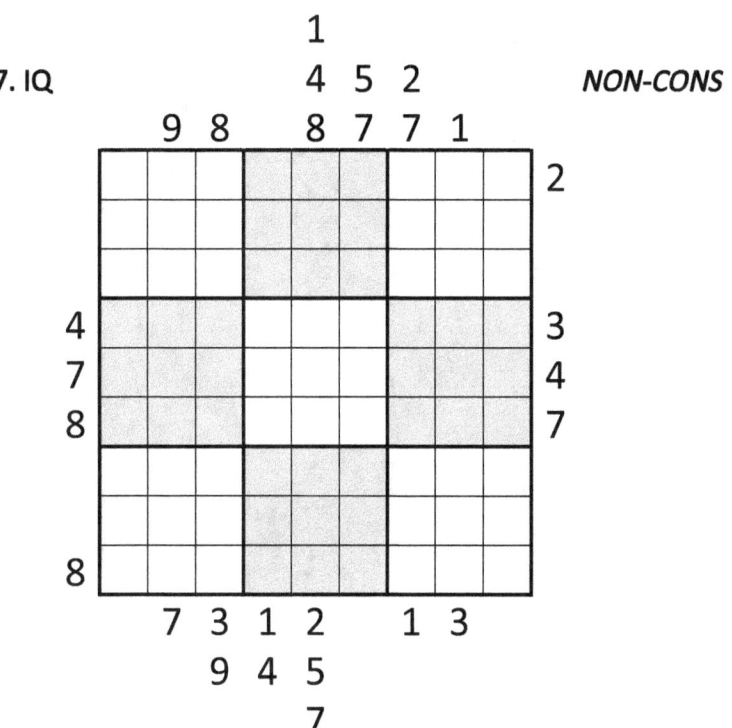

NON-CONS

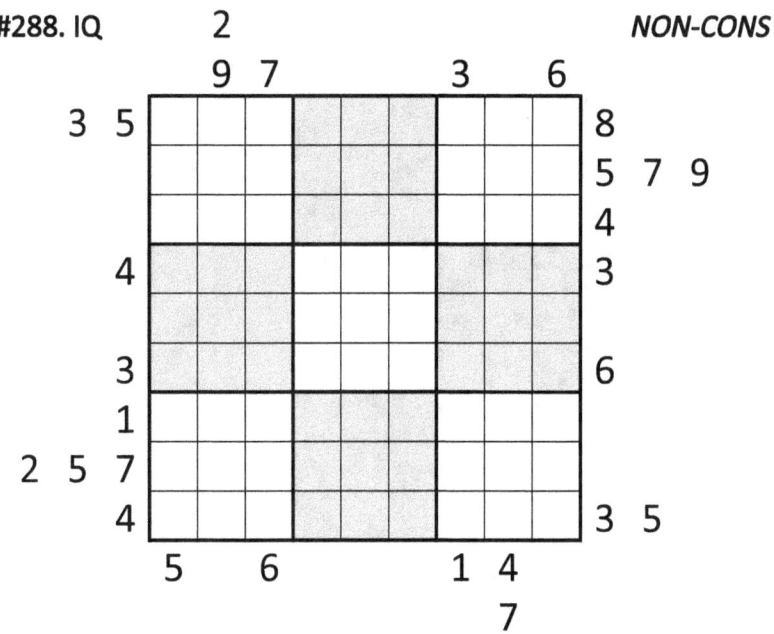

NON-CONS

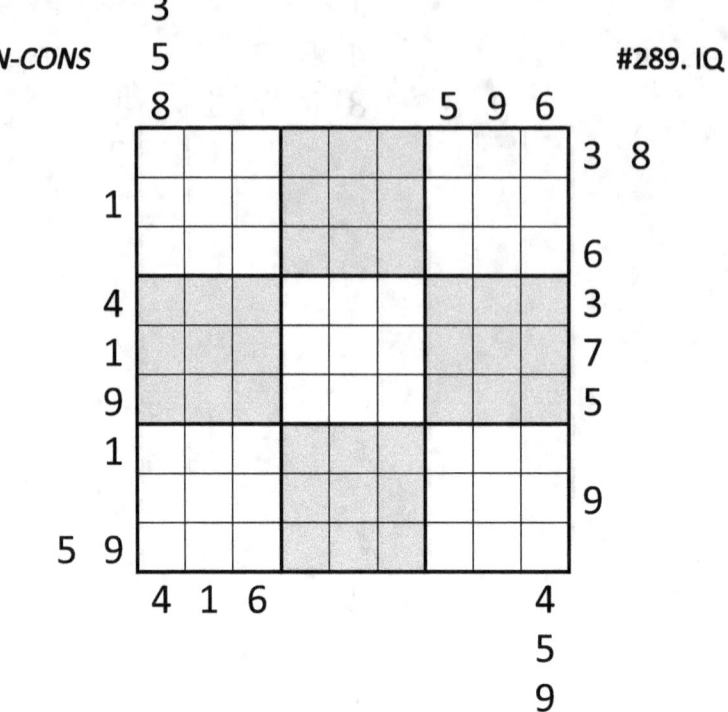

Outside Sudoku by amazon.com/djape, page 150

NON-CONS

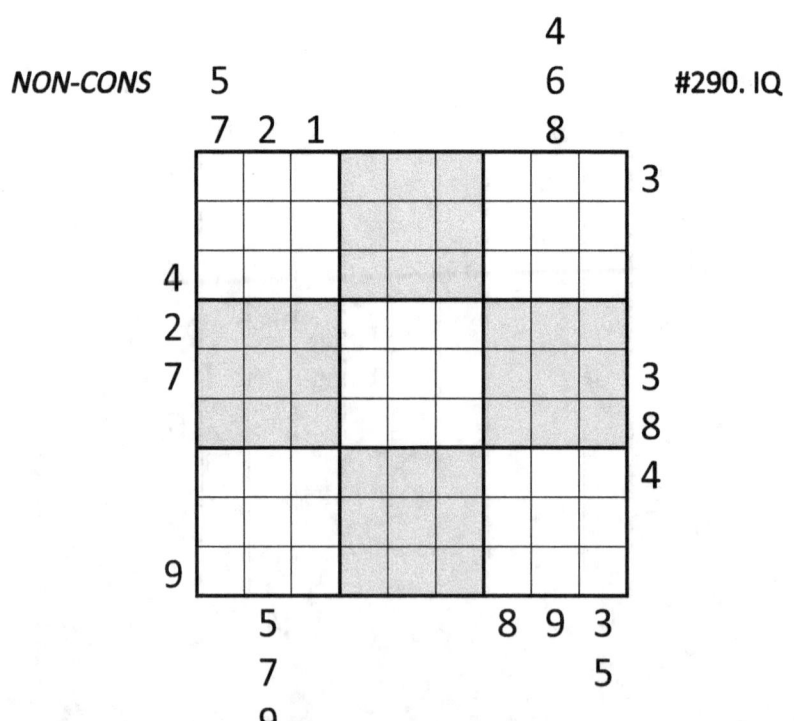

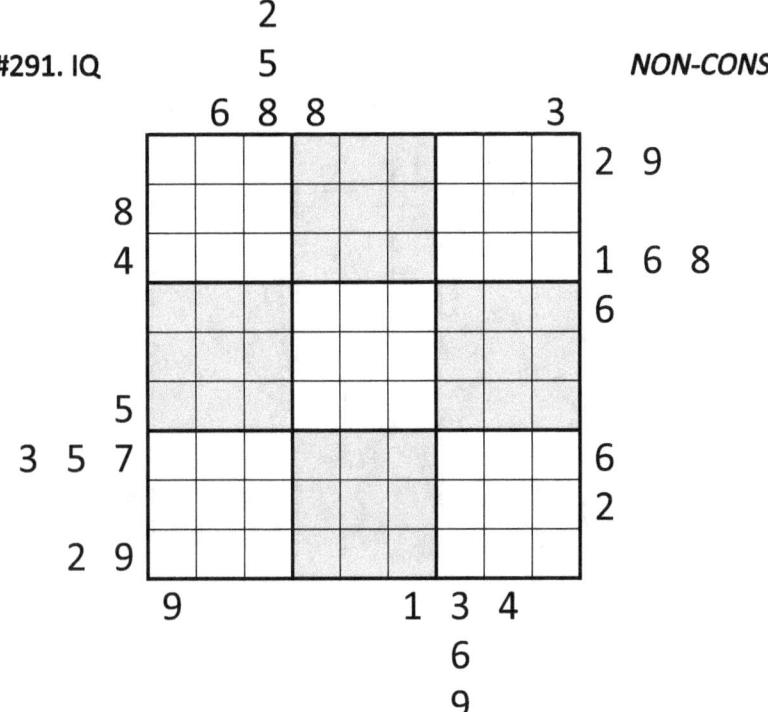

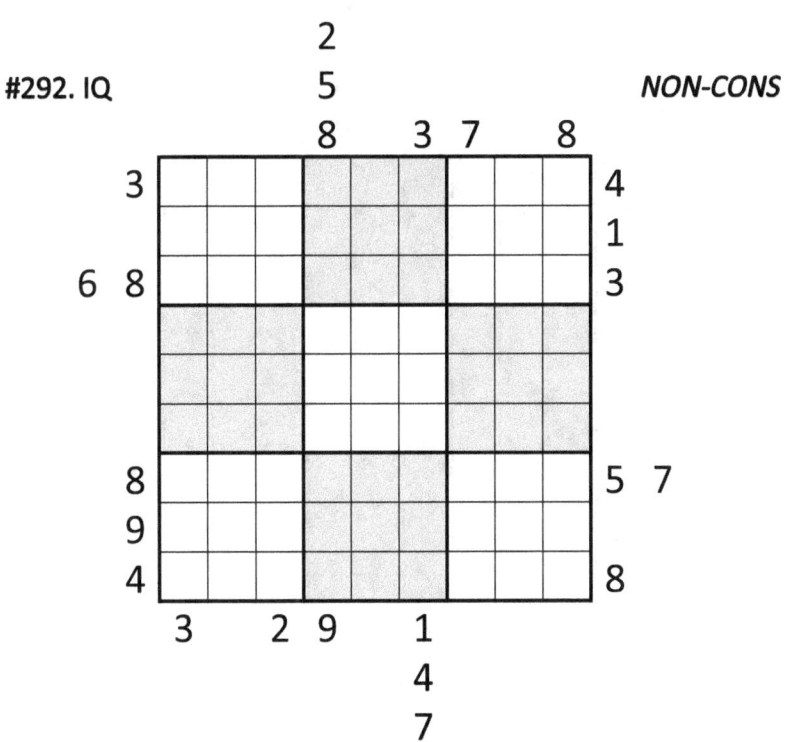

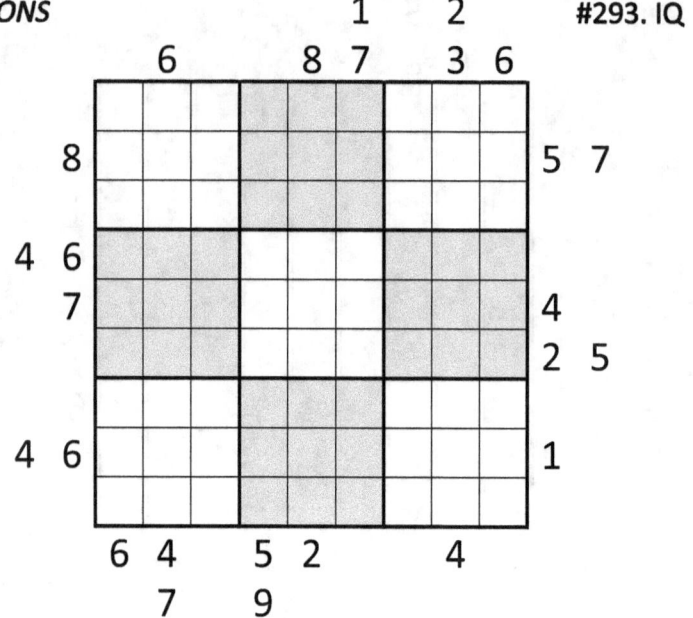

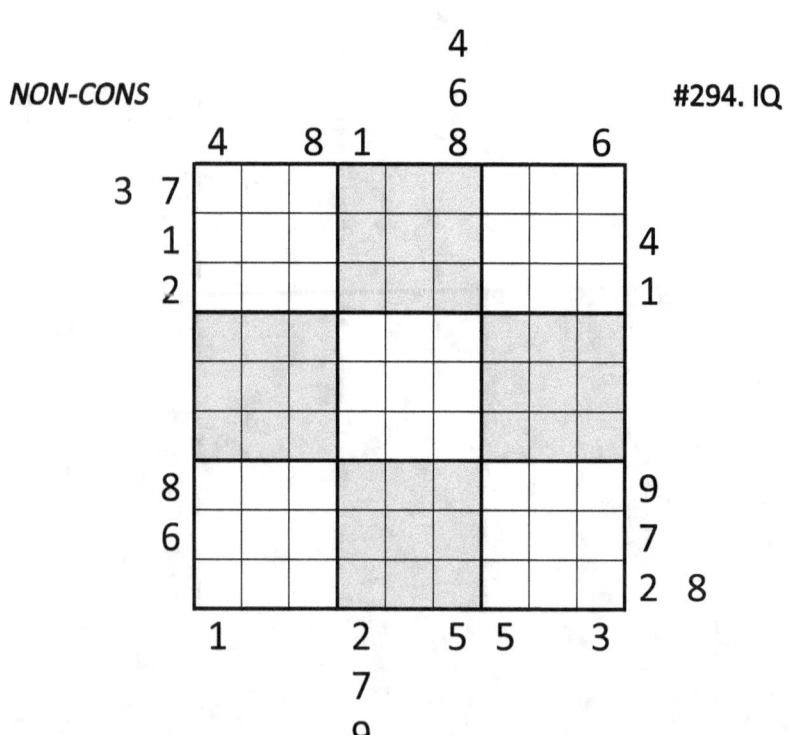

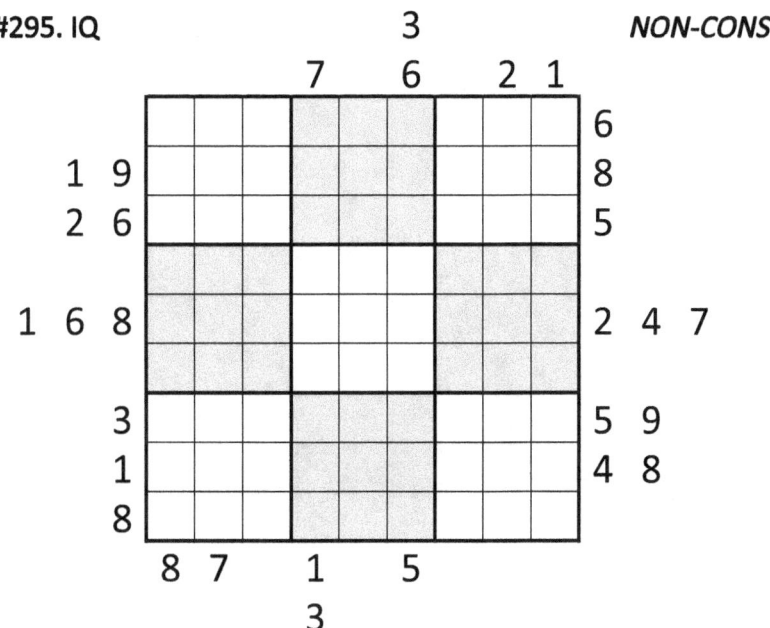

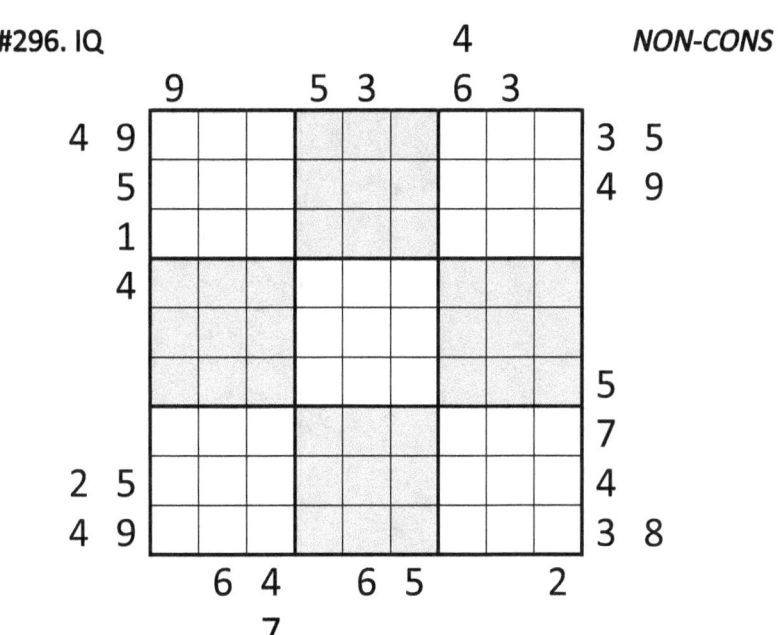

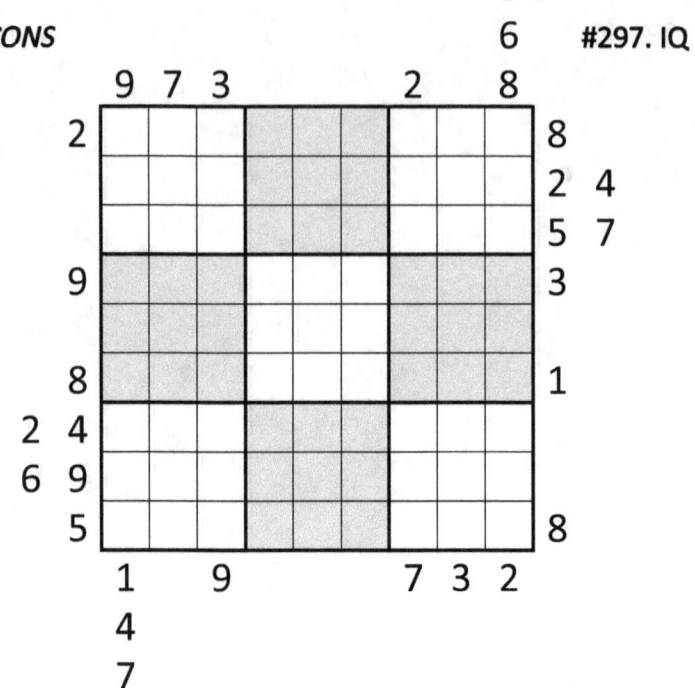

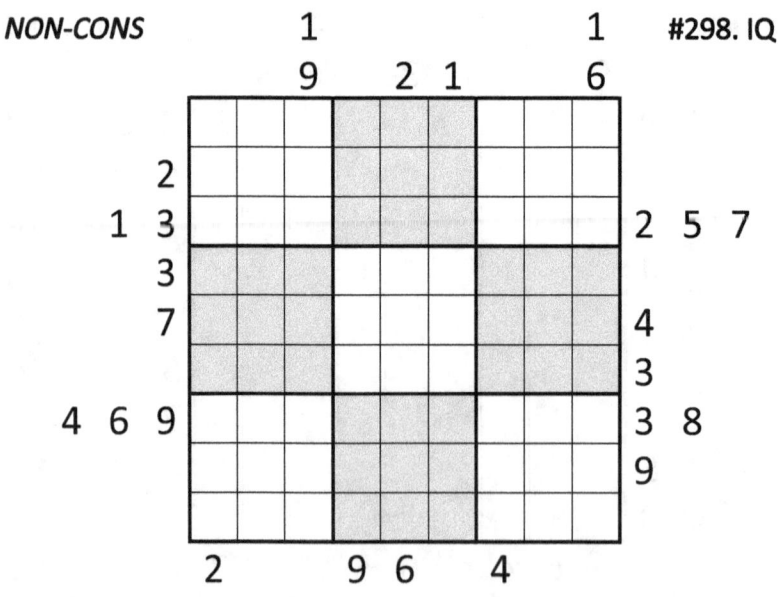

NON-CONS

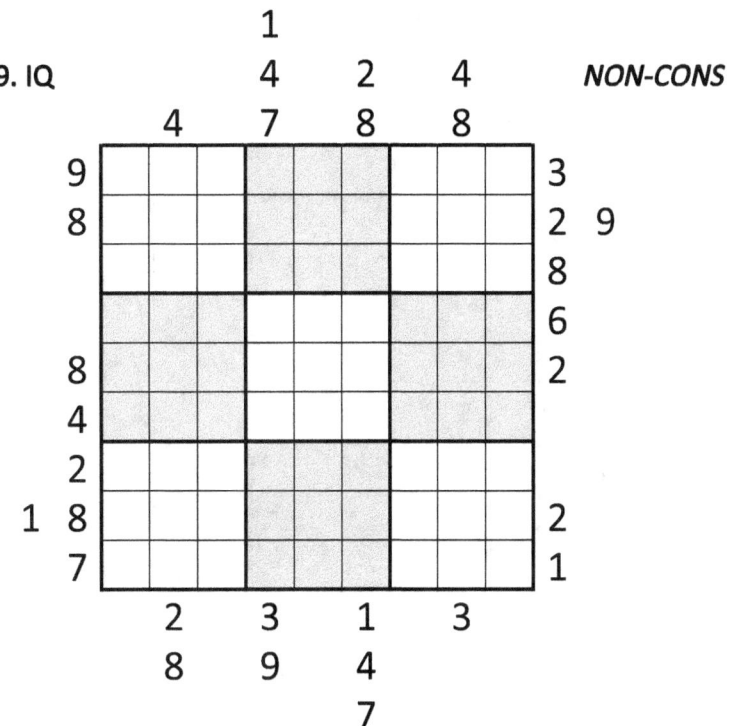

NON-CONS

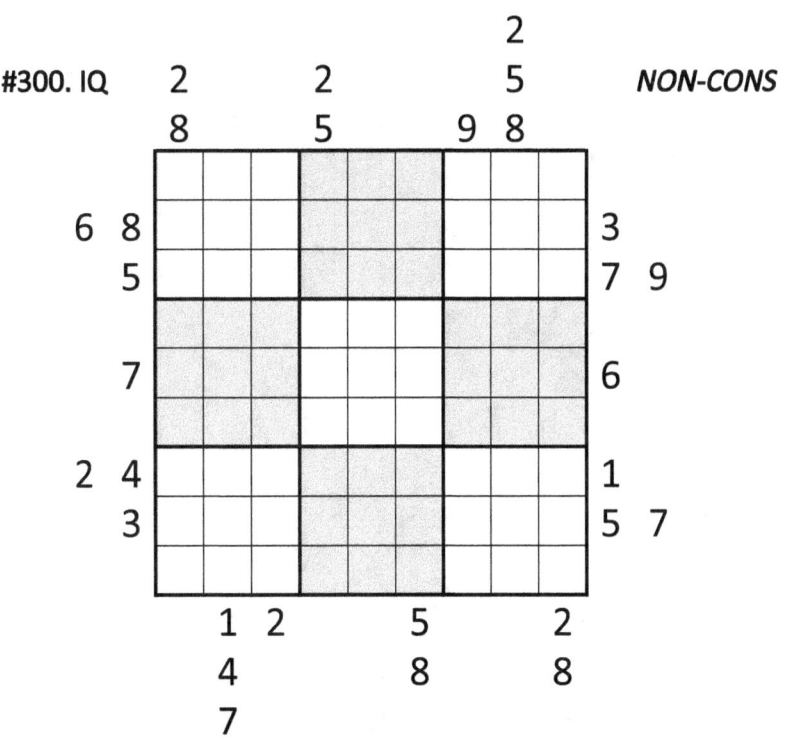

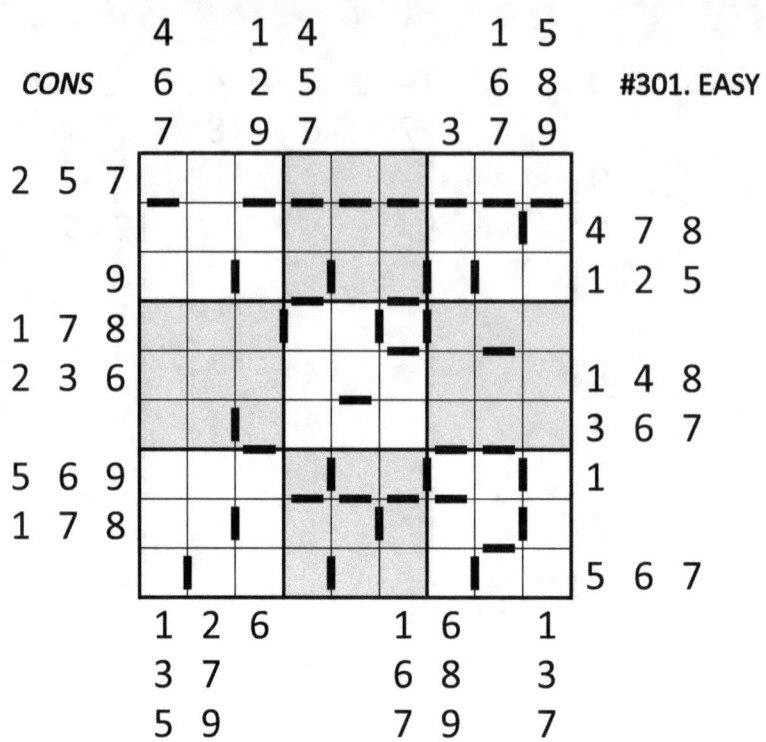

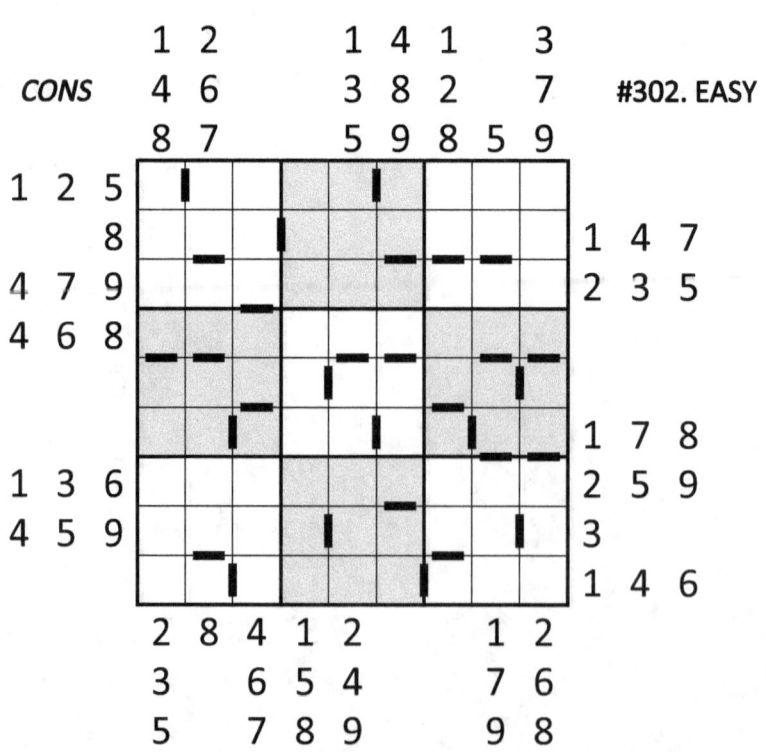

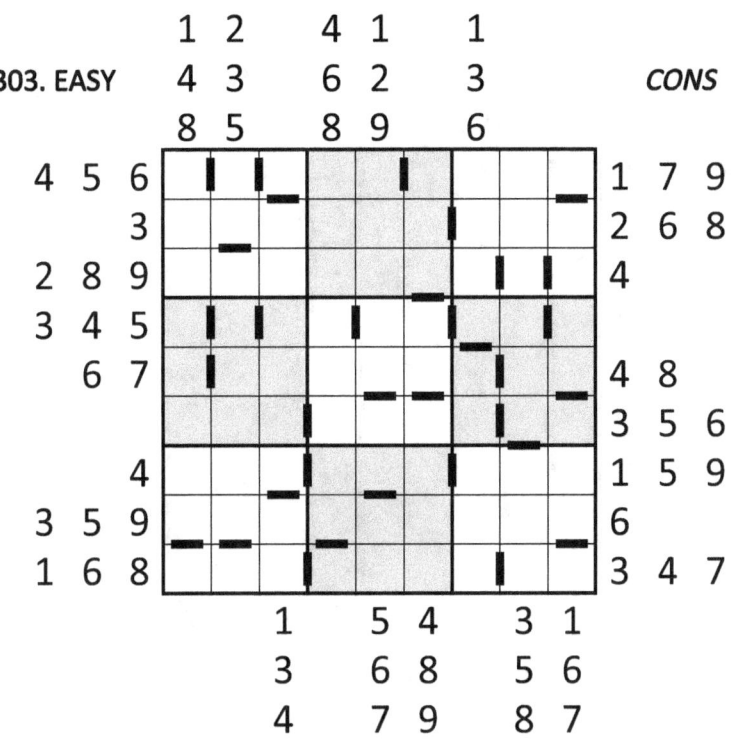

#303. EASY — CONS

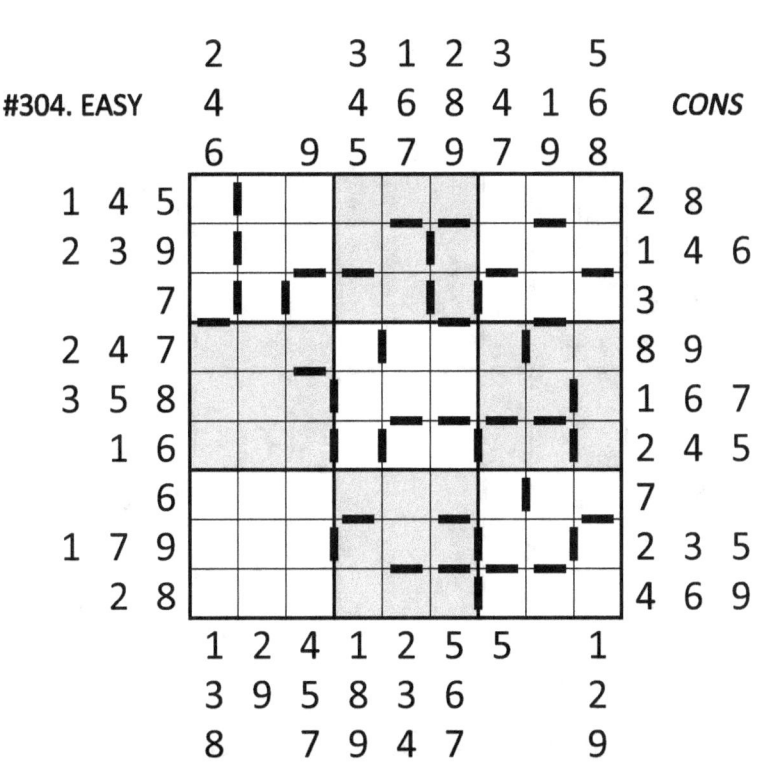

#304. EASY — CONS

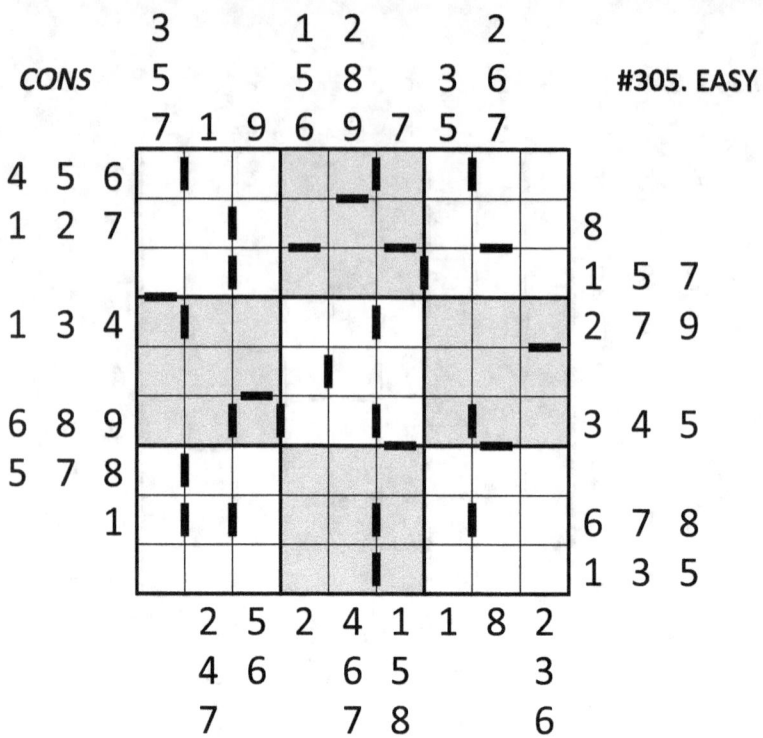

#305. EASY

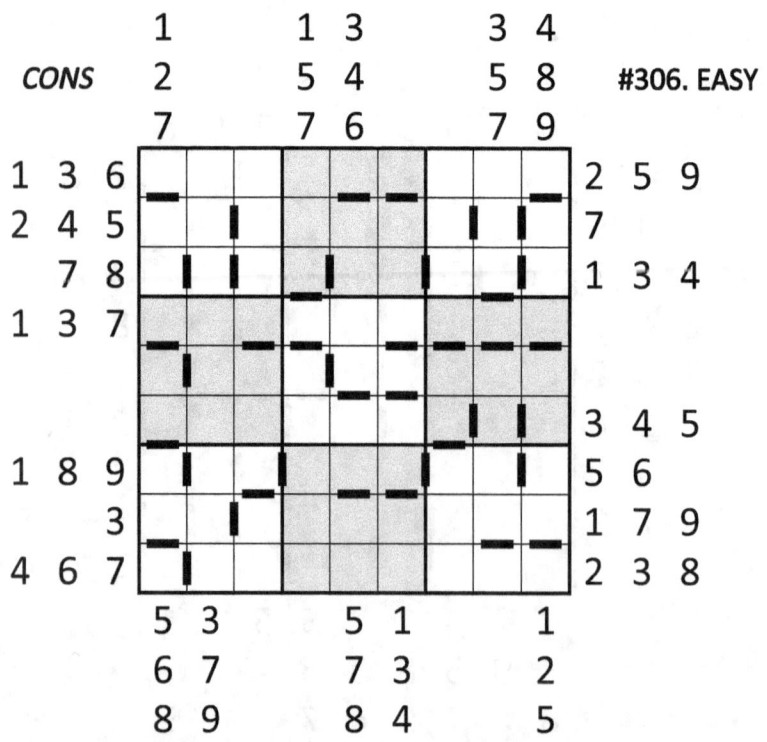

#306. EASY

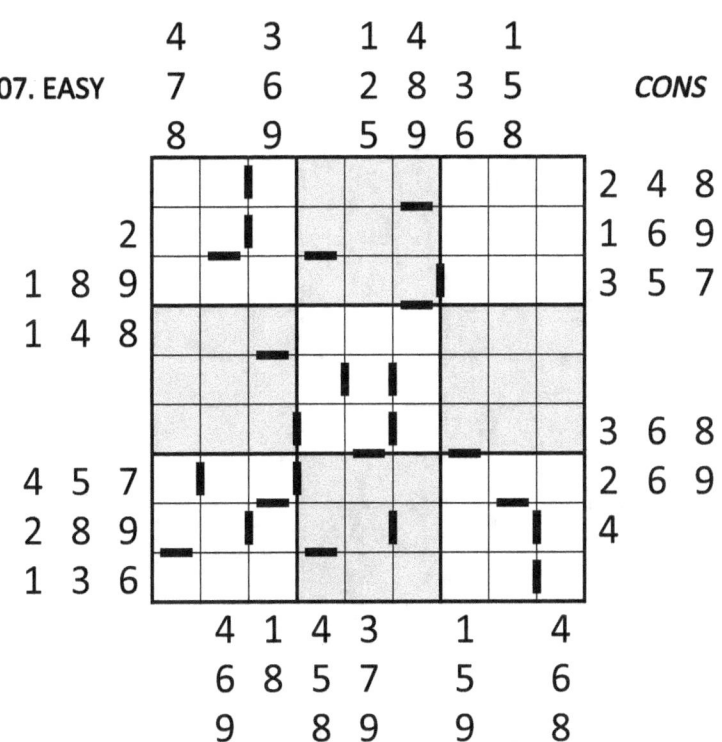

#308. EASY

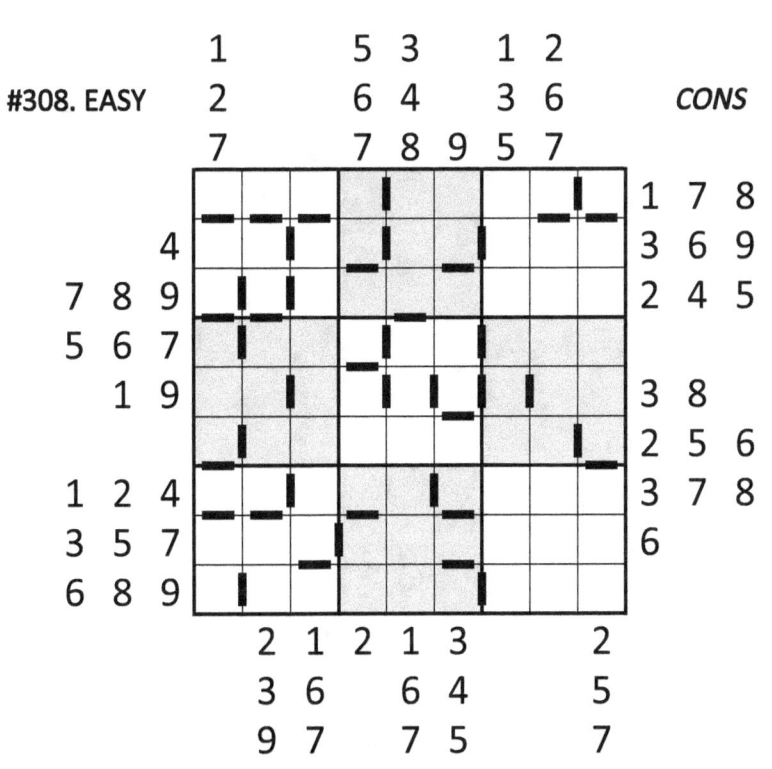

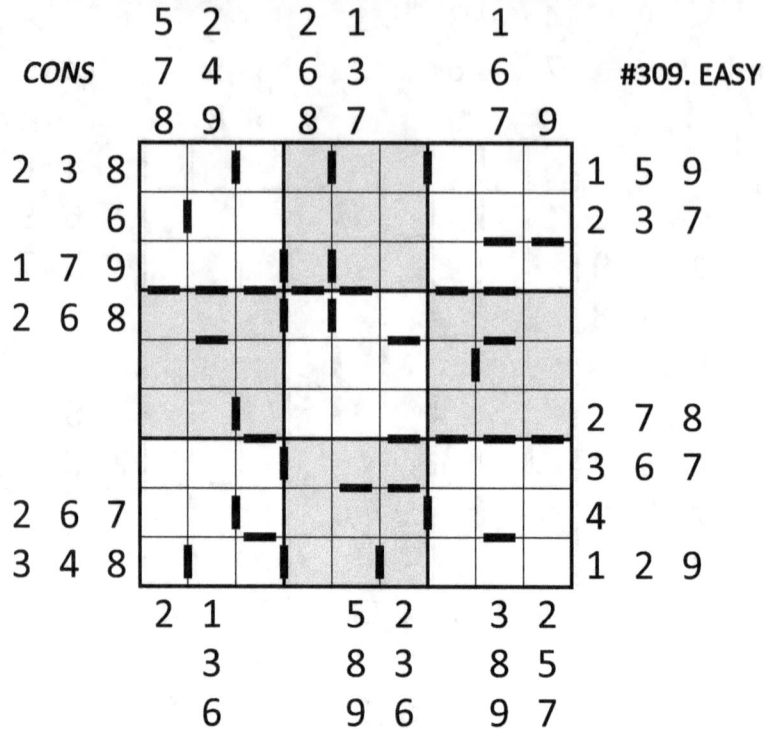

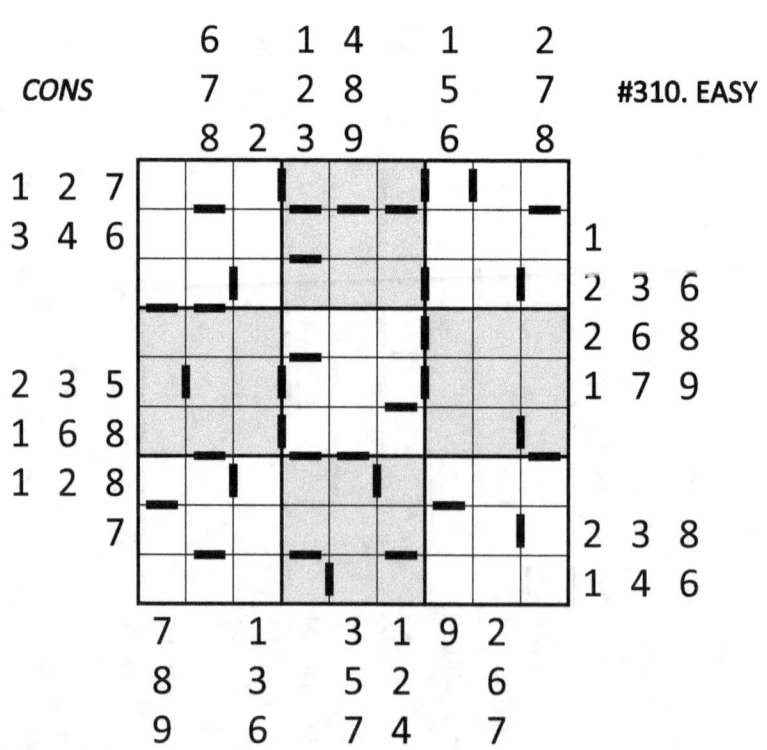

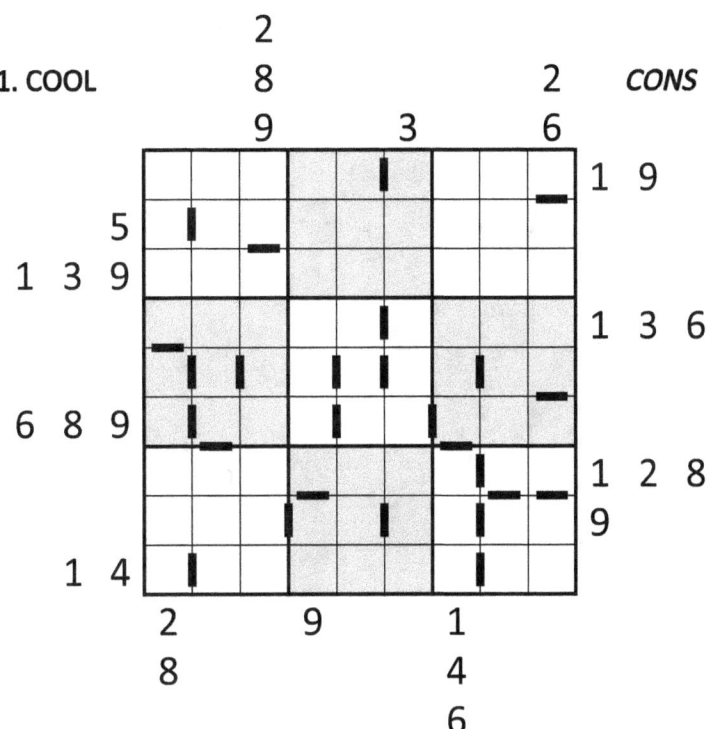

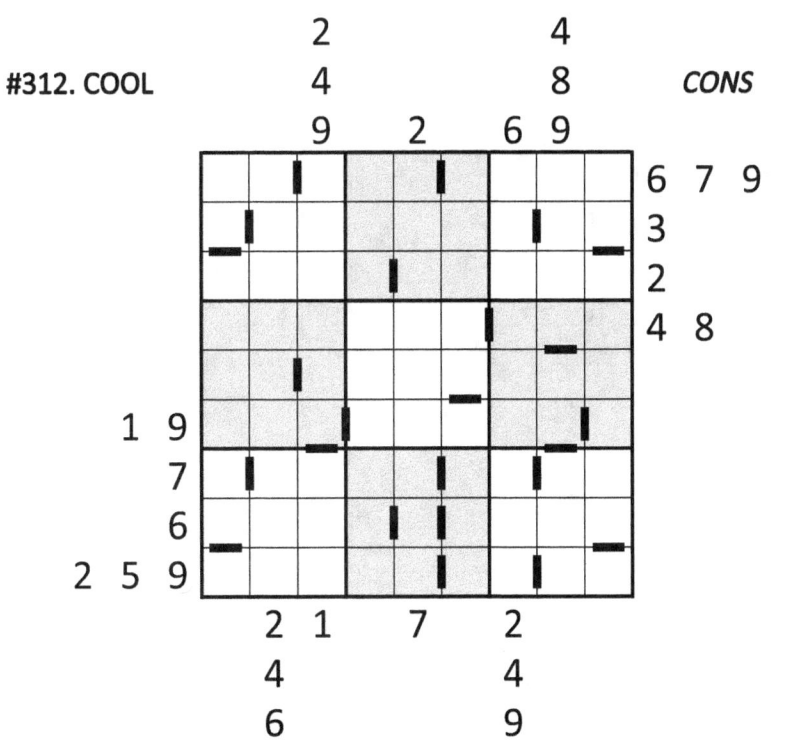

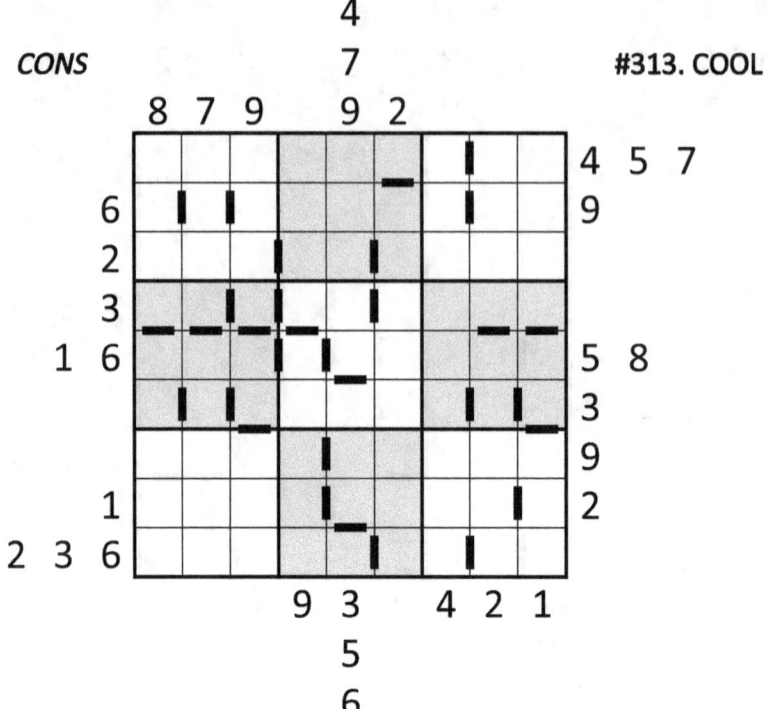

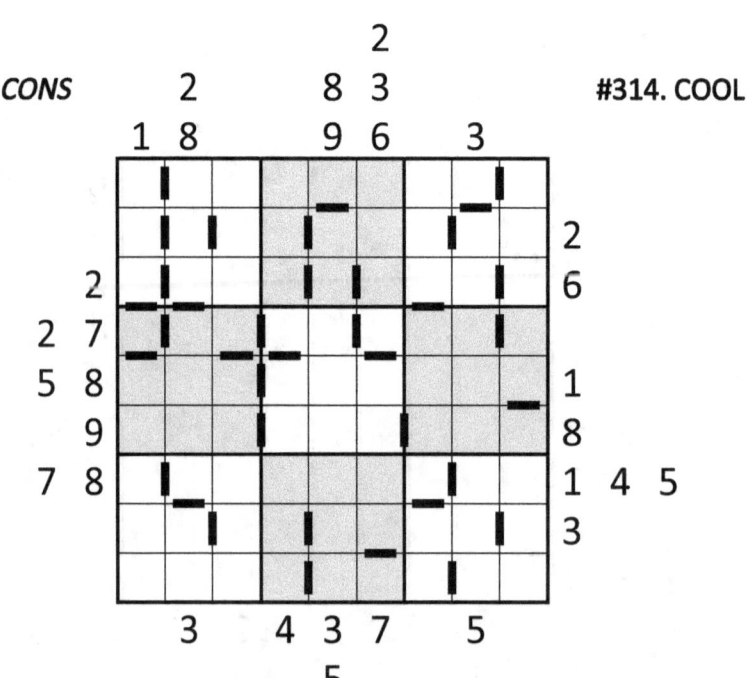

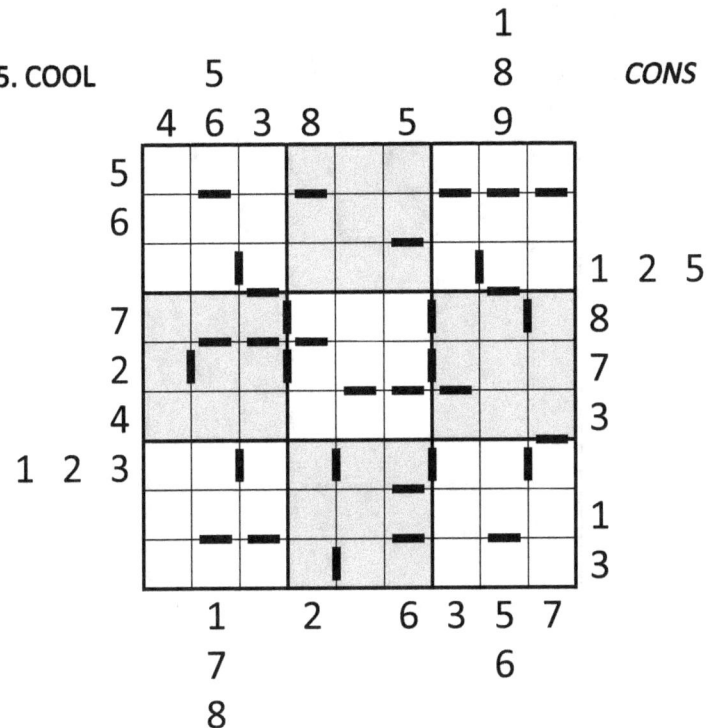

#315. COOL CONS

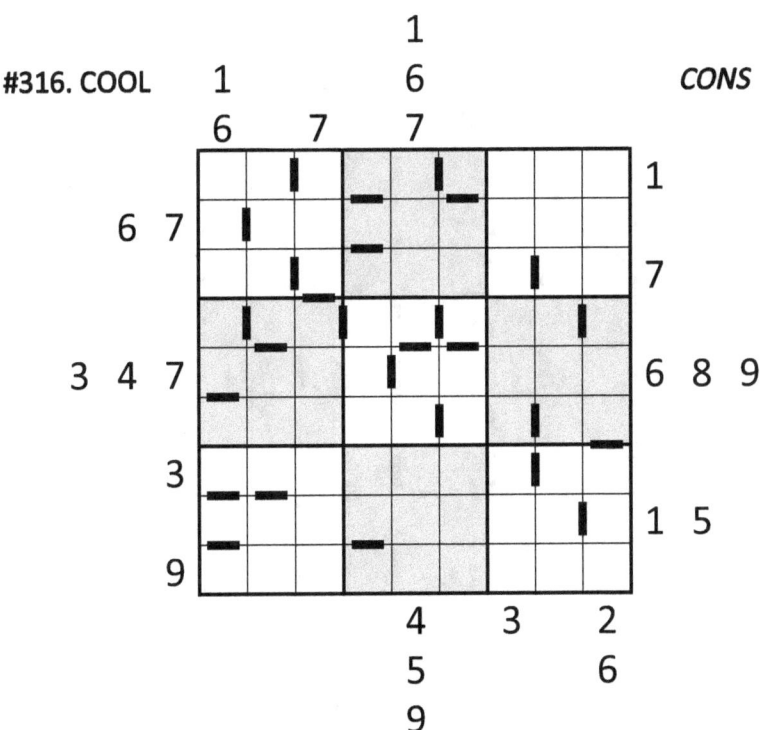

#316. COOL CONS

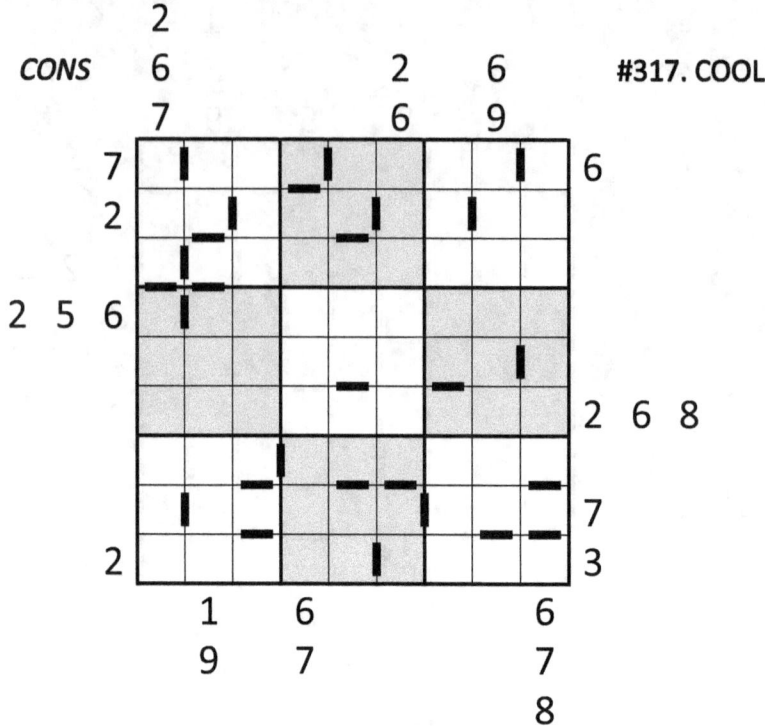

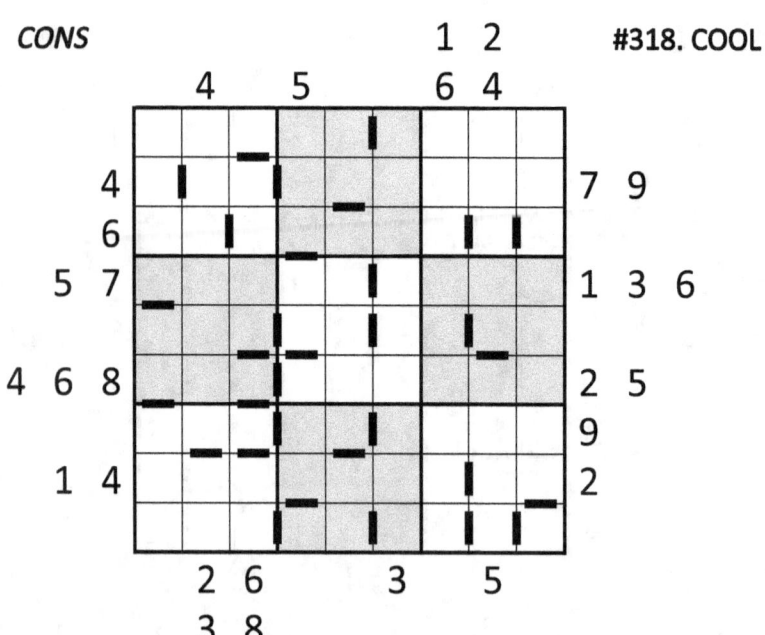

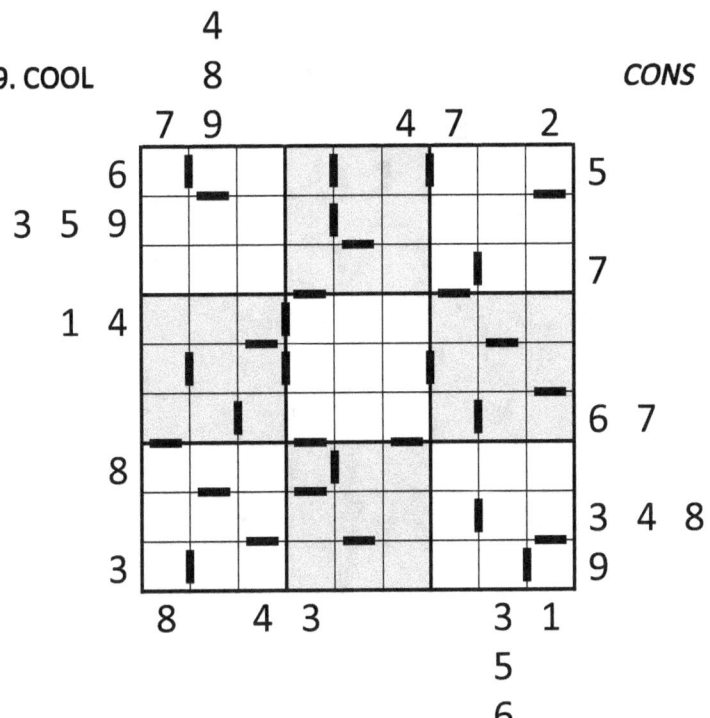

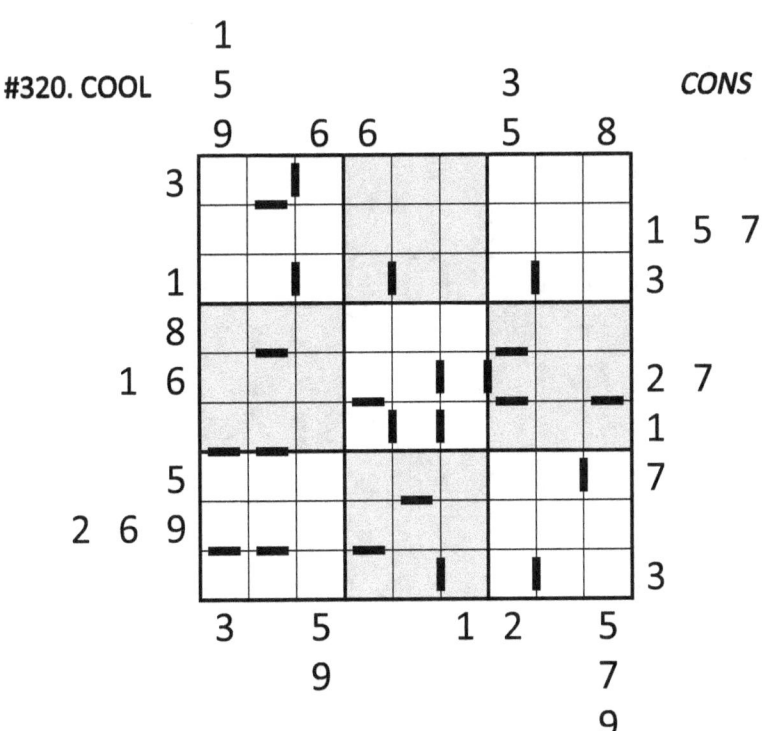

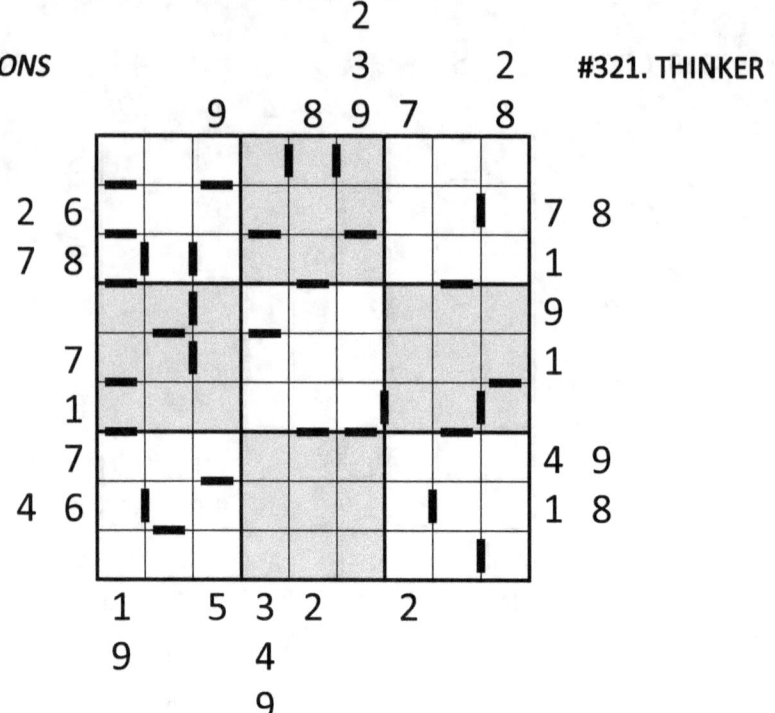

CONS

#321. THINKER

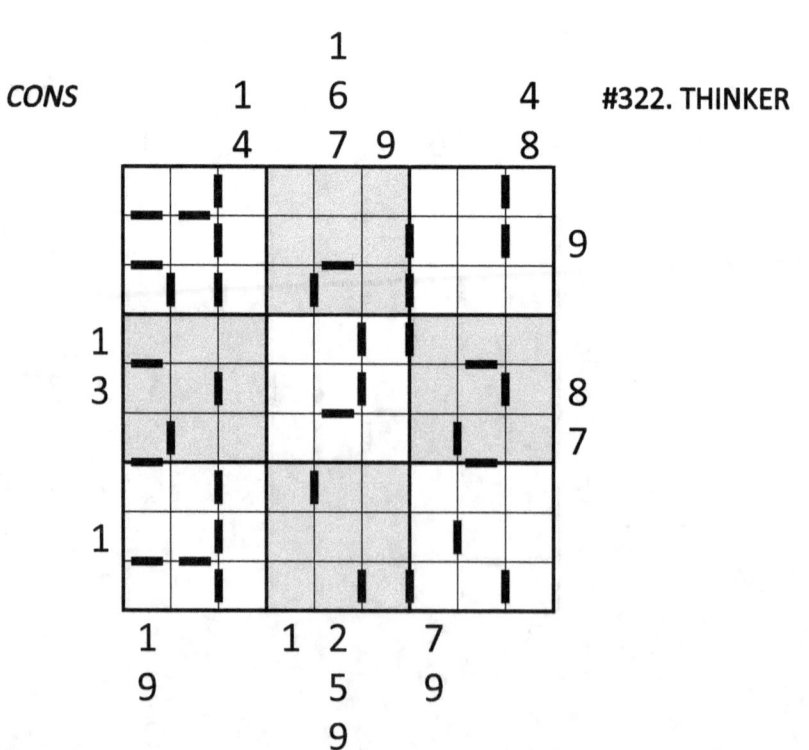

CONS

#322. THINKER

#323. THINKER

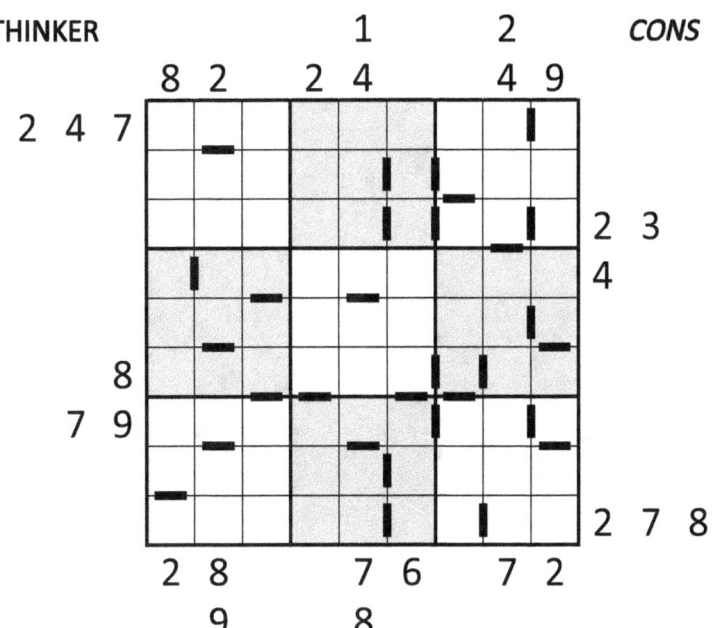

#324. THINKER

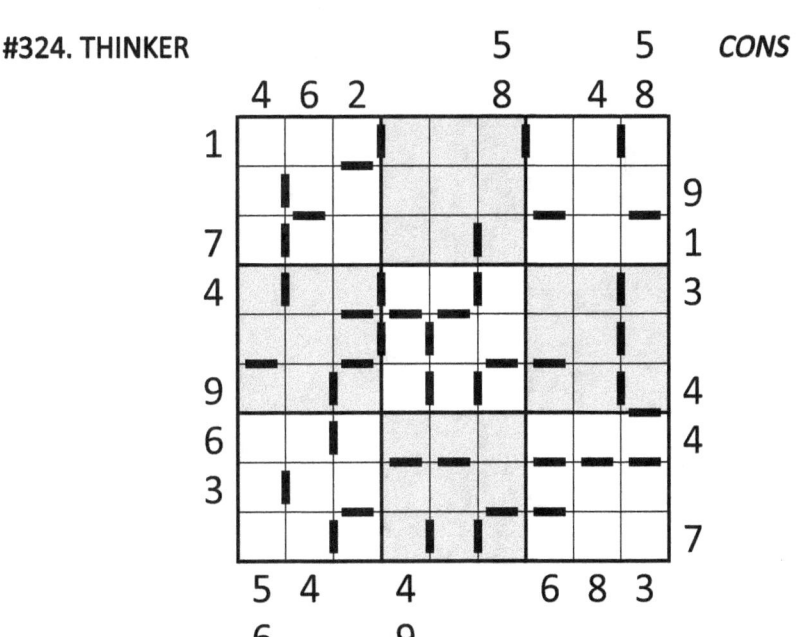

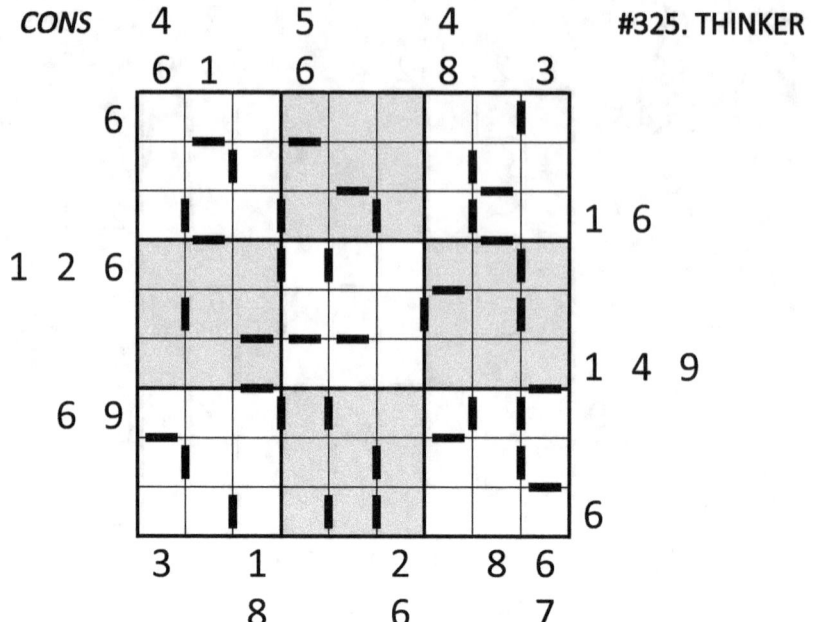

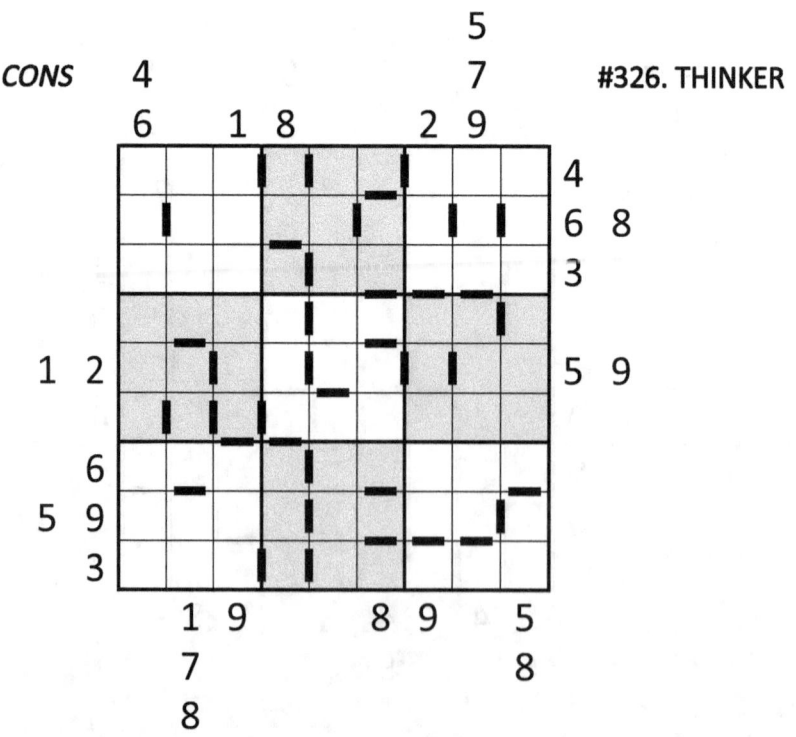

#327. THINKER

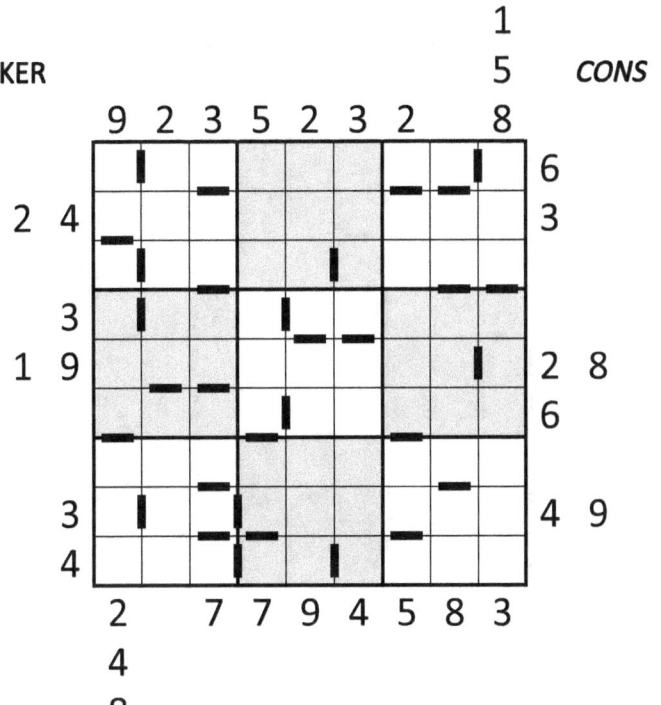

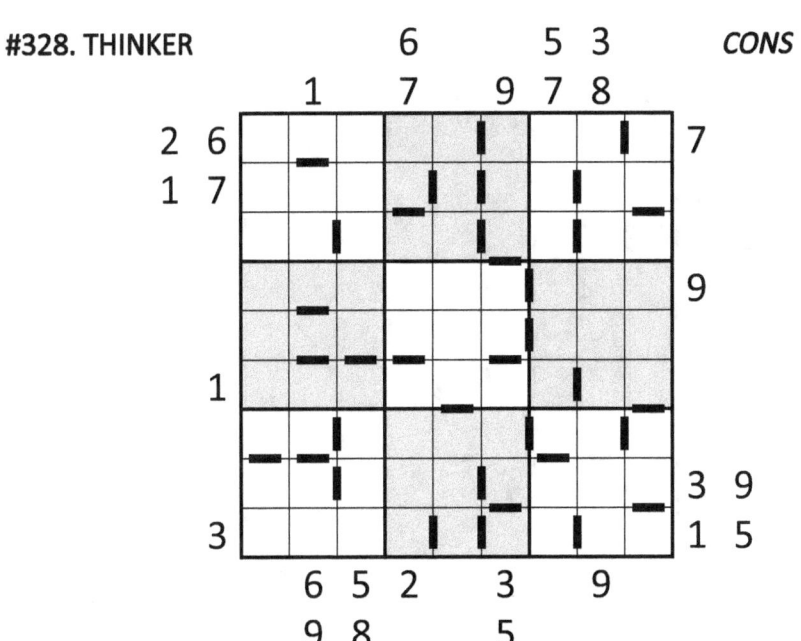

#328. THINKER

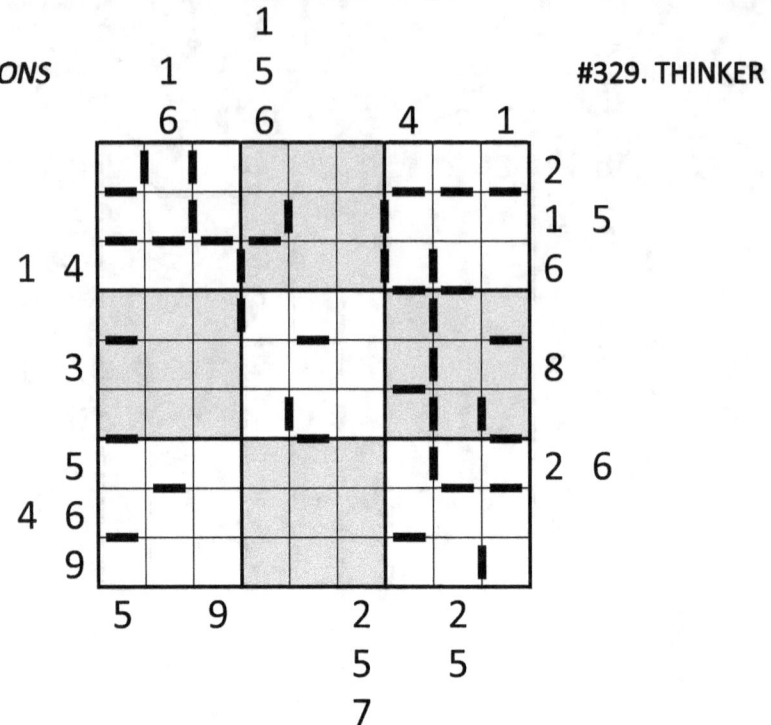

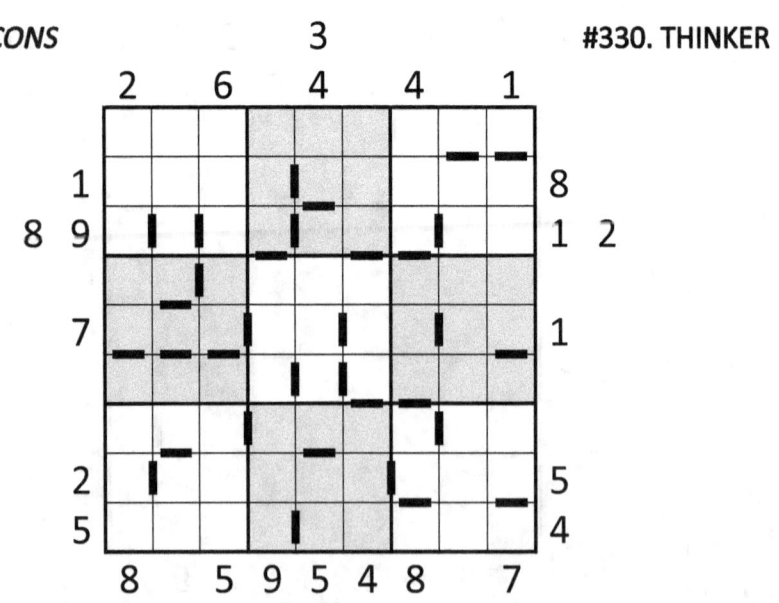

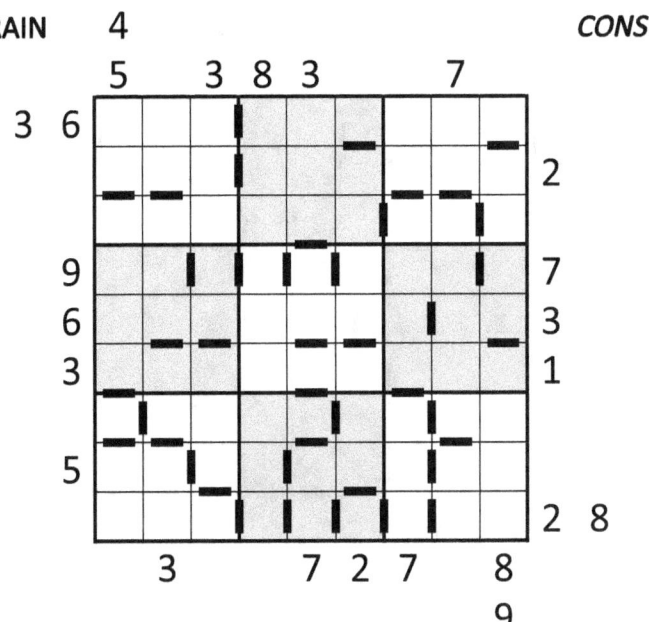

#332. BRAIN

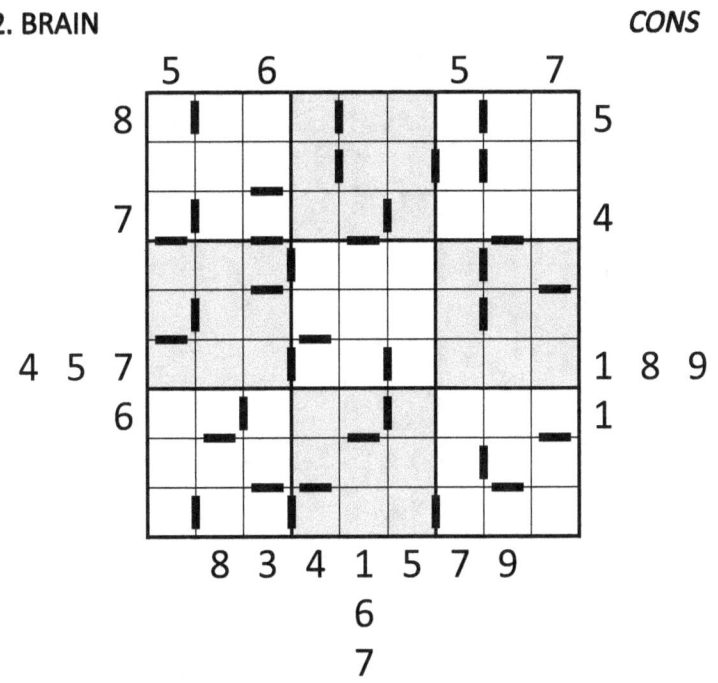

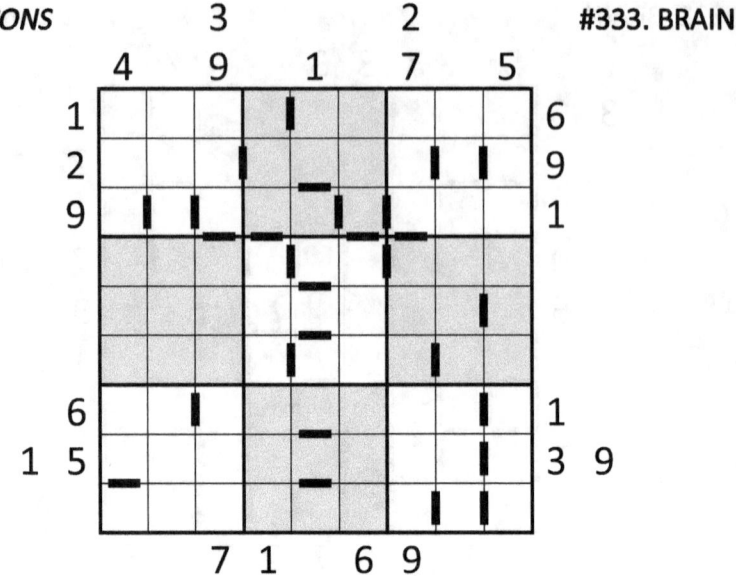

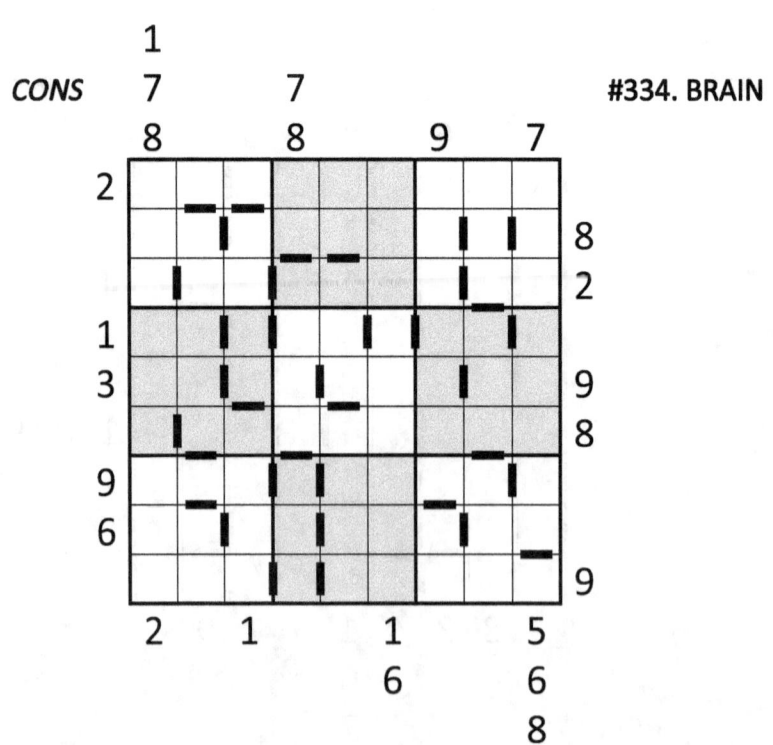

#335. BRAIN

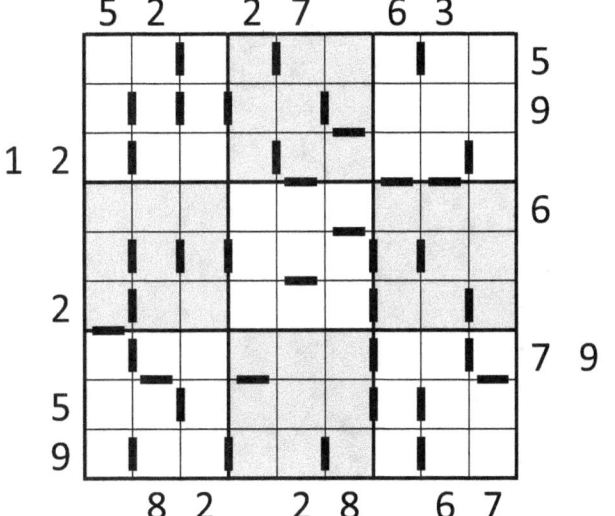

#336. BRAIN

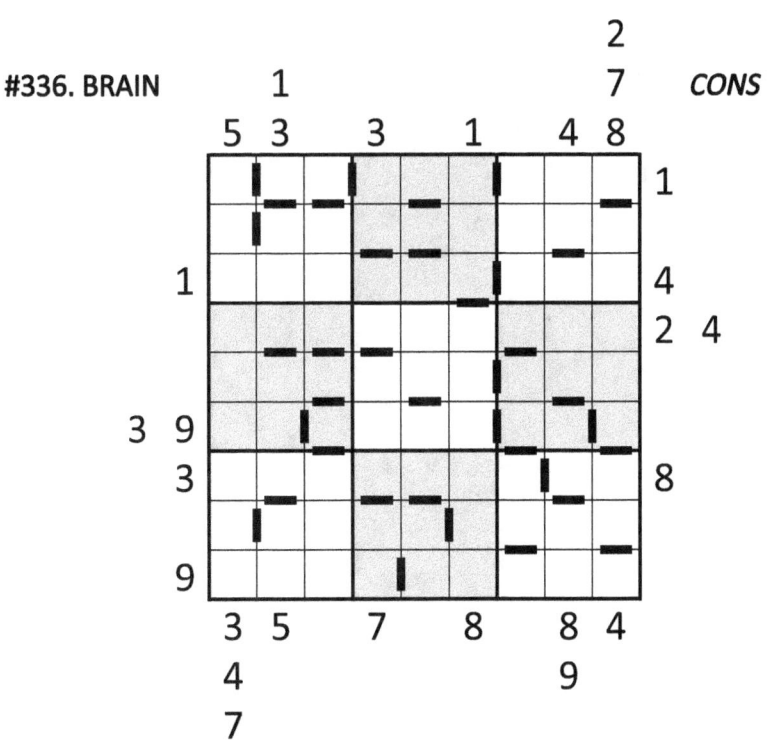

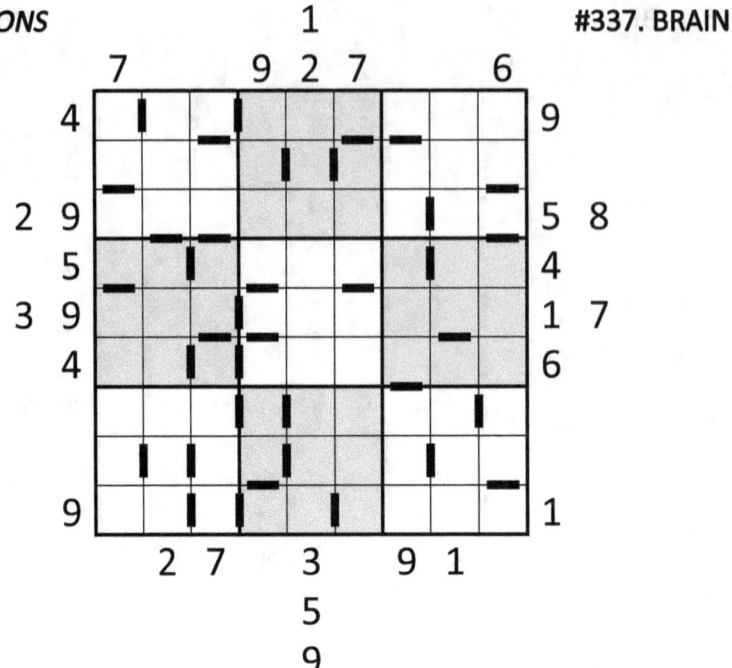

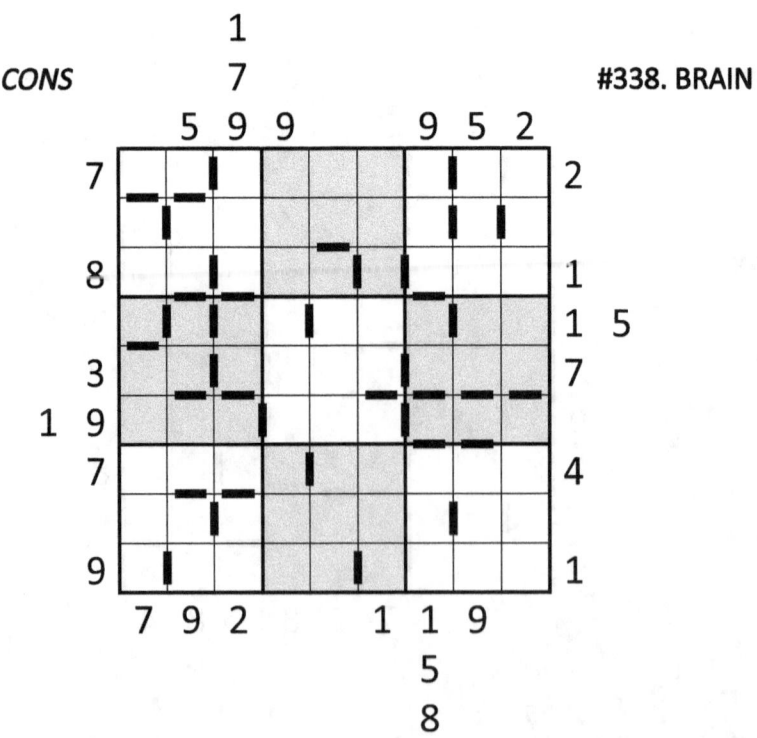

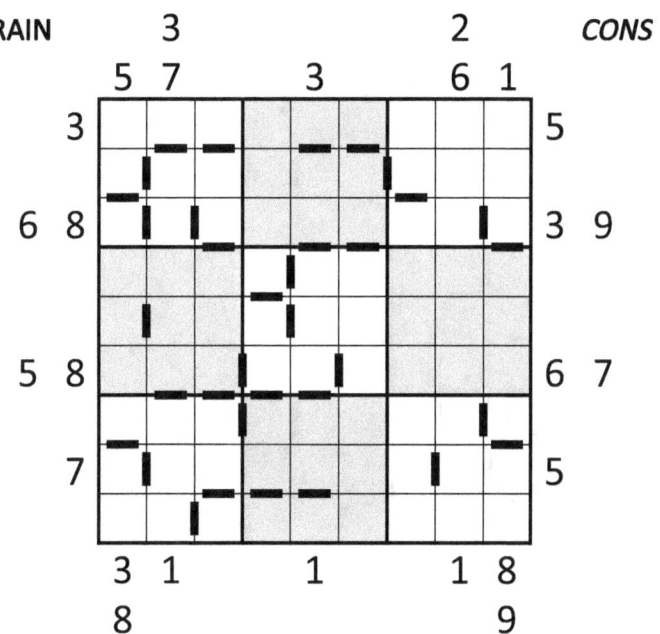

#340. BRAIN

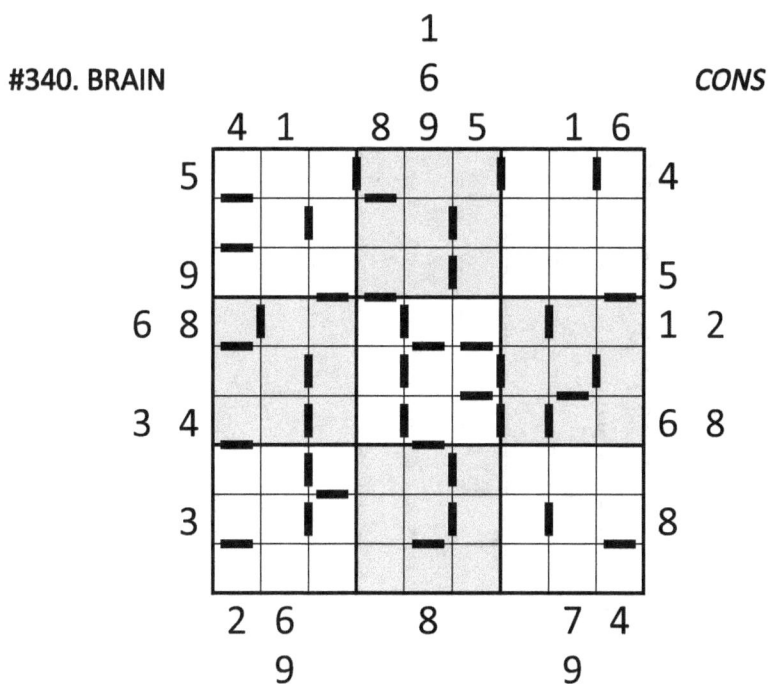

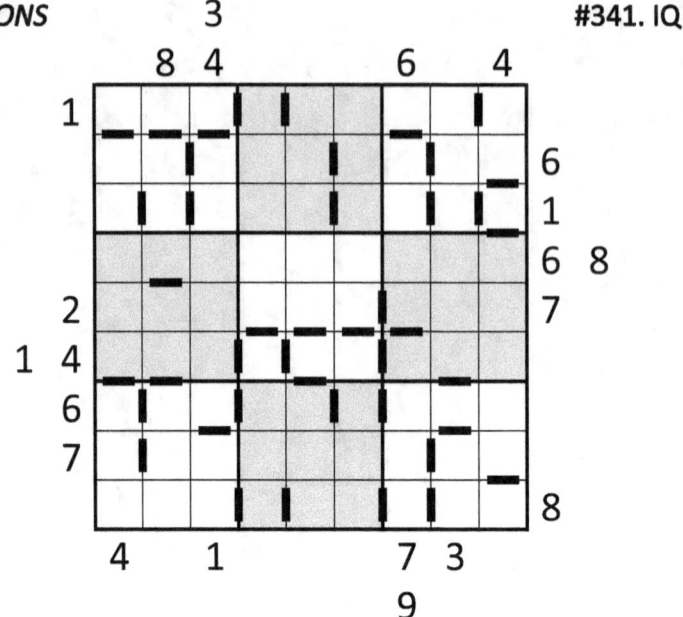

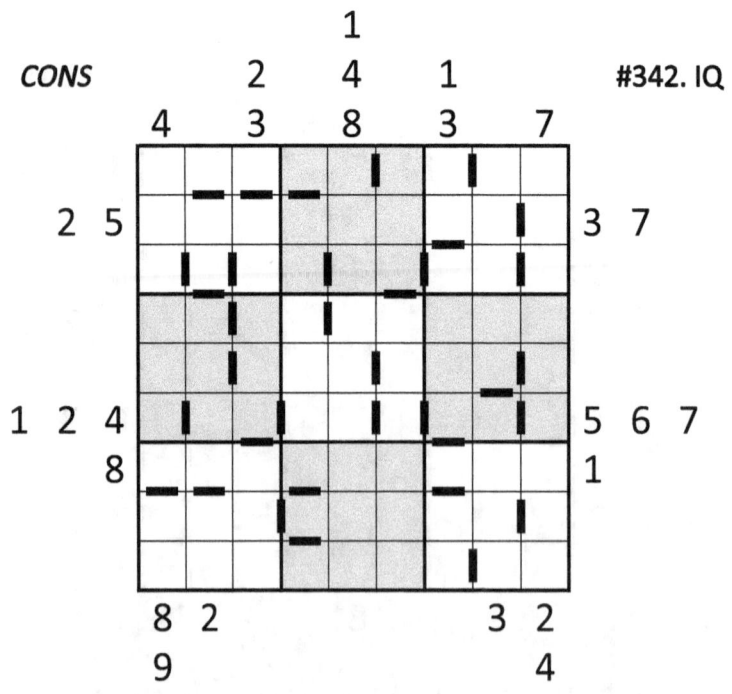

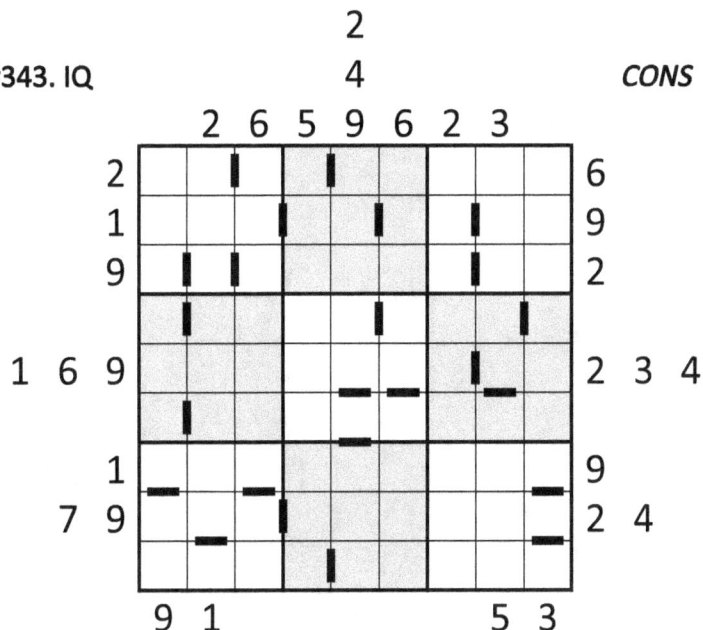

#344. IQ

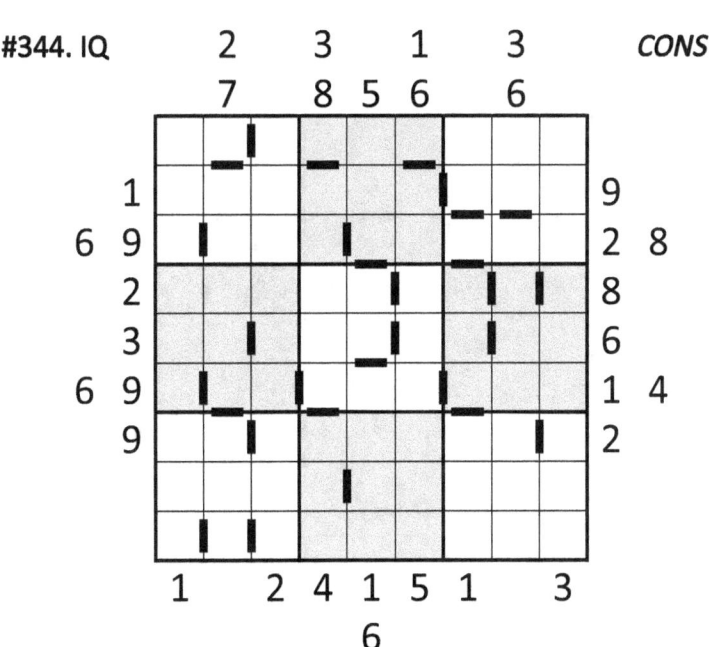

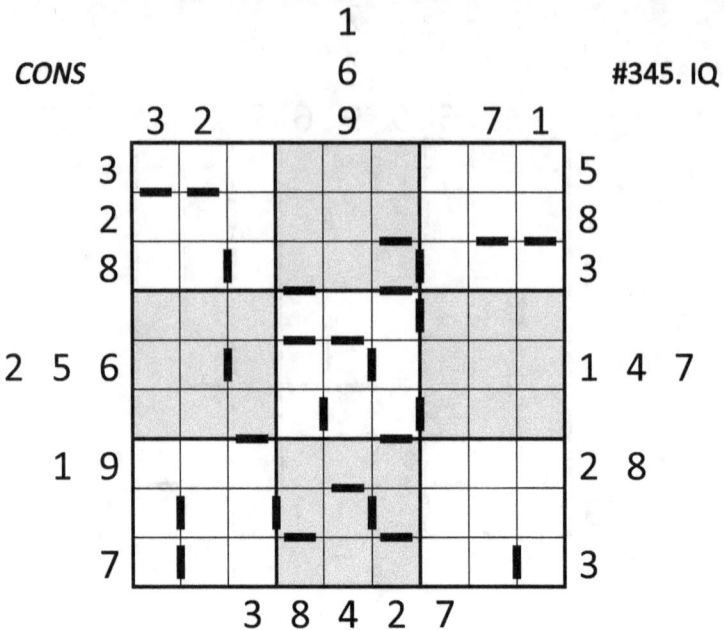

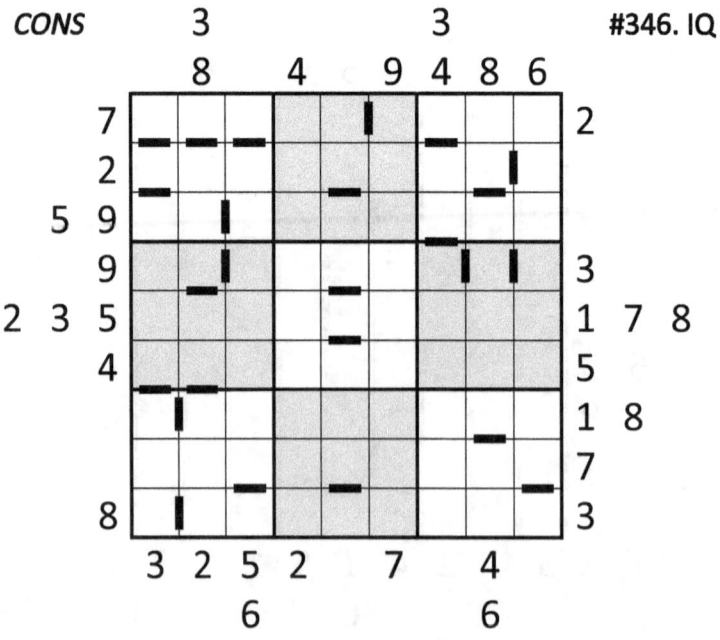

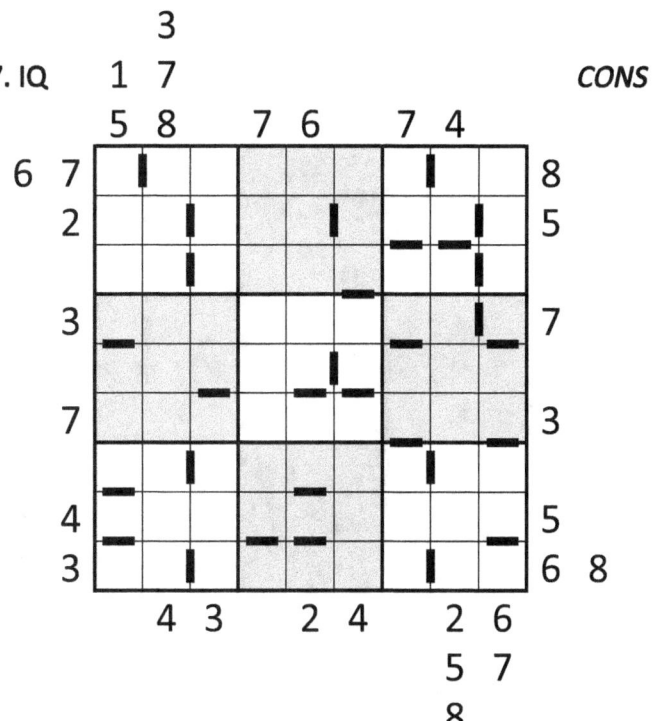

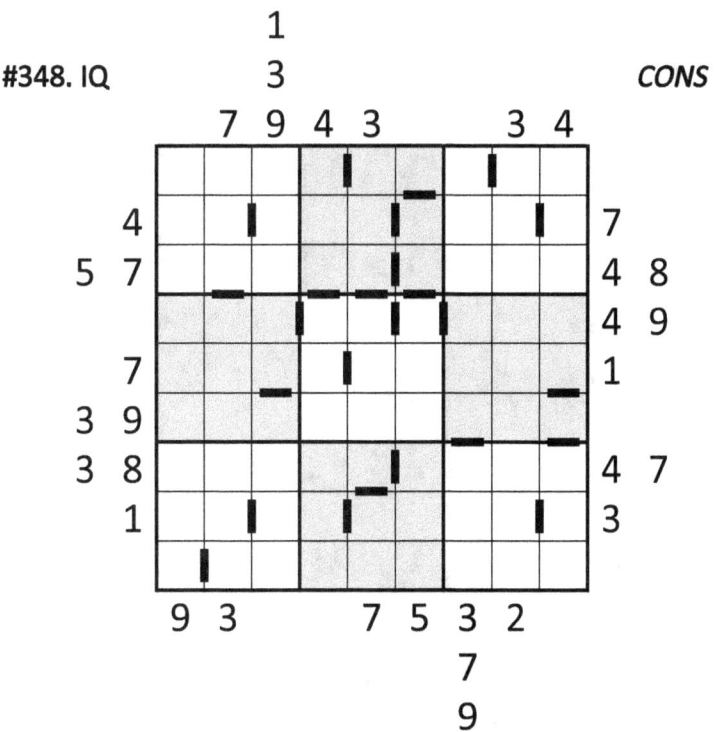

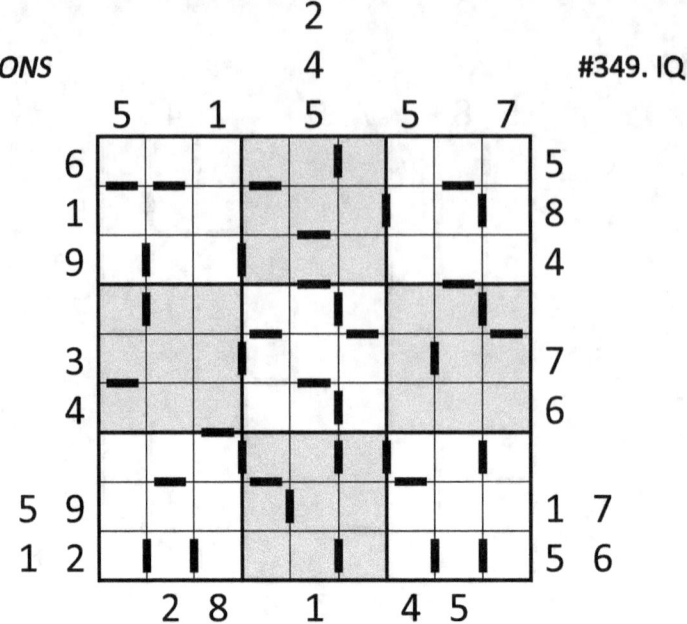

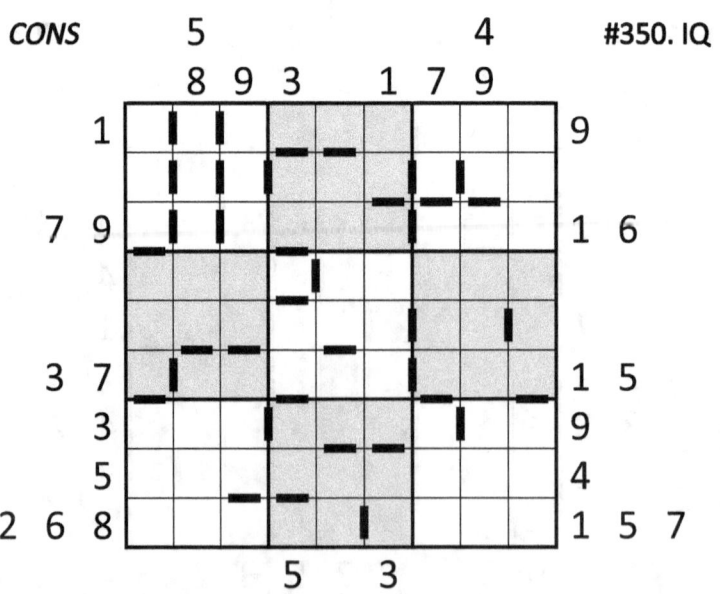

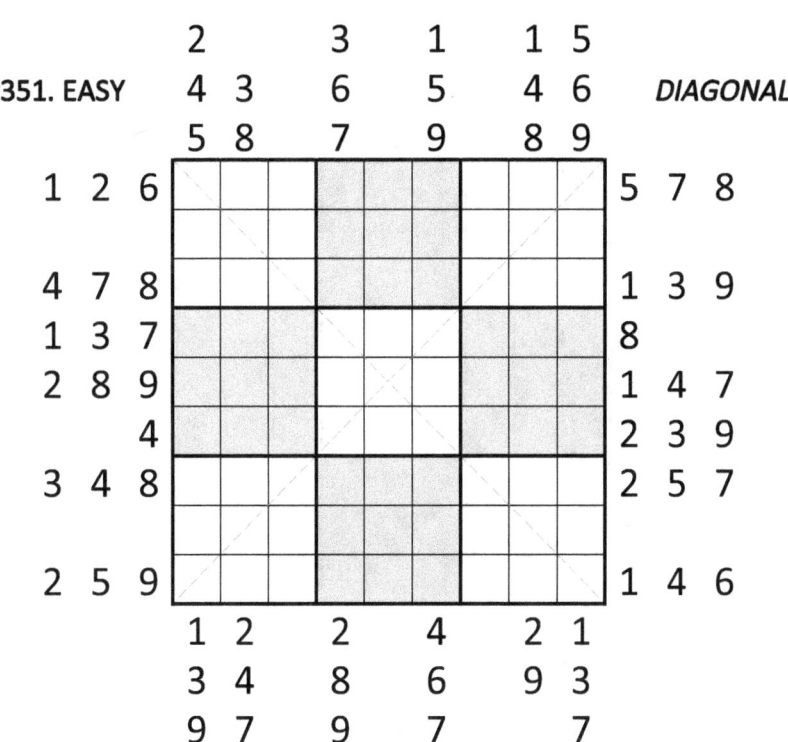

#351. EASY — DIAGONAL

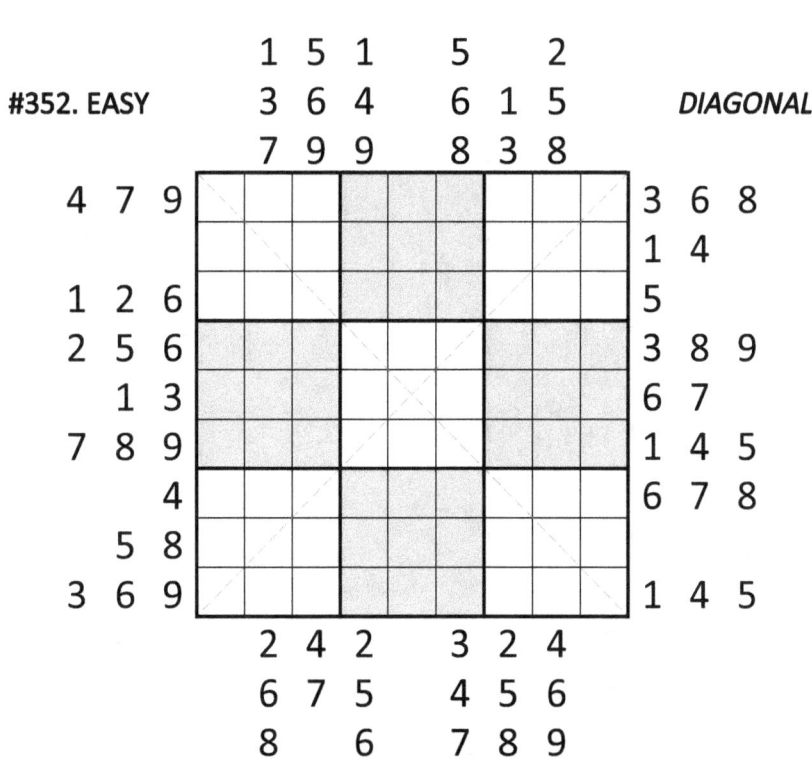

#352. EASY — DIAGONAL

Outside Sudoku by amazon.com/djape, page 181 _____

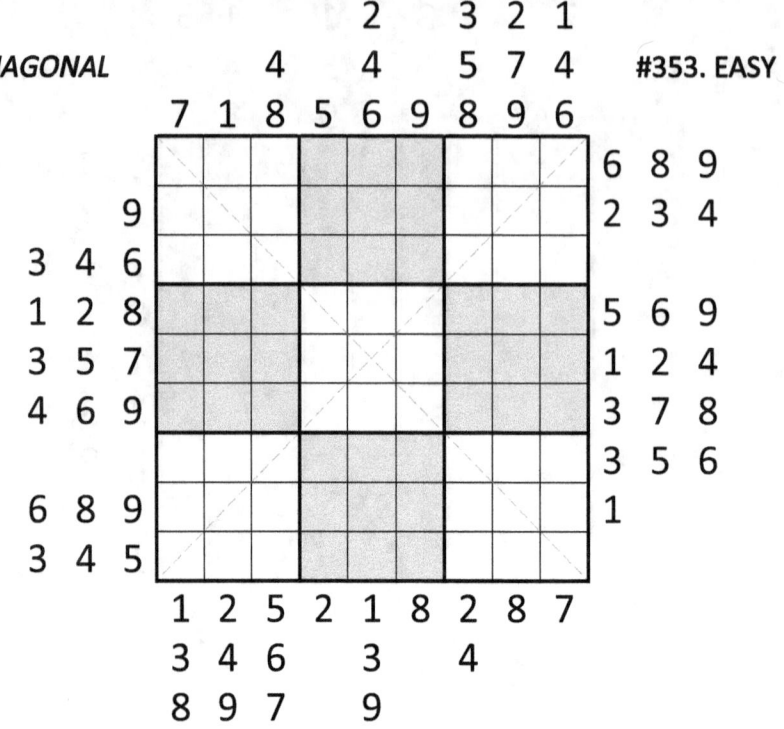

DIAGONAL   #353. EASY

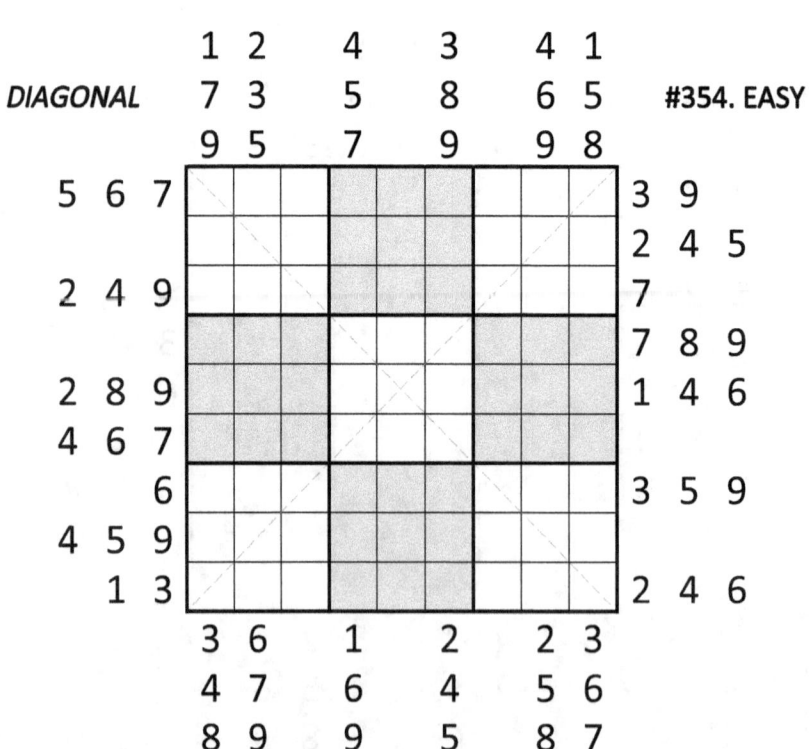

DIAGONAL   #354. EASY

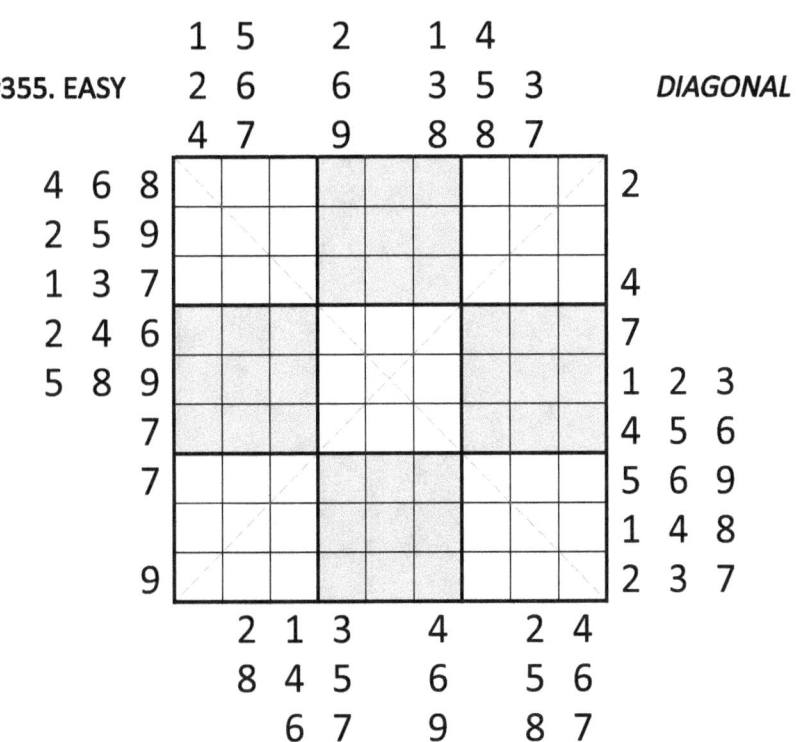

#355. EASY   DIAGONAL

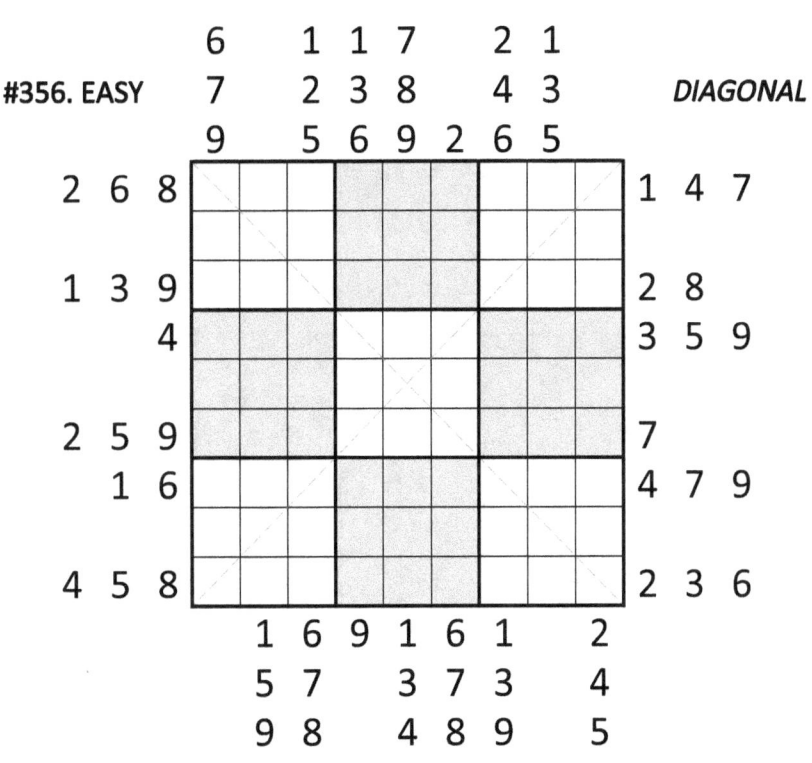

Outside Sudoku by amazon.com/djape, page 183 _____

#356. EASY   DIAGONAL

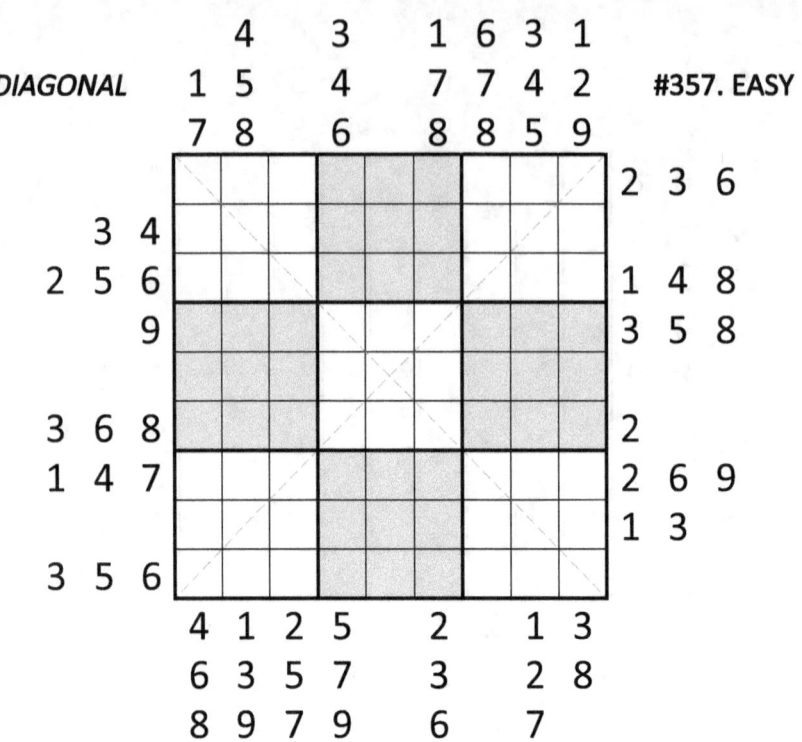

DIAGONAL   #357. EASY

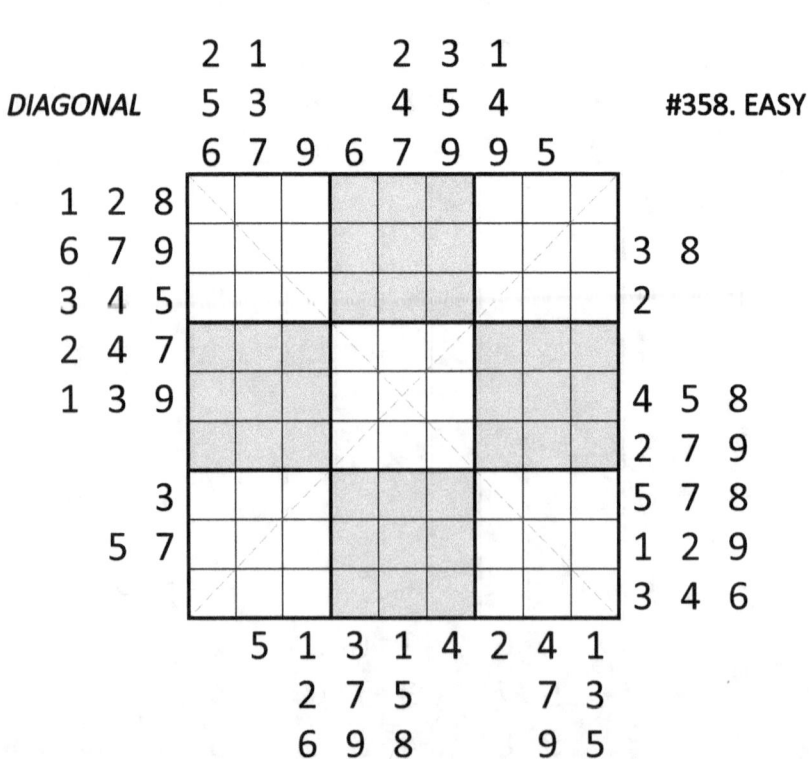

DIAGONAL   #358. EASY

_____ *Outside Sudoku by amazon.com/djape, page 184*

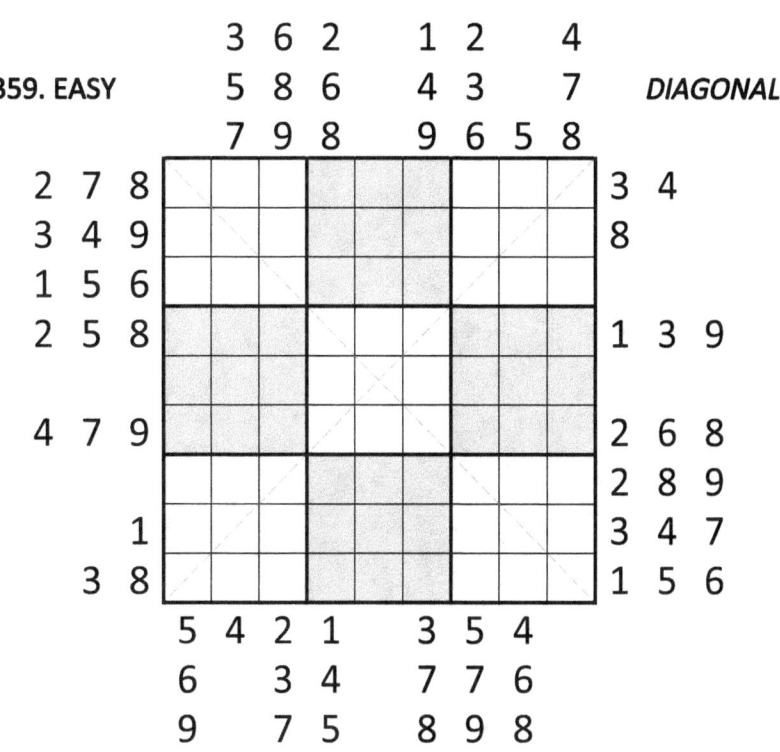

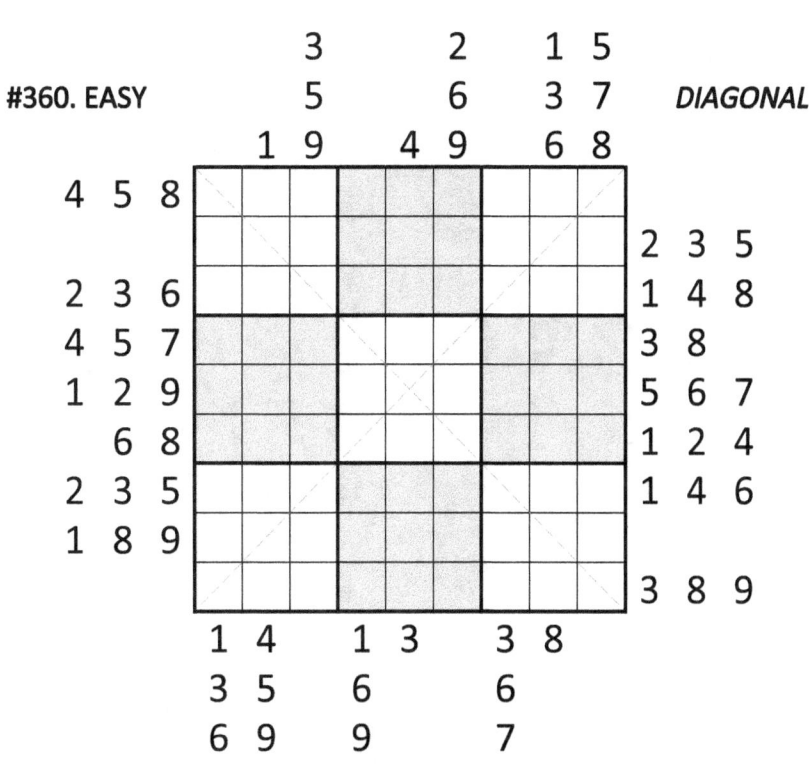

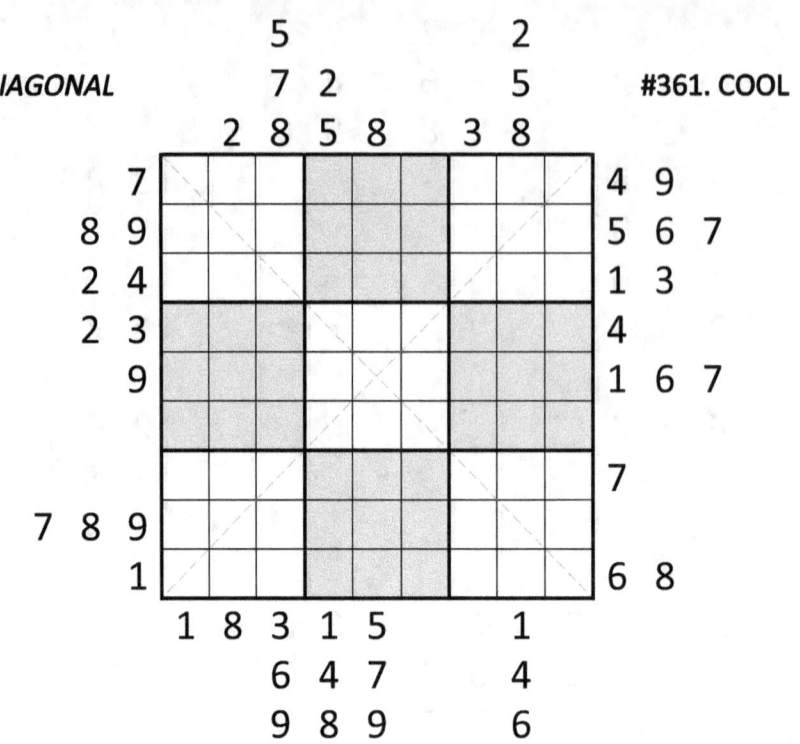

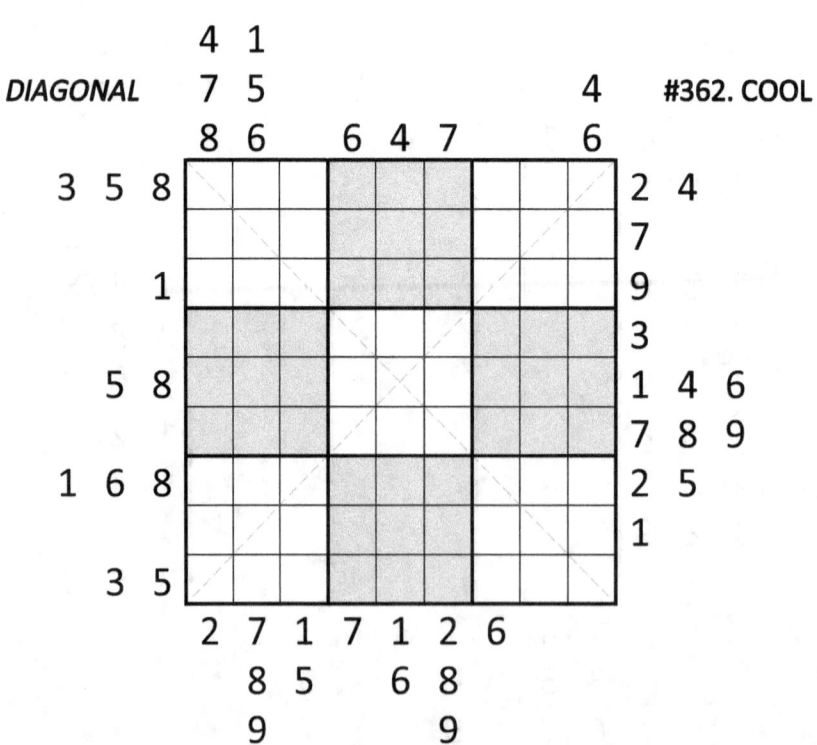

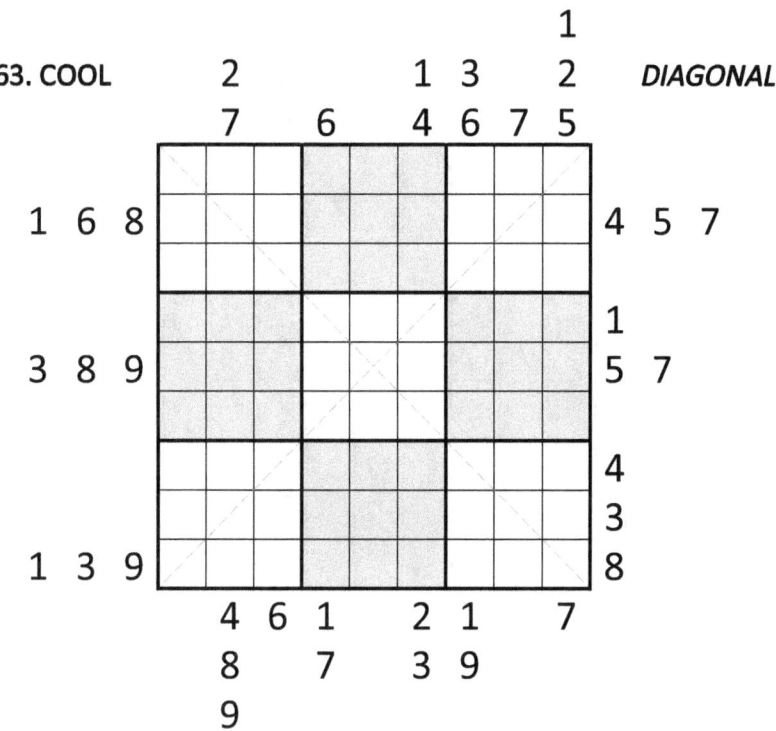

#363. COOL — DIAGONAL

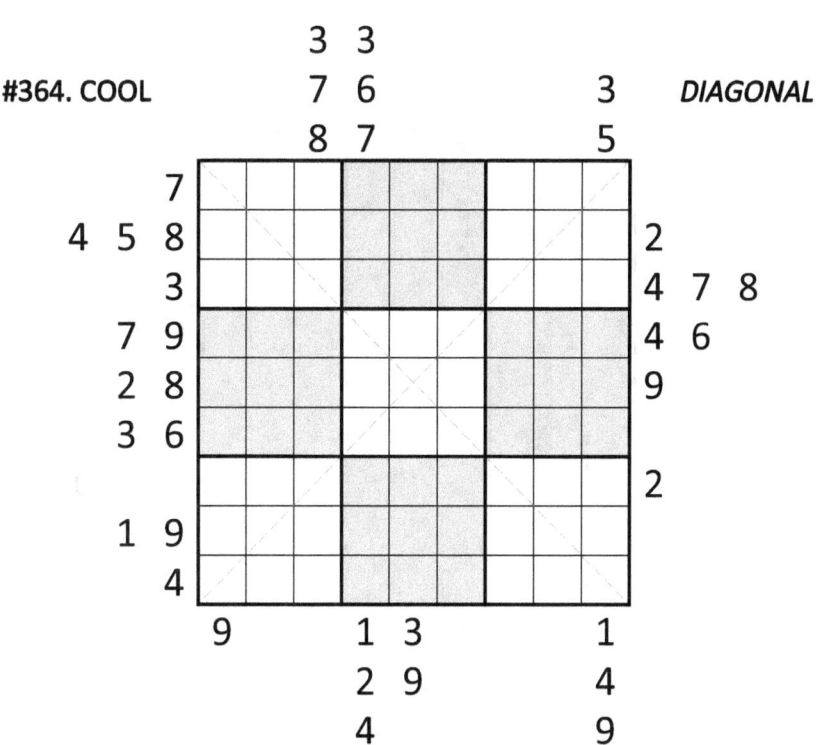

#364. COOL — DIAGONAL

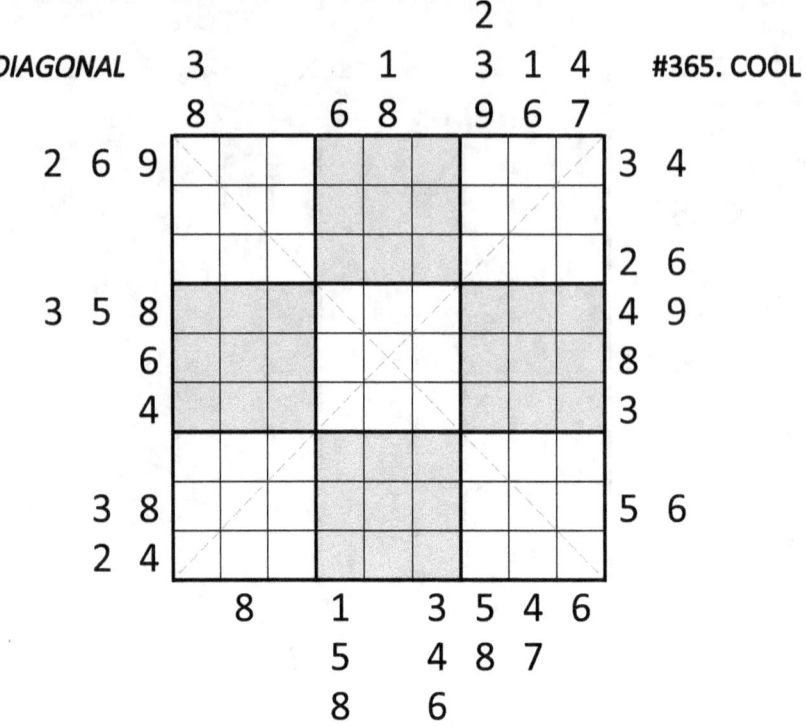

#365. COOL

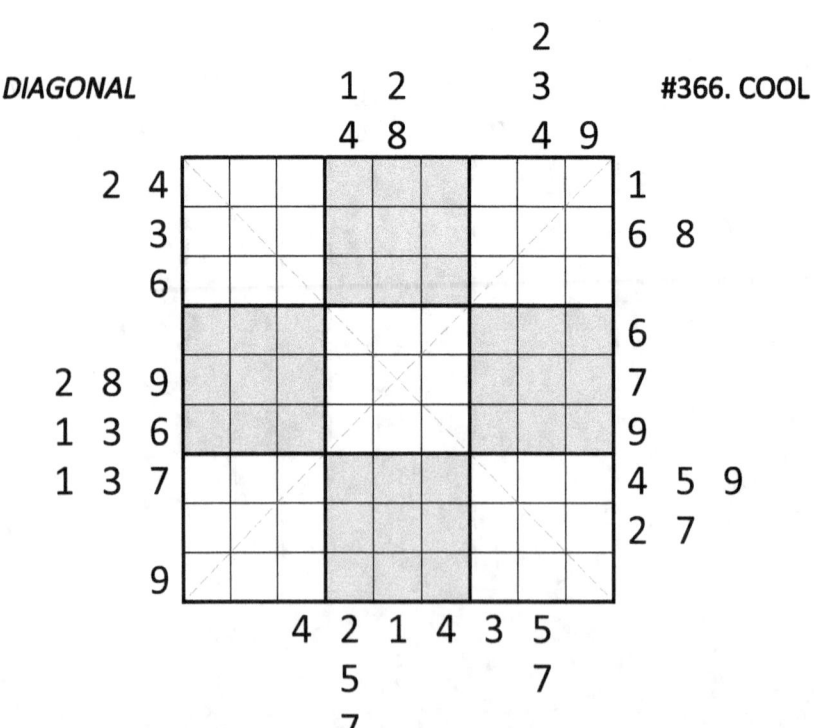

#366. COOL

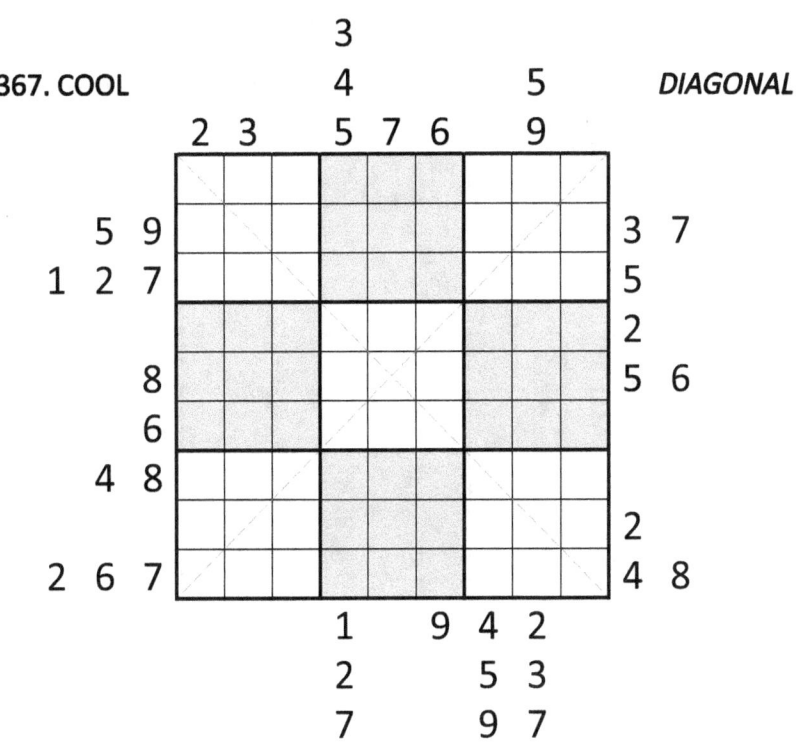

#367. COOL   DIAGONAL

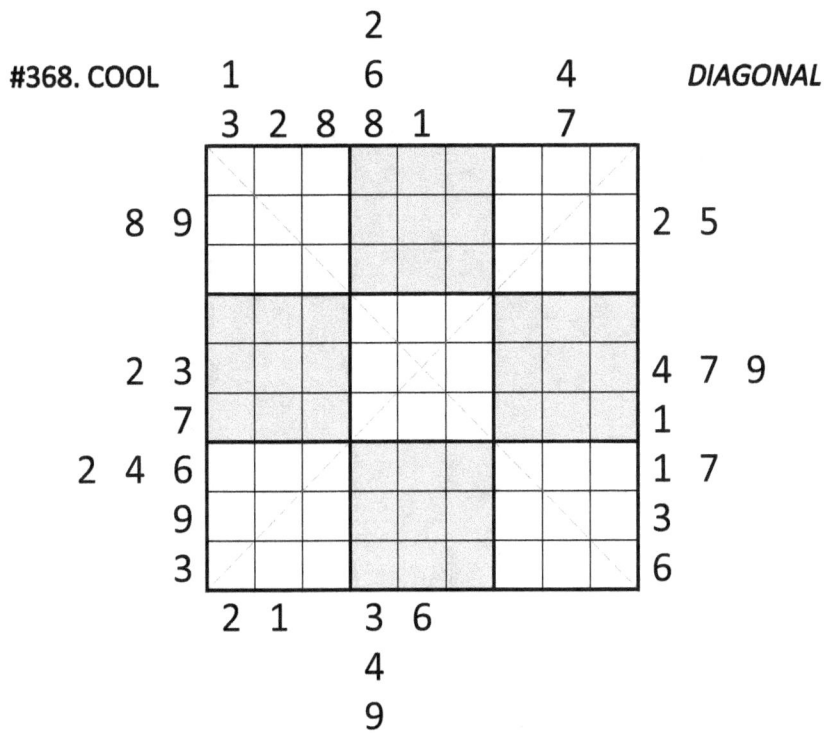

Outside Sudoku by amazon.com/djape, page 189 _____

#368. COOL   DIAGONAL

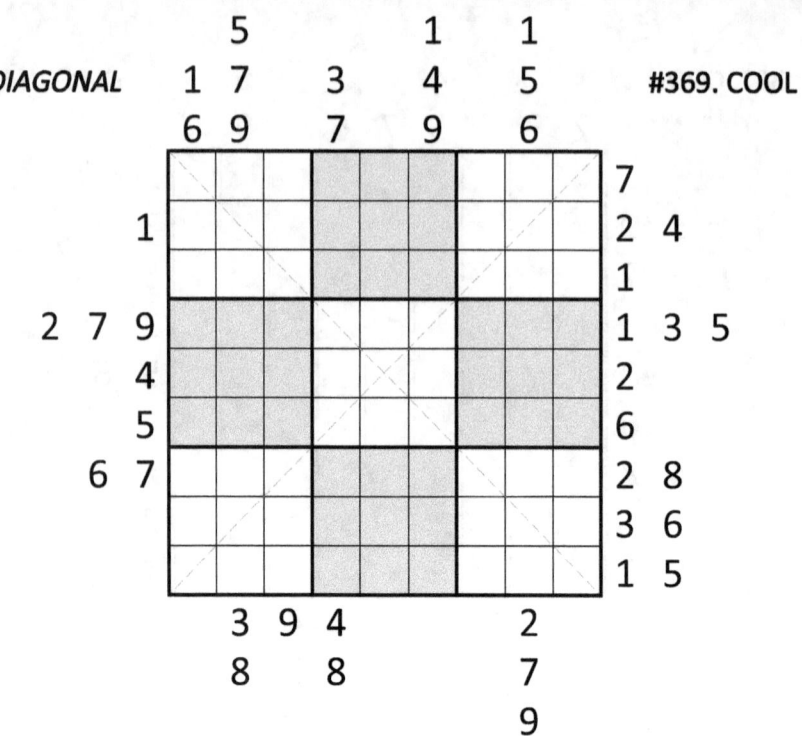

DIAGONAL   #369. COOL

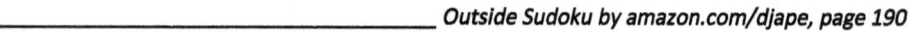

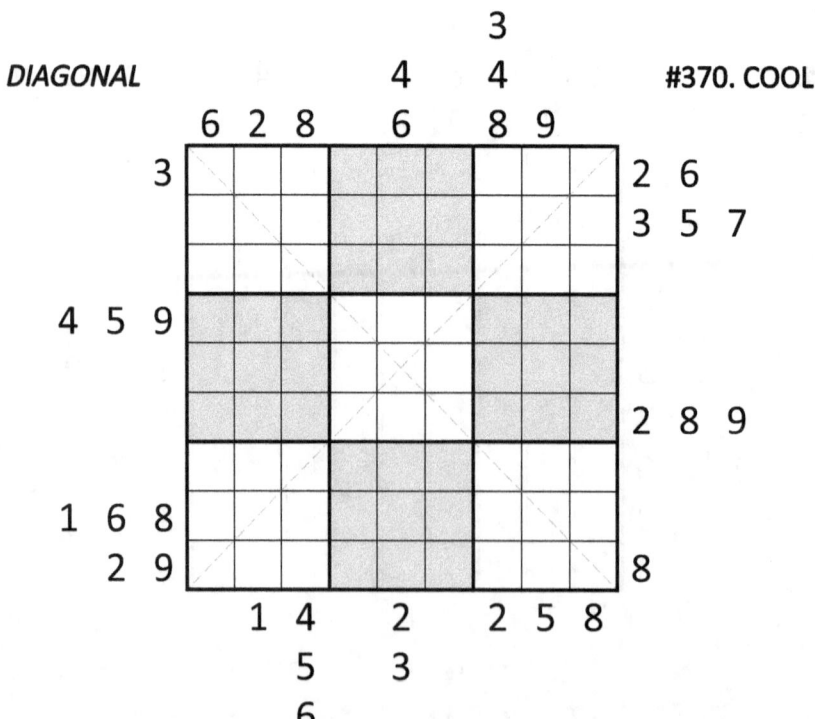

DIAGONAL   #370. COOL

#371. THINKER

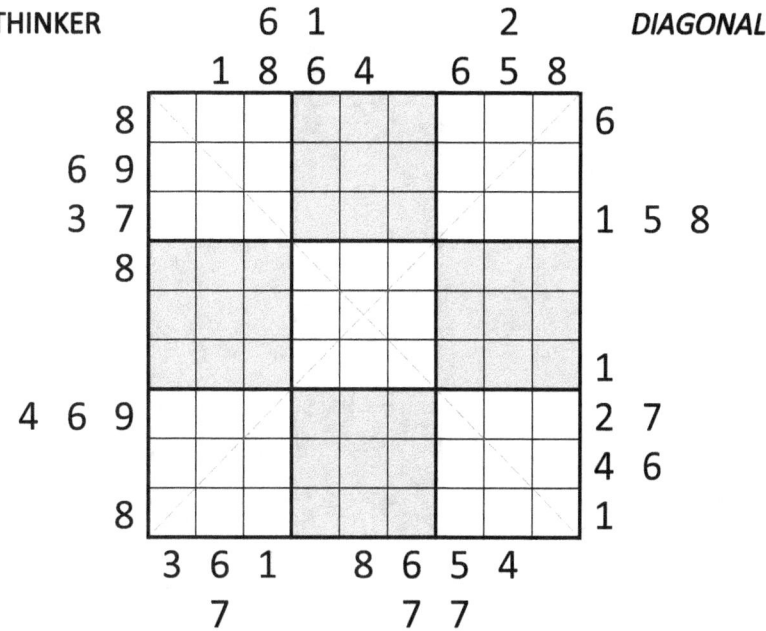

DIAGONAL

#372. THINKER

DIAGONAL

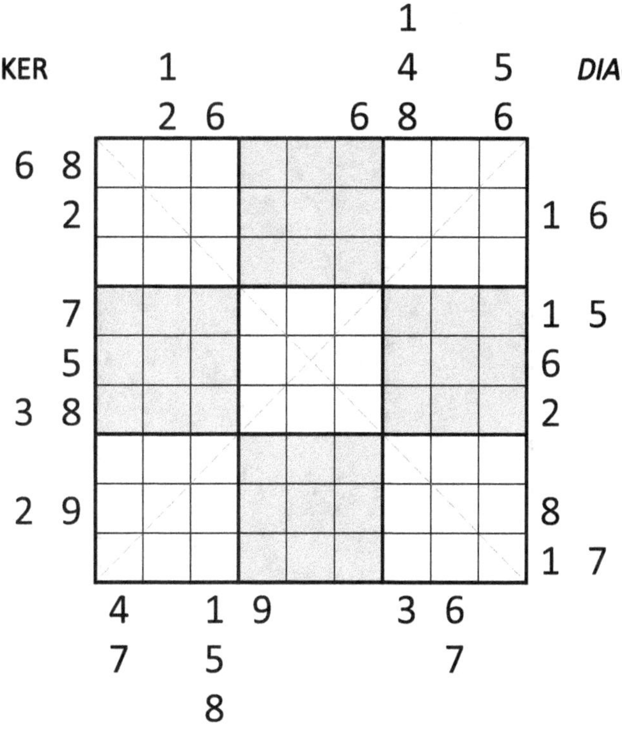

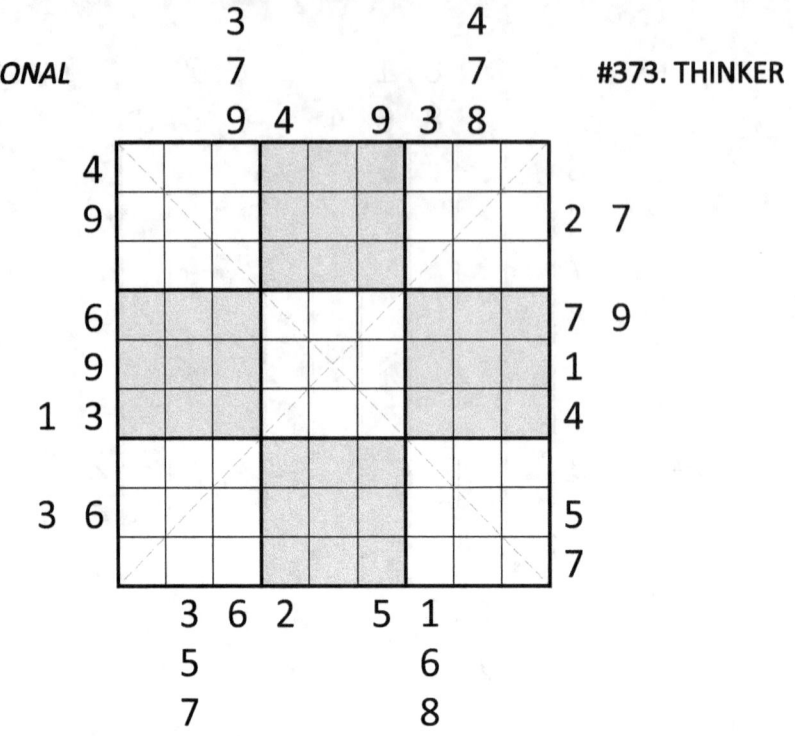

DIAGONAL   #373. THINKER

DIAGONAL   #374. THINKER

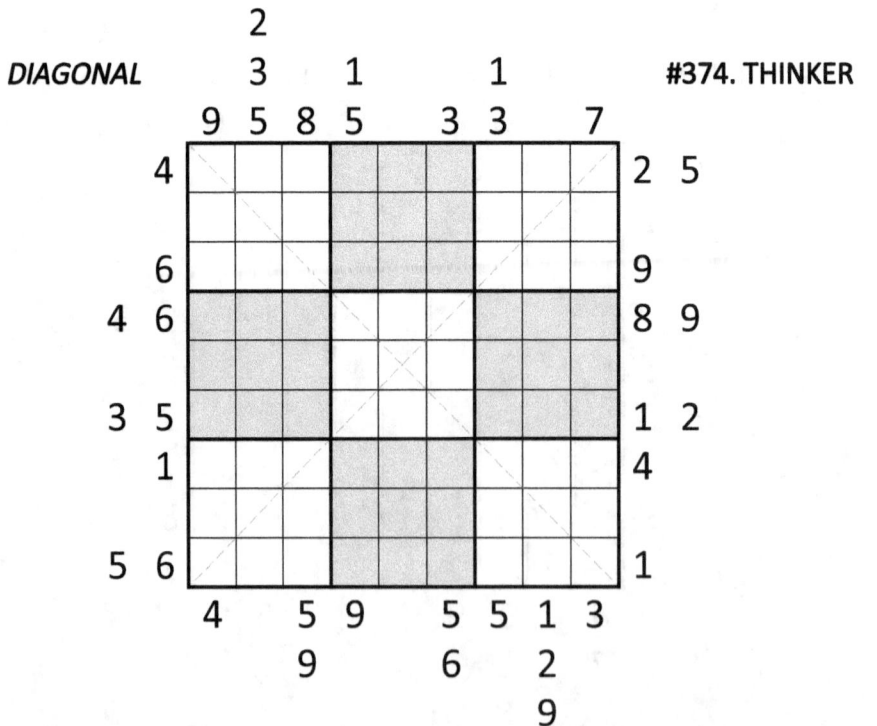

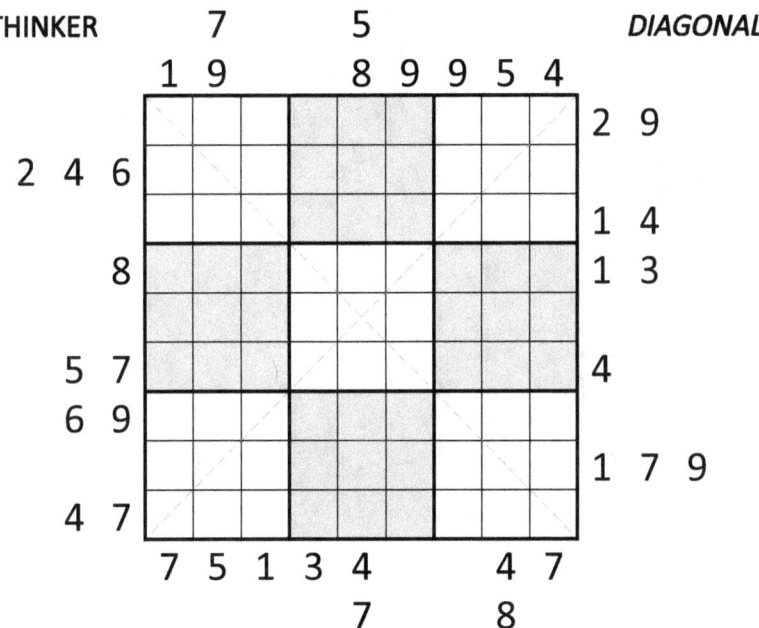

#376. THINKER

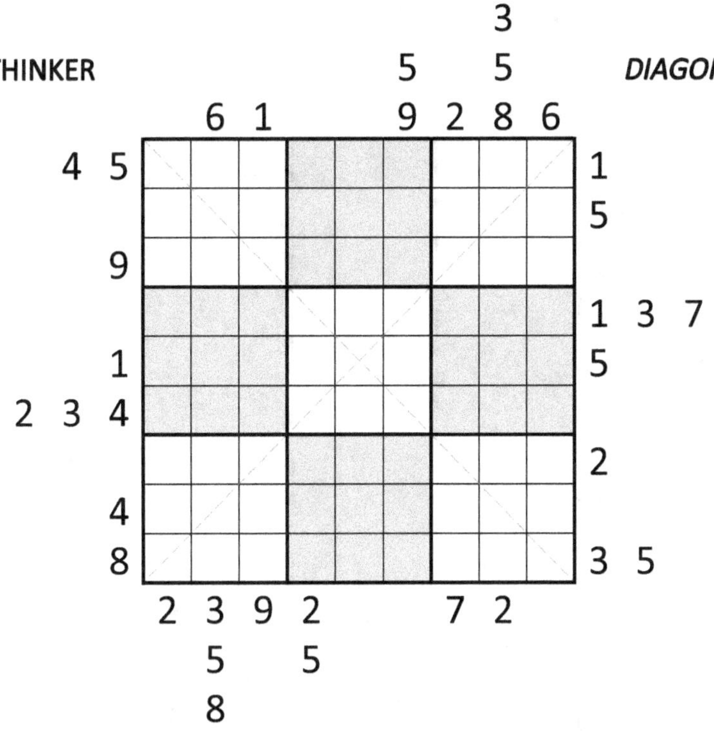

DIAGONAL   1   **#377. THINKER**

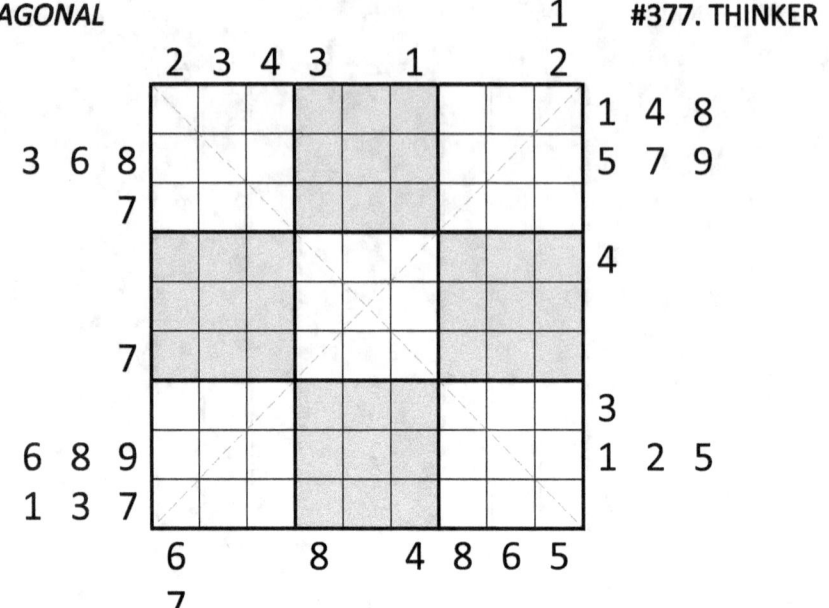

DIAGONAL   **#378. THINKER**

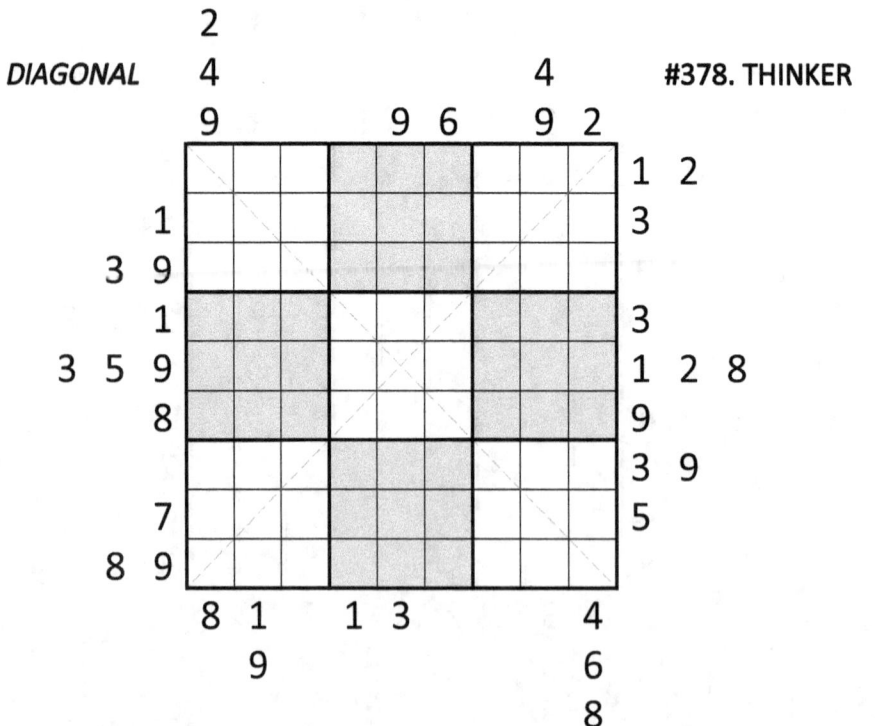

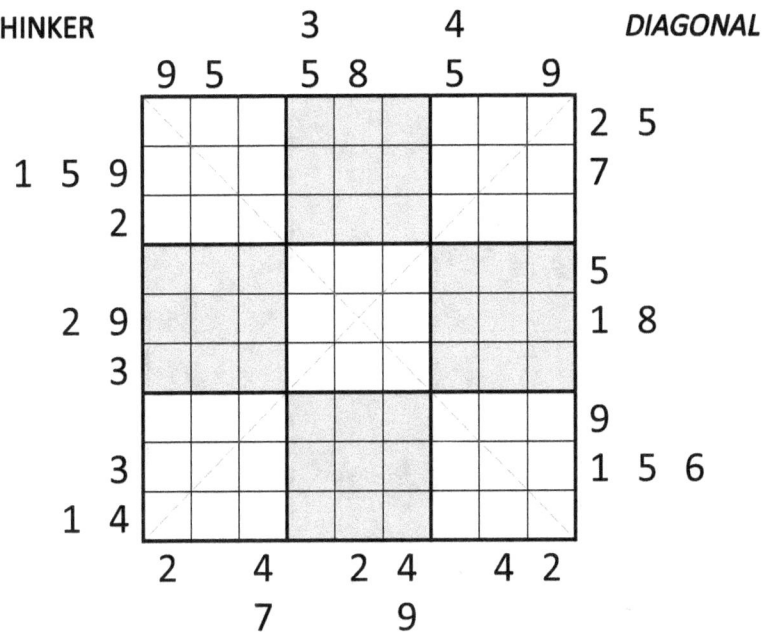

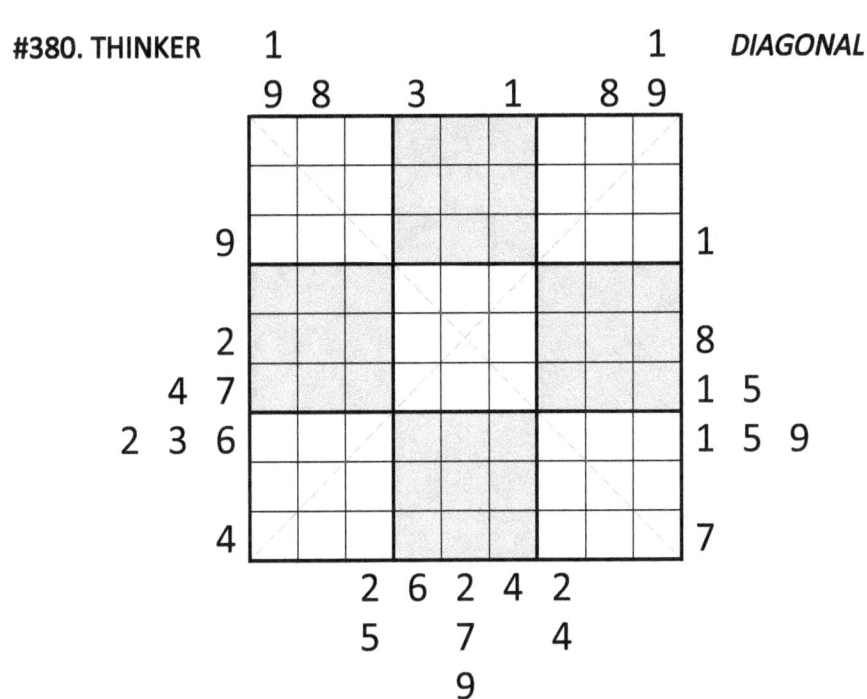

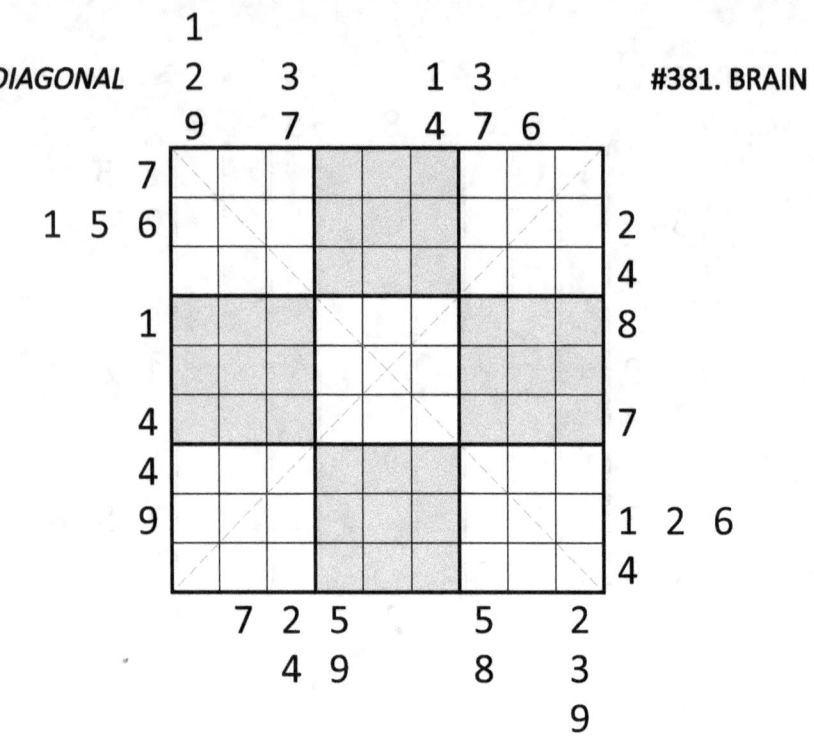

DIAGONAL   #381. BRAIN

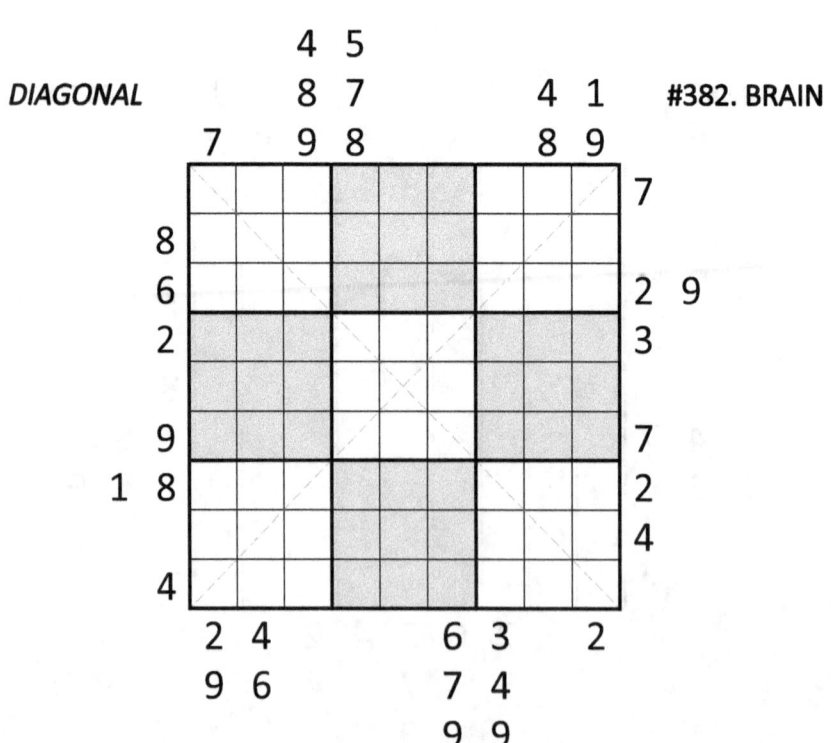

DIAGONAL   #382. BRAIN

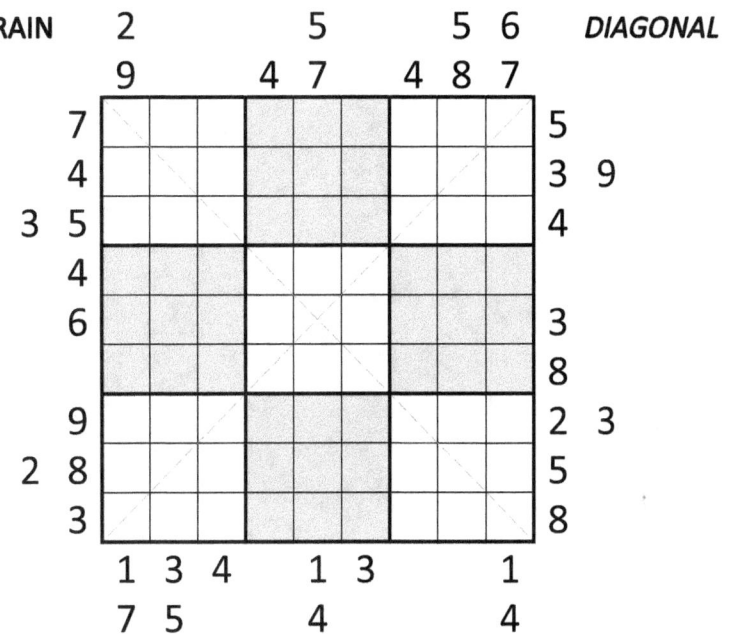

#383. BRAIN — DIAGONAL

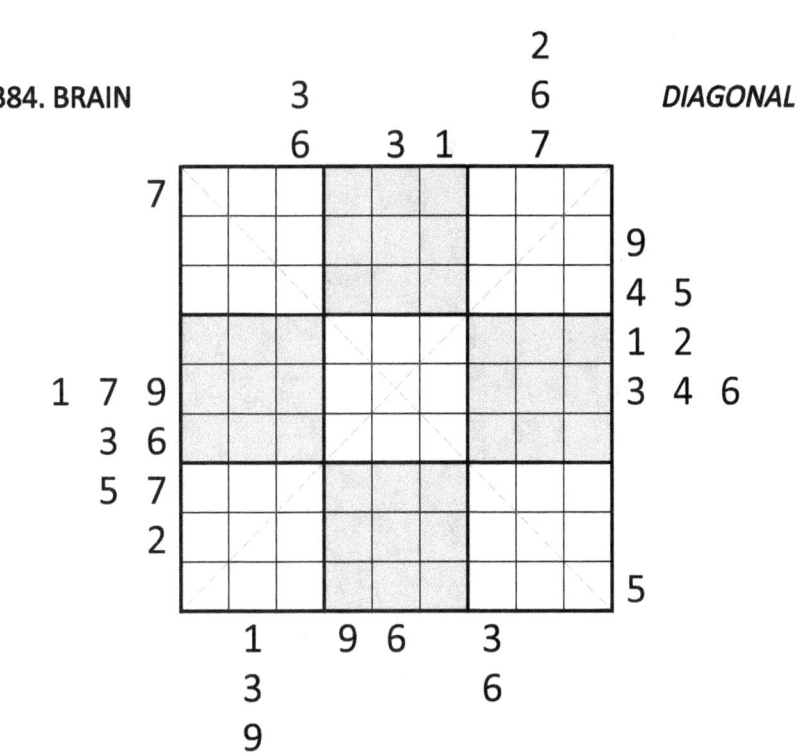

#384. BRAIN — DIAGONAL

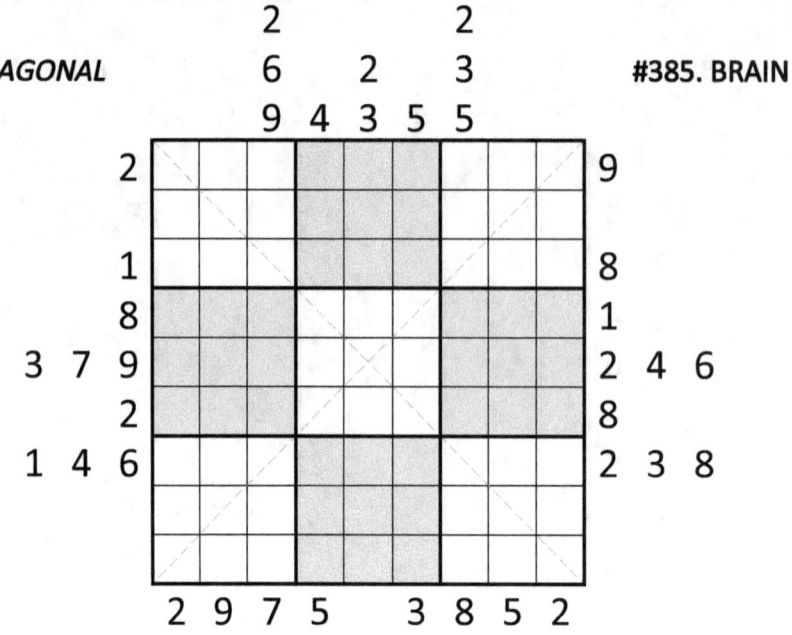

#385. BRAIN

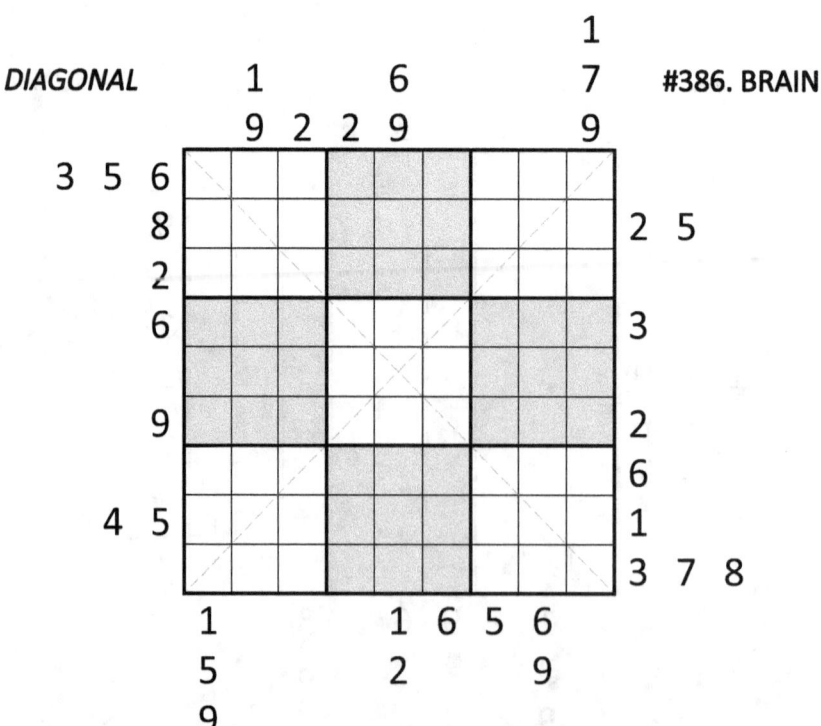

#386. BRAIN

#387. BRAIN

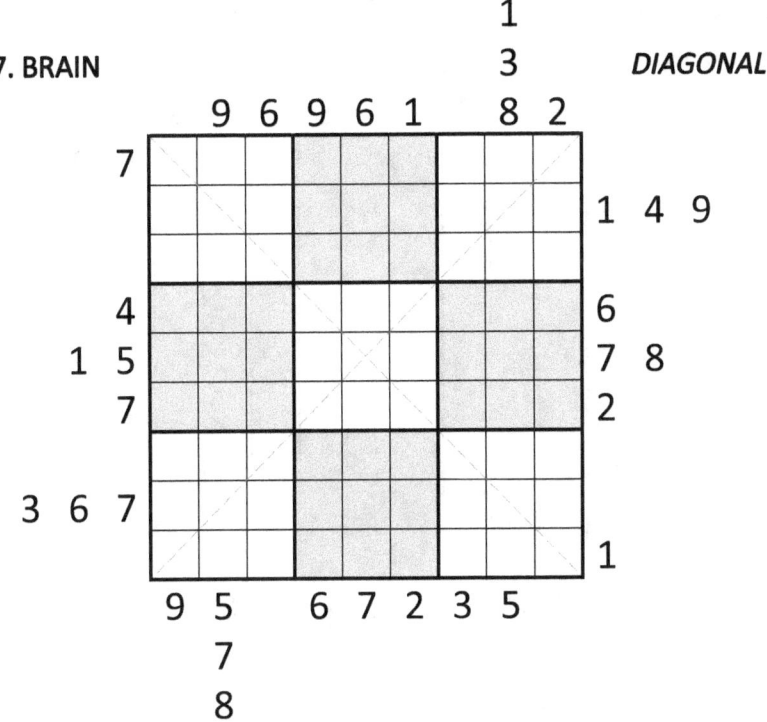

DIAGONAL

#388. BRAIN

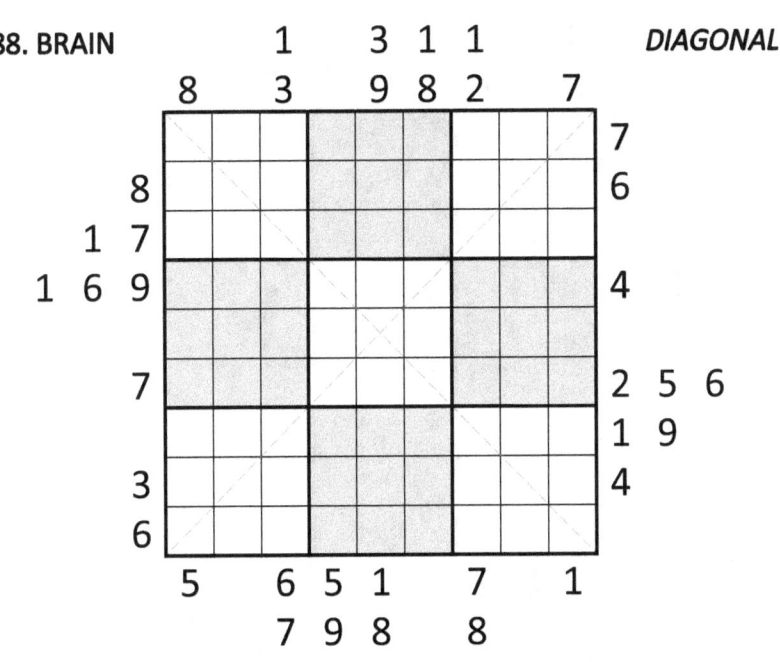

DIAGONAL

DIAGONAL

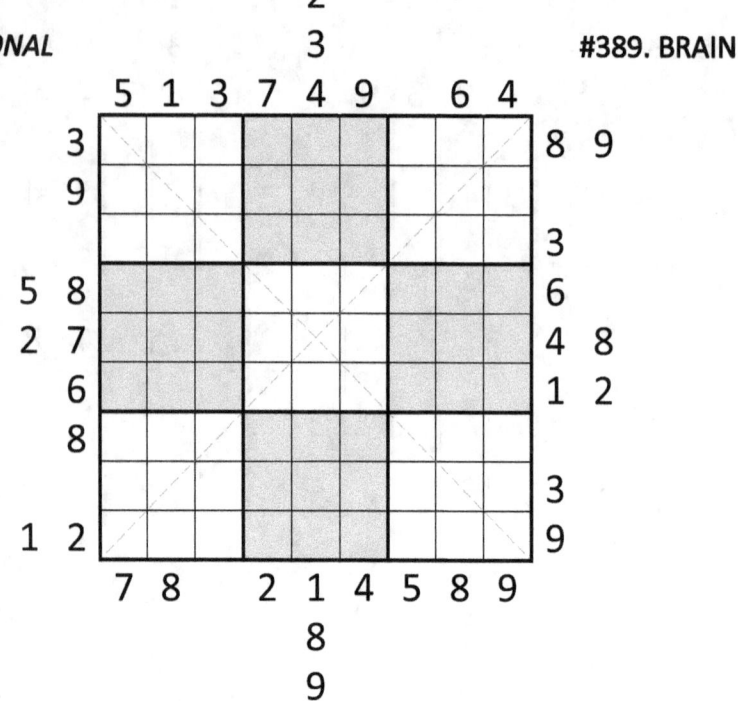

#389. BRAIN

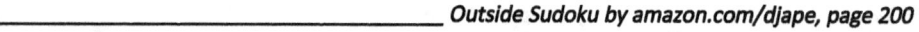

DIAGONAL

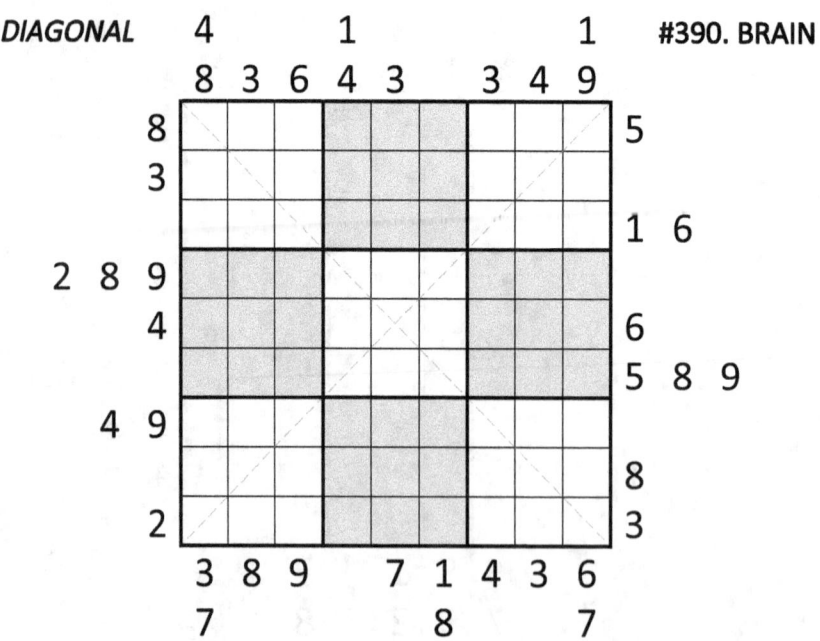

#390. BRAIN

DIAGONAL

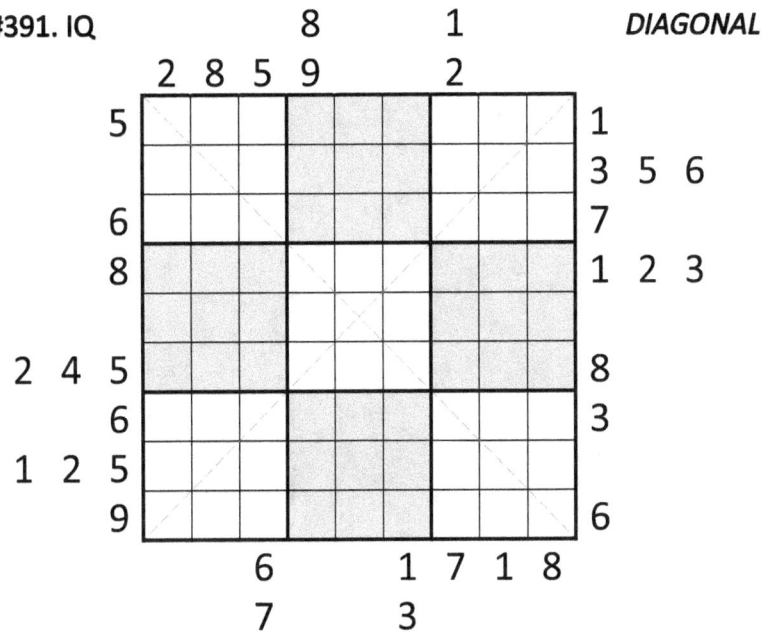

#392. IQ DIAGONAL

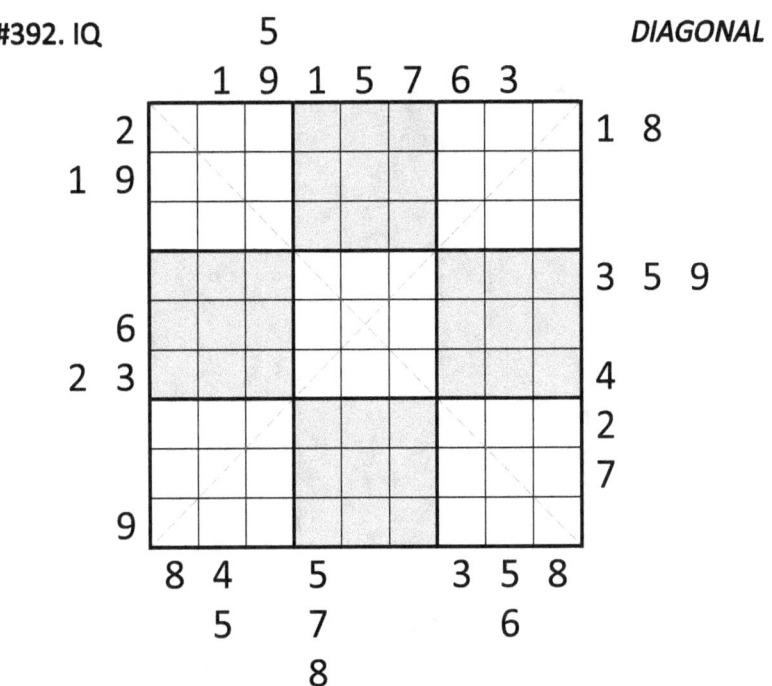

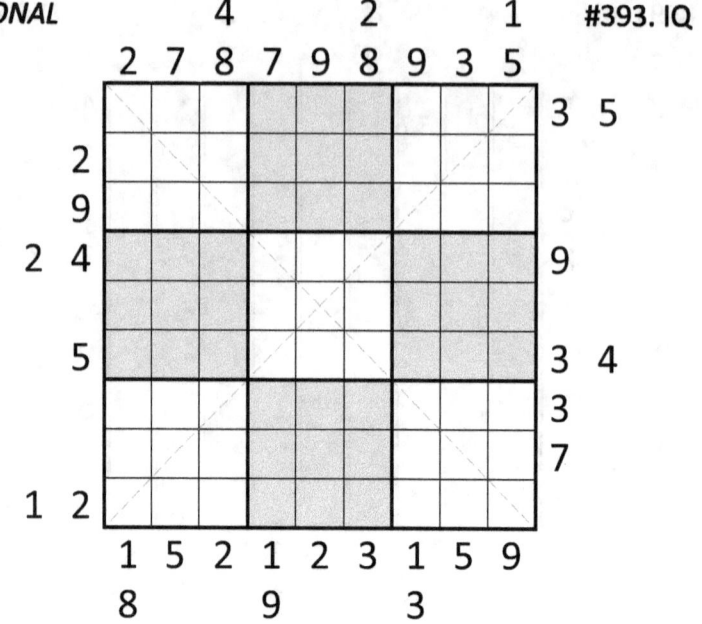

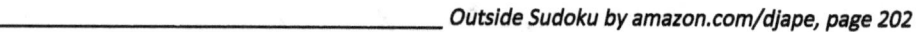

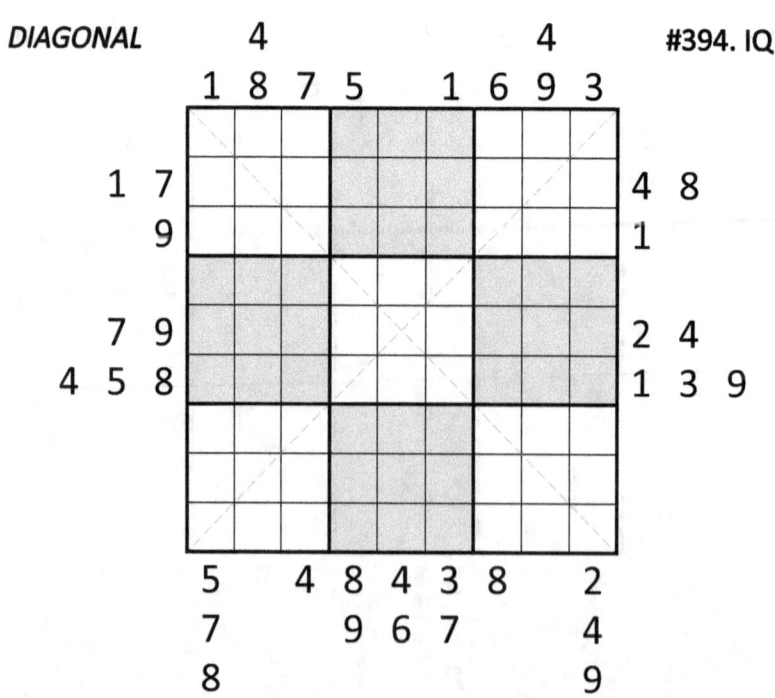

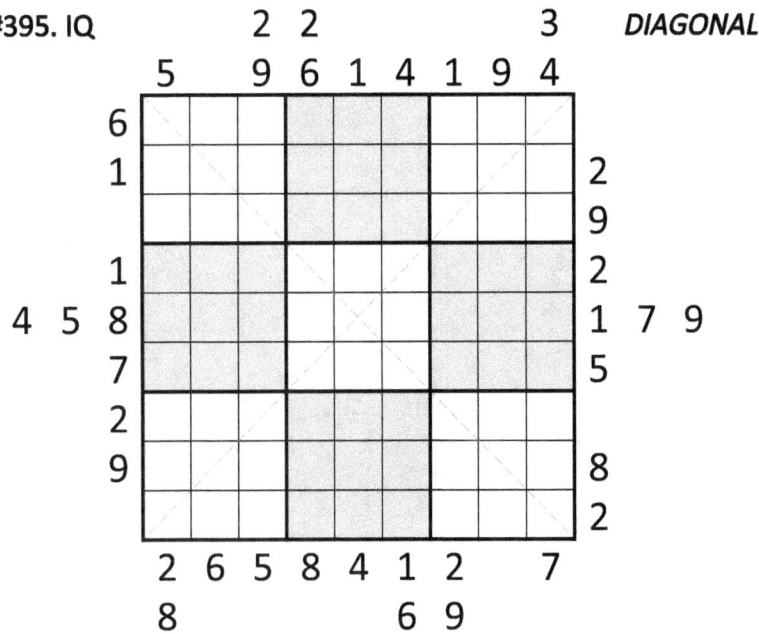

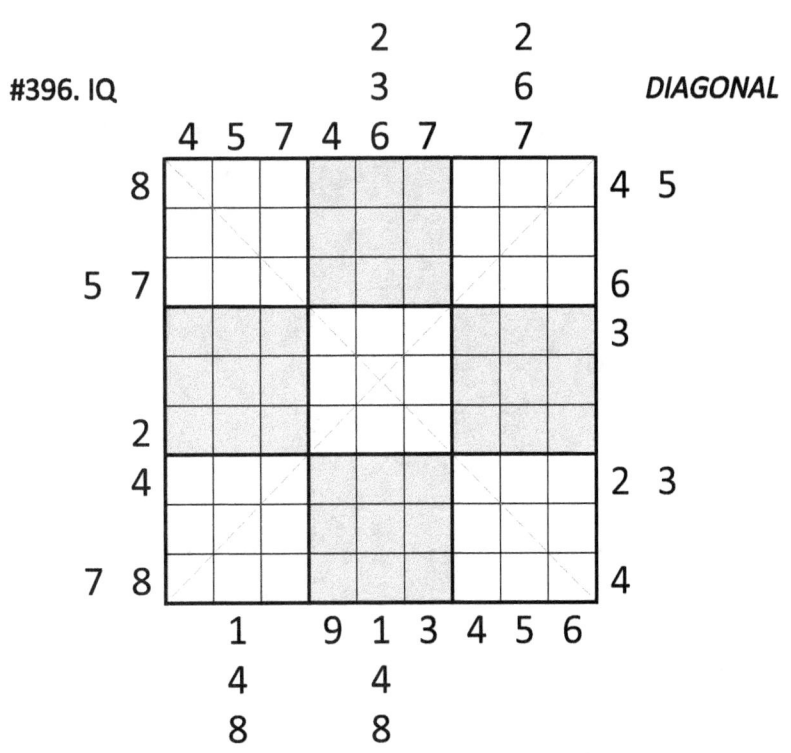

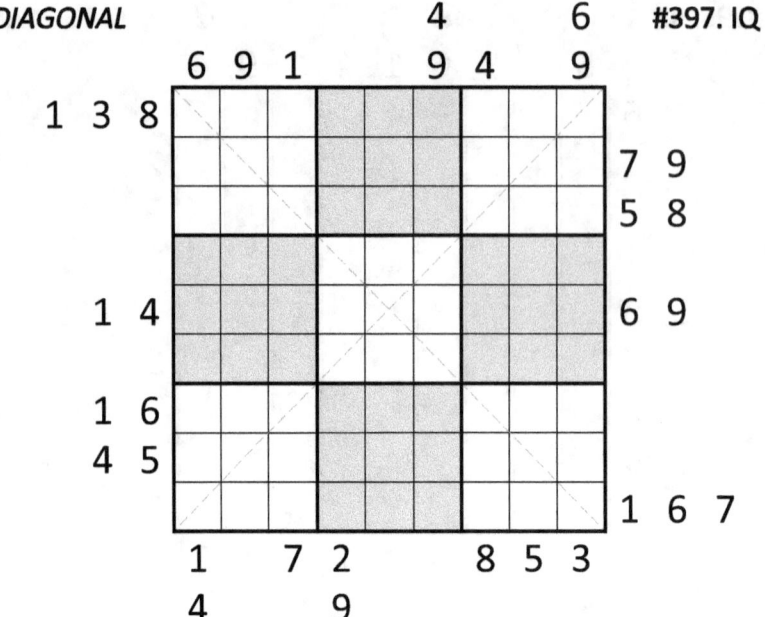

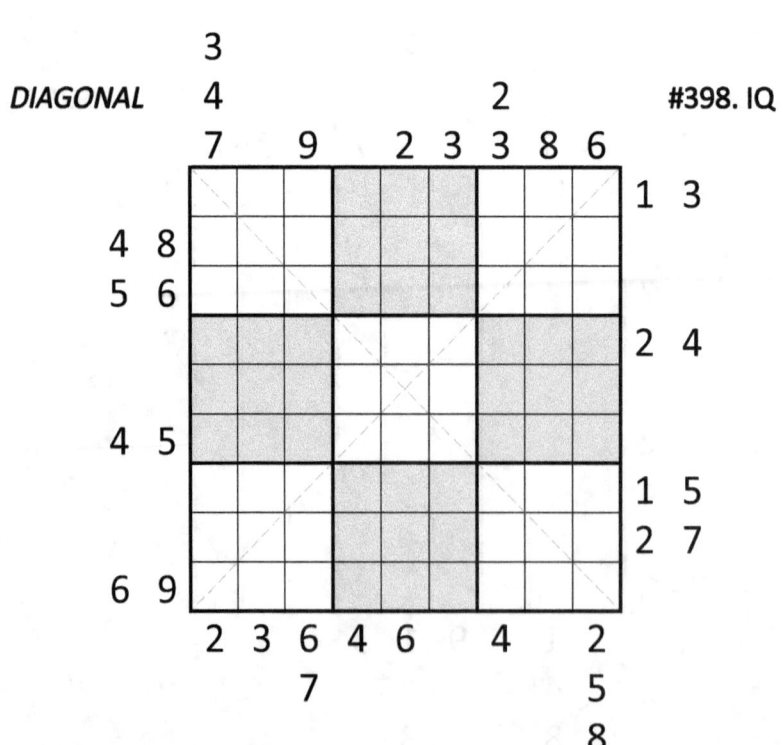

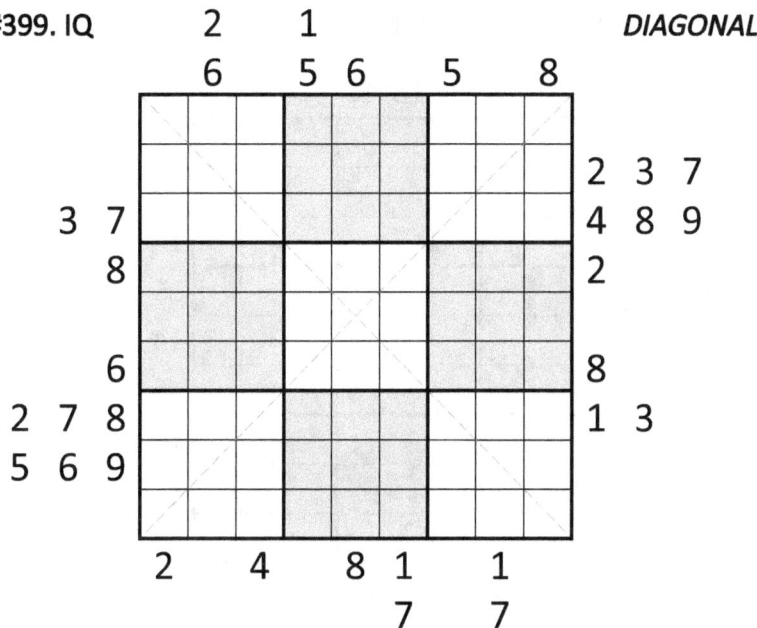

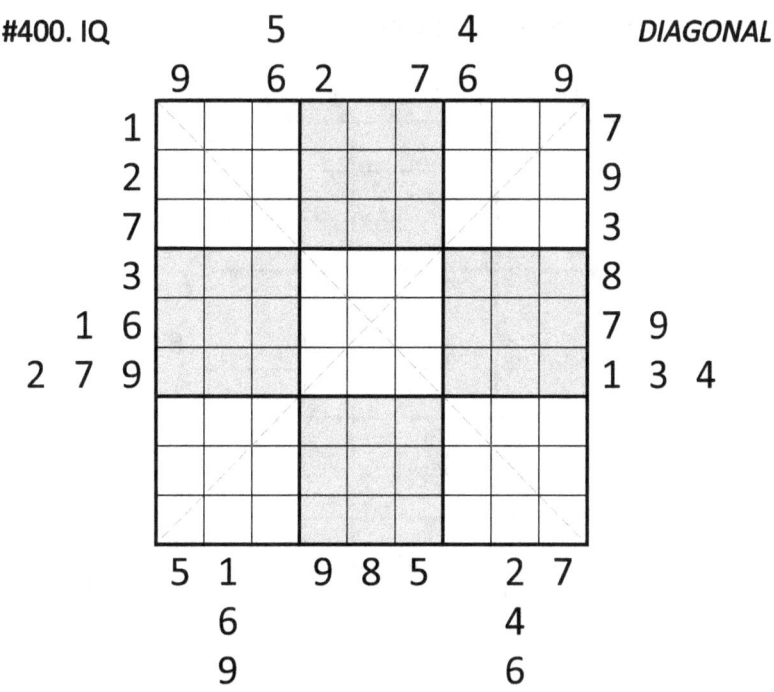

#401. HARD - FRAME

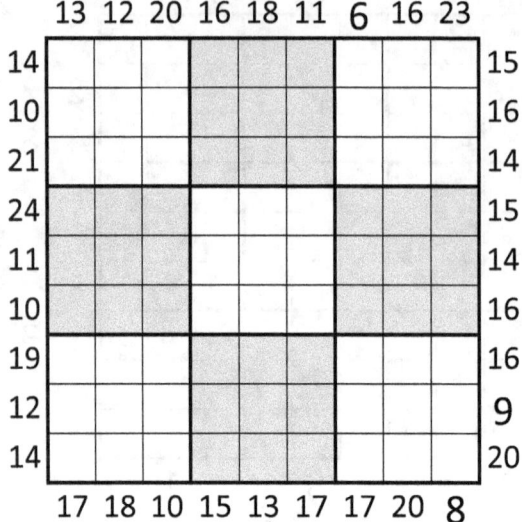

#402. HARD - FRAME

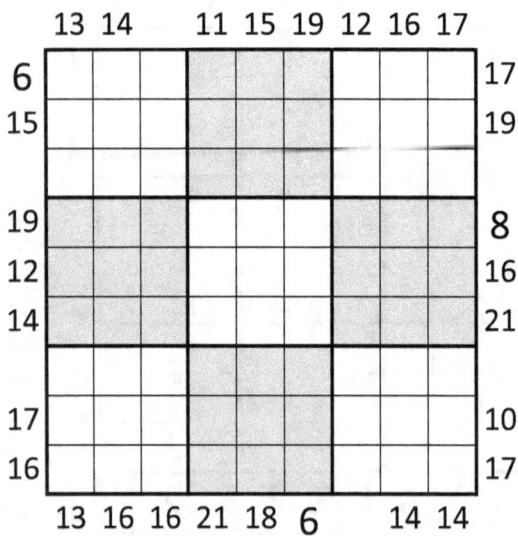

#403. HARD - FRAME - NON-CONS

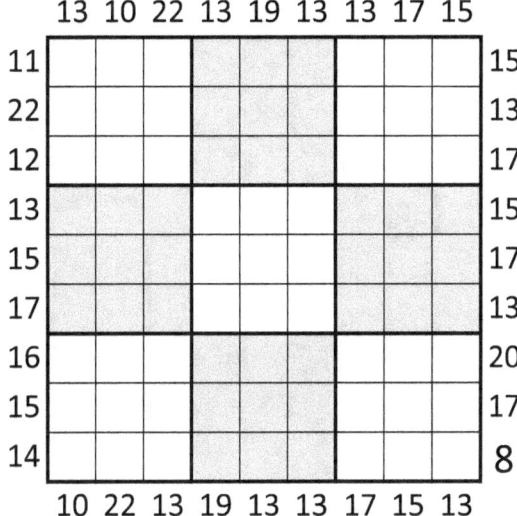

#404. HARD - FRAME - NON-CONS

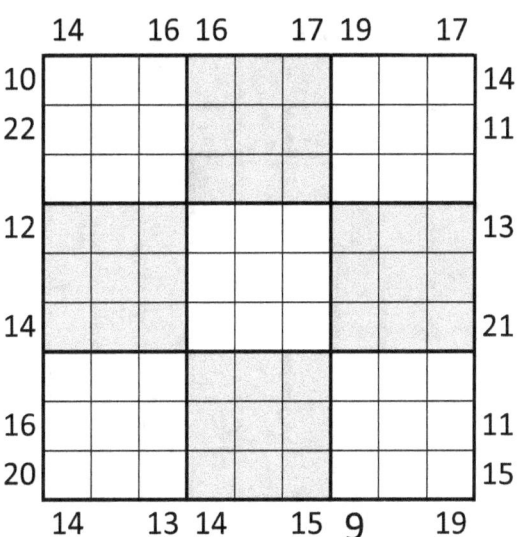

#405. HARD - FRAME - JIGSAW

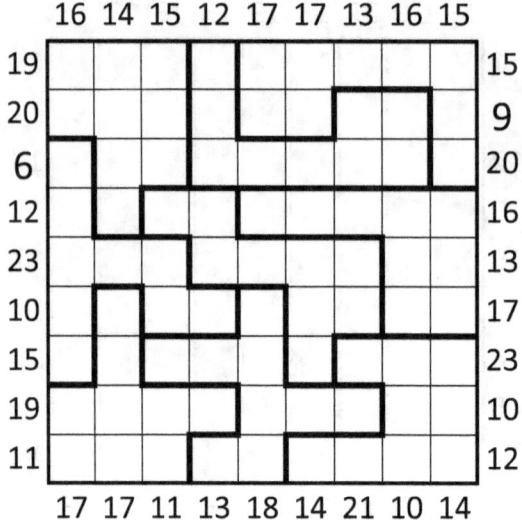

#406. HARD - FRAME - JIGSAW

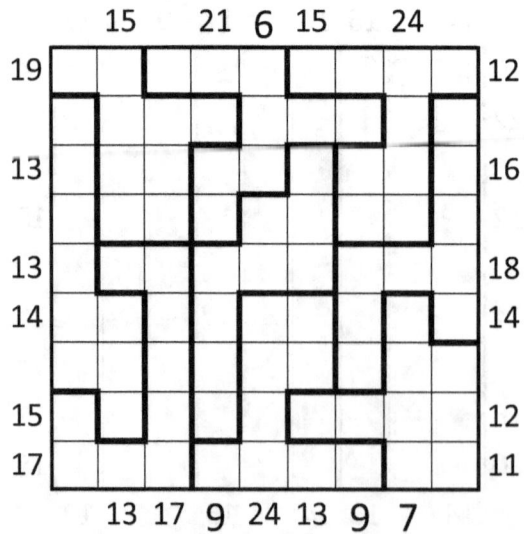